Disaster Risk Reduction

Methods, Approaches and Practices

Series editor

Rajib Shaw, Graduate School of Media and Governance, Keio University, Fujisawa, Japan

About the Series

Scope of the Series

Disaster risk reduction is a process that leads to the safety of communities and nations. After the 2005 World Conference on Disaster Reduction, held in Kobe, Japan, the Hyogo Framework for Action (HFA) was adopted as a framework for risk reduction. The academic research and higher education in disaster risk reduction has made, and continues to make, a gradual shift from pure basic research to applied, implementation-oriented research. More emphasis is being given to multi-stakeholder collaboration and multi-disciplinary research. Emerging university networks in Asia, Europe, Africa, and the Americas have urged process-oriented research in the disaster risk reduction field. With this in mind, this new series will promote the output of action research on disaster risk reduction, which will be useful for a wide range of stakeholders including academicians, professionals, practitioners, and students and researchers in related fields. The series will focus on emerging needs in the risk reduction field, starting from climate change adaptation, urban ecosystem, coastal risk reduction, education for sustainable development, community-based practices, risk communication, and human security, among other areas. Through academic review, this series will encourage young researchers and practitioners to analyze field practices and link them to theory and policies with logic, data, and evidence. In this way, the series will emphasize evidence-based risk reduction methods, approaches, and practices.

More information about this series at http://www.springer.com/series/11575

Bhaswati Ray · Rajib Shaw
Editors

Urban Drought

Emerging Water Challenges in Asia

Springer

Editors
Bhaswati Ray
Sivanath Sastri College
University of Calcutta
Kolkata, West Bengal, India

Rajib Shaw
Graduate School of Media and Governance
Keio University
Fujisawa, Japan

ISSN 2196-4106 ISSN 2196-4114 (electronic)
Disaster Risk Reduction
ISBN 978-981-10-8946-6 ISBN 978-981-10-8947-3 (eBook)
https://doi.org/10.1007/978-981-10-8947-3

Library of Congress Control Number: 2018961196

This Springer imprint is published by the registered company Springer Nature Singapore Pte Ltd.
The registered company address is: 152 Beach Road, #21-01/04 Gateway East, Singapore 189721, Singapore

Preface

The world has witnessed unprecedented urban growth since the beginning of the twenty-first century. Much of this growth is concentrated in the less-developed countries of Asia. Asia is home to 54% of the world's urban population, and the percentage share is expected to increase to 66% by 2050. Ninety-three percentage of the growth is expected to be in the less-developed countries with more than 60% in the cities of Asia. It is also the million-plus cities that are witnessing the most phenomenal urban growth under the impact of rural-to-urban migration. Along with climate change impacts, the rapid pace of mass urbanization is putting cities under increasing water insecurity and water stress. There is also an increased demand for water in order to maintain the intense pace of activities and a high standard of living in urban areas. Cities also concentrate on anthropogenic carbon emissions that accelerate global warming and impact the hydrological cycle to further increase water stress in cities. Increase in impervious surface and consequent depletion of groundwater reserves, increased water pollution from city waste, and industrial affluent decrease the availability of finite water resources. Choked water bodies and urban drainage systems increase the vulnerability to floods. Cities thus spatially concentrate the water demand of millions into a small area and also increase the frequency and intensity of water-related disasters like floods and droughts under climate change impacts, hindering development prospects.

Confronted with multifaceted and complex challenges, conventional water management practices have proved to be inadequate to address urban water insecurity and water stress. Transforming urban water systems into more resilient and hence more sustainable systems would require innovative approaches. Urban communities are seeking resilience in the urban water system and to future uncertainties in water supply, to create resilient, livable, productive, and sustainable cities.

This book in 24 chapters deals with the various aspects of urban water insecurity, concepts, and relevance with special reference to urban water insecurity in Asian cities. The first section discusses the concepts of urban water insecurity and the implications of climate change on urban drought and develops an urban water security index. The second section deals with various case studies from Asia covering varied dimensions of urban water insecurity and the policy responses to manage urban water crisis. The selected cities range from megacities in developing countries like Delhi and Kolkata in India and Dhaka in Bangladesh to million-plus cities and even small towns in developed countries like Japan.

Covering all aspects of urban drought and water insecurity, this book is intended for students, researchers, academia, policy makers, and development practitioners in the fields of water resource management, urban planning, and disaster management, especially in the Asia-Pacific region. It will help to better understand the complex scenario of urban drought caused by rapid urbanization, unplanned urban growth, and climate change and to evolve policies for sustainable water systems and resilient cities.

Fujisawa, Japan
January 2019

Bhaswati Ray
Rajib Shaw

About This Book

It is already evident that urban areas are threatened by water insecurity under the impacts of rapid urbanization, competing and increased water use, and climate-change-induced water stress particularly in the cities of Asia. Yet, nearly half of the world population lives in urban areas and the percentage is expected to reach 60 in the next two decades. Ensuring water security is a defining global challenge and involves, in addition to having enough water resources, the mitigation of water-related risks, such as flood and drought, addressing conflicts over shared water resources and resolving stress among stakeholders that compete for increased water use. It is embedded in all aspects of development, poverty reduction, food security, health, social equity, and environmental sustainability. It also involves increasing the economic efficiency.

This book attempts to study the various concepts and dimensions of urban water insecurity and to analyze the threats to urban water security with case studies from Asia. Inefficient water use by households and industries, fragmented management of water resources between sectors and institutions, climate-induced water shortage, environmental degradation of water sources, and inadequate use of alternate sources are issues of major concern. While developing a water security index for urban areas, the book also explores the mitigation measures for increased urban water resilience that would require innovative approaches, both infrastructure-based and system-based. The 24 chapters include overview articles on the theoretical framework of water security and lessons learnt from Asian cities including the analysis of problems and best practices adopted in urban water governance.

The book is principally targeted to study the concepts and dimensions of urban water insecurity and the implications of climate change on urban drought and to explore the parameters for developing urban water security index. The book also deals with various case studies from Asian cities covering varied dimensions of urban water insecurity and explores the innovative approaches to water development and management in these cities.

Covering all aspects of urban drought and water insecurity, this book will be a valuable resource material for students, researchers, academia, policy makers, and development practitioners. The primary target groups for this book are students and researchers in the fields of water resource management, urban planning, and disaster management, especially from the countries of the Asia-Pacific region. The collective knowledge from this book will help policy planners and practitioners to better understand the complex scenario of urban drought caused by rapid urbanization, unplanned urban growth, and climate change and to evolve policies for sustainable water systems and resilient cities.

Contents

Editors and Contributors

About the Editors

Bhaswati Ray is an Associate Professor in Sivanath Sastri College, University of Calcutta, Kolkata, India. She is also a Guest Faculty at the Postgraduate Department of Geography, Vivekanandha College for Women, Kolkata. She was awarded Pan-Asian Risk Reduction (PARR) Fellowship by Kyoto University (International Environmental and Disaster Management Laboratory, Graduate School of Global Environmental Studies), Japan, in July 2015 for integrated research on water resource management with special reference to fluoride, drought, and salinity affected regions of West Bengal, India. She is a Member of the American Planning Association. She has a reference book as a sole author, more than 20 publications including academic papers and chapters and almost 25 conference presentations to her credit. An ardent researcher with immense field experience, her expertise includes GIS mapping, global and regional water stress, water insecurity and water resource management, urban and rural local governance, housing typology and built form, community participation, sustainable development, disaster risk reduction, and resilience.

Rajib Shaw is a Professor of Keio University in its Shonan Fujisawa Campus (SFC) in Japan. Earlier, he was the Executive Director of Integrated Research on Disaster Risk (IRDR), a decade-long research program co-sponsored by the International Council for Science (ICSU), the International Social Science Council (ISSC), and the United Nations International Strategy for Disaster Reduction (UNISDR). He is the Chair of the UNISDR's Science Technical Advisory Group (STAG) at the global level, as well as the Co-Chair of UNISDR's Asia Science Technology Academic Advisory Group (ASTAAG). In IRDR, he currently serves as the Science Committee Member. He is also the Senior Fellow of Institute for Global Environmental Strategies (IGES), Japan, and the Chairperson of SEEDS Asia, a Japanese NGO. Previously, he was a Professor in the Graduate School of

Global Environmental Studies of Kyoto University, Japan. His expertise includes community-based disaster risk management, climate change adaptation, urban risk management, and disaster and environmental education. He has published more than 40 books and over 300 academic papers and chapters.

Contributors

Basanta Raj Adhikari Institute for Disaster Management and Reconstruction, Sichuan University-The Hong Kong Polytechnic University, Chengdu, China; Pulchowk Campus, Institute of Engineering, Tribhuvan University, Kirtipur, Nepal

Wijitbusaba (Ann) Marome Faculty of Architecture and Planning, Urban Futures Research Unit of the Thammasat University, Bangkok, Thailand

Ali Ardalan Department of Climate Change and Health, Institute of Environmental Research, Tehran University of Medical Sciences, Tehran, Iran; Harvard Humanitarian Initiative, Harvard University, Cambridge, USA

A. J. M. K. K. Aviruppola University of Peradeniya, Peradeniya, Sri Lanka

Emily Ying Yang Chan Division of Global Health and Humanitarian Medicine, Jockey Club School of Public Health and Primary Care, Chinese University of Hong Kong, Hong Kong, China; International Centre of Excellence for Collaborating Centre for Oxford University and CUHK for Disaster and Medical Humanitarian Response (ICoE-CCOUC), Integrated Research on Disaster Risk (IRDR), Hong Kong, China; WHO Thematic Platform for Health Emergency and Disaster Risk Management Research Network, World Health Organization, Geneva, Switzerland

Faye Cruz Manila Observatory, Quezon City, Philippines

Serjmyadag Dalai Institute of Disaster Research, National Emergency Management Agency, Ulaanbaatar, Mongolia

Oyunbaatar Dambaravjaa Information and Research Institute of Meteorology, Hydrology and Environment, Ulaanbaatar, Mongolia

Jessica Dator-Bercilla Manila Observatory, Quezon City, Philippines; Christian Aid, London, India

Sameer Deshkar Department of Architecture and Planning, Visvesvaraya National Institute of Technology, Nagpur, India

Kapil Gupta Indian Institute of Technology Bombay, Mumbai, India

Mahdi Hadi Center for Water Quality Research, Institute for Environmental Research, Tehran University of Medical Sciences, Tehran, Iran

Noorul Haq Department of Geography, University of Peshawar, Peshawar, Pakistan

Mahfuzul Haque Department of Development Studies, Adjunct Faculty, University of Dhaka, Dhaka, Bangladesh

Janice Ying-en Ho Division of Global Health and Humanitarian Medicine, Jockey Club School of Public Health and Primary Care, Chinese University of Hong Kong, Hong Kong, China

Shimpei Iwasaki Department of Environmental Science, International College of Arts and Sciences, Fukuoka Women's University, Higashi-Ku, Fukuoka, Japan

Mona Khaleghy Rad Department of Climate Change and Health, Institute of Environmental Research, Tehran University of Medical Sciences, Tehran, Iran

Attaullah Khan Department of Geography, Government College Lund Khwar, Mardan, Pakistan

Mohammad Golam Kibria Department of Soil Science, Bangladesh Agricultural University, Mymensingh, Bangladesh

Md. Anwarul Abedin Department of Soil Science, Bangladesh Agricultural University, Mymensingh, Bangladesh

Bijon Kumer Mitra Institute for Global Environmental Strategies, Hayama, Japan

Tarun Kumar Mondal Department of Geography, University of Kalyani, Nadia, West Bengal, India

Gemma Narisma Manila Observatory, Ateneo de Manila University, Quezon City, Philippines

Vinay Nikam EnviroTech Consultants, Thane, India

Parmakope Urban Development Planning Agency, Banda Aceh, Indonesia

Emma Porio Manila Observatory, Ateneo de Manila University, Quezon City, Philippines

Gomboluudev Purevjav Information and Research Institute of Meteorology, Hydrology and Environment, Ulaanbaatar, Mongolia

Alfi Rahman Department of International Environment and Resources Policy, International Post Graduate Program in Human Security, Graduate School of International Cultural Studies, Tohoku University, Sendai, Miyagi, Japan; Faculty of Social and Political Sciences, Tsunami and Disaster Mitigation Research Center (TDMRC), Syiah Kuala University, Banda Aceh, Indonesia

Atta-ur Rahman Department of Geography, University of Peshawar, Peshawar, Pakistan

Bhaswati Ray Sivanath Sastri College, University of Calcutta, Kolkata, India

K. W. G. Rekha Nianthi University of Peradeniya, Peradeniya, Sri Lanka

Paramita Roychowdhury Department of Geography, Krishnagar Government College, Nadia, West Bengal, India

Aiko Sakurai Department of Social Sciences, Toyo Eiwa University, Tokyo, Japan

Samiullah Department of Geography, University of Peshawar, Peshawar, Pakistan

Narendra Man Shakya Pulchowk Campus, Institute of Engineering, Tribhuvan University, Kirtipur, Nepal

Vinod K. Sharma Indian Institute of Public Administration, New Delhi, India

Rajib Shaw Graduate School of Media and Governance, Keio University, Fujisawa, Japan

Suresh Das Shrestha Central Department of Geology, Tribhuvan University, Kirtipur, Nepal

Shyamli Singh Indian Institute of Public Administration, New Delhi, India

Stephen Anthony Sutton Department of Psychological and Clinical Sciences, Charles Darwin University, Darwin, Australia; Tsunami and Disaster Mitigation Research Center (TDMRC), Syiah Kuala University, Banda Aceh, Indonesia; Bushfire and Natural Hazards Cooperative Research Centre, East Melbourne, Australia

Tomoko Takeda Institute for Global Environmental Strategies, Hayama, Japan

Deepthi Wickramasinghe Department of Zoology and Environment Sciences, University of Colombo, Colombo, Sri Lanka

Wanglin Yan Keio University, Fujisawa, Japan

Antonia Yulo-Loyzaga Manila Observatory, Quezon City, Philippines

Chapter 1
Defining Urban Water Insecurity: Concepts and Relevance

Bhaswati Ray and Rajib Shaw

Abstract Nearly half of the world's total population resides in towns and cities, and the percentage is projected to reach 60 in the next two decades. The United Nations predict that by 2050, the urban population of the world will reach 6.3 billion from 3.5 billion. 93% of the growth would be in the developing countries with more than 80% in the cities of Asia and Africa. The rapid increase in population and the fast pace of urbanization along with climate change are putting cities under increasing water stress. The two main challenges facing urban areas are the inadequate and inequitable access to safe water and improved sanitation facilities on one hand and increased frequency and intensity of disasters including floods and droughts with adverse consequences on economic growth, health and well-being on the other. Water insecurity can be defined as the lack availability of sufficient water of good quality to meet basic human requirements, livelihoods and ecosystem functions, and an increased risk of water-linked disasters. Water security thus involves, in addition to having enough water resources, reduction of disaster risks associated extreme weather events floods, resolving conflicts over shared water systems, reducing stress among different stakeholders and competing uses of water. It is thus embedded in various development issues including poverty alleviation, food security, social equity and environmental sustainability. It also involves increasing the economic efficiency. Based primarily on literature review, the chapter intends to include the various concepts and threats to urban water security and measures to improve the same.

Keywords Urbanization · Climate change · Water stress · Water security

B. Ray (✉)
Sivanath Sastri College, University of Calcutta, Kolkata, India
e-mail: bhaswati173@gmail.com

R. Shaw
Graduate School of Media and Governance, Keio University, Fujisawa, Japan

B. Ray and R. Shaw (eds.), *Urban Drought*, Disaster Risk Reduction,
https://doi.org/10.1007/978-981-10-8947-3_1

1.1 Introduction

The world has witnessed unprecedented urban growth since the beginning of the twenty-first century. The world's urban population, which stood at just 10% of the global total at the onset of the century, reached an unprecedented 50% by the beginning of the twenty-first century (United Nations Department of Economics and Social Affairs 2010), when the total urban population was 3.6 billion. It is the million plus cities that witnessed the most phenomenal urban growth. In 1900, there were 16 'million cities' across the globe with population exceeding one million or more, most of them being located in the developed nations. By 2000, the number of million cities increased to 400, with three quarters of the cities located in developing countries (United Nations Department of Economics and Social Affairs 2010). The United Nations predicts that by 2050, urban population of the world will increase from 3.6 billion to 6.3 billion.

The next few decades will witness the most rapid urban growth in human history, with 2.6 billion additional urban dwellers expected by 2050 (McDonald et al. 2014; United Nations Population Division 2011). The world's urban population would then reach 6.2 billion, an increase from the global population of 3.6 billion (United Nations Population Division 2011). 93% of the growth is expected to be concentrated in the less developed countries with more than 80% in the cities of Asia and Africa. 37% of this growth would be contributed by just three countries—India, China and Nigeria. The rapid increase in population and the fast pace of urbanization along with climate change are putting cities under increasing water insecurity.

Water scarcity is the inability to meet the demand for water in an area due to the unavailability of sufficient amount of water. Water scarcity can be both physical water scarcity and socio-economic scarcity of water. Physical scarcity is denoted by demand for water exceeding supply and is a consequence of inadequacy of available water resources to fulfil the demand for water. This occurs due to overexploitation of water resources. A socio-economic scarcity, on the other hand, results from insufficient investment and political apathy to develop water resources. It is often a result of poor governance rather than insufficient availability. Water scarcity is further accentuated by climate change and altered patterns of flood and droughts, increased pollution, depletion and overuse. It thus emerges from a combination of hydrological variability and high human use, which may in part be mitigated by storage infrastructure (United Nations 2016). Water stress, a component of water scarcity, is defined as the difficulty of obtaining fresh water for use during a specific period of time and may result in further depletion and deterioration of water resources, resulting in increased water scarcity (Science Daily 2017). According to the World Bank, roughly 1.6 billion people already live in countries with water scarcity, and that number is likely to increase in future. Large tracts of Central and West Asia suffer from physical water scarcity, while South Asian countries, particularly India, Nepal and Bangladesh, suffer from socio-economic scarcity. China and India, where urban growth is expected to be maximum, suffer most from water scarcity—both physical and socio-economic (Fig. 1.1).

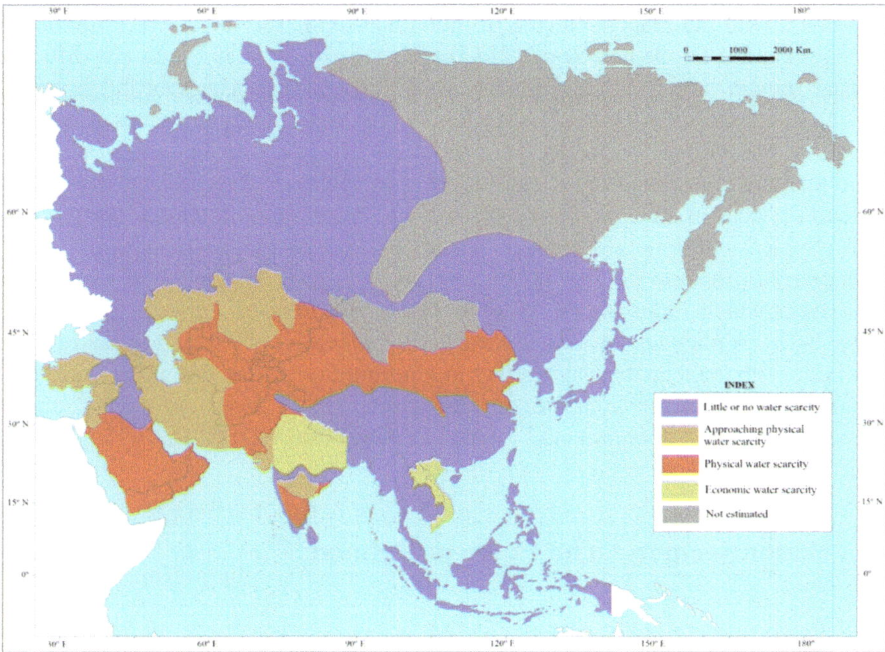

Fig. 1.1 Water scarcities in Asia

Water insecurity, on the other hand, can be defined as the lack of availability of adequate water of good quality to meet basic human requirements, livelihoods and ecosystem functions, and an increased risk of water-linked disasters. The two main challenges facing urban areas are the inequitable access to safe water and sanitation facilities and the increased frequency and intensity of disasters like flood and droughts, with adverse consequences on economic growth, health and well-being. It prevents both accessibility and affordability of the resource to the vulnerable sections of the population and is more a governance issue than absolute shortage of water availability. Water insecurity is an effect, rather than a cause, of sociopolitical domination and infrastructural exclusion. While policy experts often respond to water insecurity by focusing on the issue of physical scarcity and estimates of water availability to explain water insecurity, it is often more than a technocratic response. Both geographical limitations like location in dry climates or far from water sources as well as financial limitations, with the poor cities unable to construct robust urban water infrastructure increases their vulnerability to urban water insecurity, at a time when the rationale for delivery of water services in a non-discriminatory manner is well documented. Institutional, operational and financial causes of water insecurity dictate the response path that differs in the rural and urban contexts.

Exponential population growth is a major challenge in the Asian cities. Growth in urban population is accompanied by a trend towards smaller, and therefore more, households with the associated cost of individual piped service delivery provision and maintenance (Hope and Rouse 2013). The most in need and the most difficult to reach remain largely unserved (Hope and Rouse 2013). The shortfall is being made up of other improved supplies at a higher cost highlighting the water insecurity issues in cities already suffering from financial gaps. Significant loss of revenue due to finances channelized to alternate safe water sources often reduces the ability of urban services to meet escalating demands in Asian cities. Increased investment needs coupled with low-cost recovery results in deteriorating services. It increases the inability of the service providers to extend services to unplanned or expanding urban areas which limits access to low-income and vulnerable groups (Hope and Rouse 2013). A major risk to water security is the neglect of existing infrastructure associated with inadequate cost recovery (Hope and Rouse 2013), high leakage loss, rationing water supply by hours per day that accelerates deterioration and introduces the risk of ingress of contaminated water (Hope and Rouse 2013; Rouse 2007).

1.2 Water Security: The Concept Defined

While the concept of water security is not new, the term appears to have gained greater profile recently, judging from a range of reports and conferences that have considered water security in isolation or in relation to the security of other resources, notably energy and food/land (Mason and Calow 2012; Martin-Nagle et al. 2012; National Intelligence Community 2012; Oxford University Water Security Network 2012; World Economic Forum 2011). The Global Water Partnership (GWP) (2000) first defined water security as an overarching goal where every person has access to enough safe water at affordable cost to lead a clean, healthy and productive life, while ensuring the environment is protected and enhanced. Water security was also defined as the availability of water in adequate quantity and quality in perpetuity to meet domestic, agricultural, industrial and ecosystem needs (Cook and Bakker 2012; Swaminathan 2001). The emphasis was on the availability of water and on environmental protection. Later, water security went on to add new dimensions of prevention and reduction of disaster risks. The Ministerial Declaration of the Second World Water Forum in Hague in 2000 defines water security as ensuring that fresh water, coastal and related ecosystems are protected and improved; that sustainable development and political stability are promoted, that every person has access to enough safe water at an affordable cost to lead a healthy and productive life and that the vulnerable are protected from the risks of water-related hazards (Gerlak and Mukhtarov 2015). Water security has also been defined by Cheng et al. (2004) as access to safe water for healthy living and food production at affordable cost while ensuring that the environment is protected from pollution and water-related disasters like floods and droughts. The concept of water security thus entails managing the effects of both overabundance and water scarcity. It implies the availability of an acceptable quantity and quality of water for health, livelihoods, ecosystems and production, coupled with an acceptable level of water-related risks to people, environments and economies (Halmatov et al. 2017; Grey and Sadoff 2007). For the Global Water Partnership (GWP), the essence of water security is that concern for the resource base itself is coupled with concern that services which exploit the resource base for human survival and well-being, as well as for agriculture and other economic enterprise, should be developed and managed in an equitable, efficient and integrated manner (Mason and Calow 2012; Global Water Partnership 2009). The World Bank defines water security as reliable availability of acceptable quantity and quality of water for production, livelihoods and health and acceptable level of often unpredictable water related risks to society (Hutchinson and Herrmann 2008).

These definitions thus combine to focus on availability of potable water for human consumption, providing water for productive activities in the agricultural and industrial sectors, for environmental protection and conservation, and for the prevention of water-related disasters and issues of national security. Water security may increasingly be referred to in articulating water's role in national and international peace and stability due to water's strategic significance as both a fugitive resource that often traverses borders and, in its embedded or virtual form, a globally traded commodity (Mason and Calow 2012; United Nations Development Programme 2006). The competition for water will aggravate in the coming decades, and people with the weakest rights like women and the landless will see their entitlements to water threatened and challenged by the more powerful. By traversing borders through water bodies, rivers and wetlands, water is often responsible for cross-border tensions in water-stressed regions. Water security also implies the ability of a population to enjoy sustained access to adequate quantities of safe water to support livelihoods, human well-being and economic development, protection against water pollution and water-related disasters, for preserving ecosystems, ensuring peace and political stability (United Nations 2013).

The Asian Water Development Outlook (AWDO), developed in 2007 by the Asia-Pacific Water Forum (APWF) and the Asian Development Bank (ADB) to highlight the important water management issues (Asian Development Bank 2013), introduced the concept of quantitative measurement of water security. Asian Water Development Outlook 2013 by the Asian Development Bank was the first comprehensive attempt to measure progress towards a water-secure future in Asia and the Pacific region. The report provides a robust, pragmatic and the first-ever quantitative and comprehensive framework for assessing water security in the countries of Asia and the Pacific (Asian Development Bank 2013) involving five different dimensions. The national water security of any country is assessed by the composite score of the five dimensions measured on a scale of one to five (Asian Development Bank 2016), where one indicates extremely poor water security while five denotes the best.

1. Household water security: it is to be measured in terms of access to piped water supply, access to sanitation and hygiene (Asian Development Bank 2013). Providing safe and reliable safe water and sanitation is essential for efforts to eradicate poverty and support economic development (Asian Development Bank 2013).

2. Economic water security: measured across various sectors, agricultural, industrial and energy water security, it is the productive use of water to sustain economic growth in food production, industry and energy sectors of the economy (Asian Development Bank 2016).
3. Urban water security: defined in terms of water supply, wastewater treatment and drainage (Asian Development Bank 2013), the indicators measure better water management designs and services to create and sustain vibrant and liveable water-sensitive cities.
4. Environmental water security: watershed disturbance, pollution, water resource development and biotic factors (Asian Development Bank 2013).
5. Resilience to water-related disasters: it includes the exposure, vulnerability, hard coping capacities and soft coping capacities to water-linked disasters like flood, drought, storm surges and coastal floods. There is a need to build resilient communities in Asia and Pacific region that can adjust to climate variability and reduce disaster risks.

1.3 Urban Water Security in Asia: Its Current Relevance

The main reason behind a growing urban population in the developing countries of Asia is rural-to-urban migration for better economic opportunities, improved standards of living and better provision of basic services like education and health. The urban population of the world has grown rapidly from 746 million in 1950 to 3.9 billion in 2014 (United Nations 2014). The number is projected to reach 6.2 billion in 2050. Asia, despite its lower level of urbanization, is home to 53% of the world's urban population (United Nations 2014). The percentage share is projected to increase to 66% by 2050.

Urbanization in Asia is best described as pseudo-urbanization or mass urbanization, with maximum concentration of urban population in the megacities, cities with population exceeding 10 million. The total population in ten megacities was 153 million in 1990 accounting for 7% of the total urban population. By 2014, the number of megacities increased to 28 (Hussein 2017), housing a population above 450 million and accounting for almost 12% of the world's total. A total of 16 of these megacities are located in Asia, 4 megacities are in Latin America, 3 each in Africa and Europe, and 2 megacities in North America (United Nations 2014). The rapid rate of urbanization has caused an immense increase in the slum population in the less developed countries. The number of slum dwellers increased from 200 million in 1990 to 853 million in 2014. Apart from putting stress on basic urban services including housing, basic infrastructure like potable water supply, good sanitation, electricity, health and education services, mass urbanization also makes cities prone to environmental degradation and increased frequency of disasters. Increasing access to municipal water supply for the world's poor is one of the Millennium Development Goals, since municipal supply is generally cleaner and safer than other water sources (McDonald et al. 2014; Howard and Bartram 2003).

Past research has shown that as cities grow in population, the total water needed for adequate municipal supply grows as well (McDonald et al. 2011, 2014; Bradley et al. 2002; Postel et al. 1996; Falkenmark and Widstrand 1992; Falkenmark and Lindh 1974). This increase in total municipal water demand is driven not just by the increase in urban population, but also by a tendency for economic development to increase the fraction of the urban population that uses municipal supply rather than other sources such as local wells or private water vendors (McDonald et al. 2014; Bartlett 2003; Bhatia and Falkenmark 1993). Urbanization and a higher standard of living further increase per capita water use, as new technologies such as showers, washing machines and dishwashers increase residential use of water (McDonald et al. 2011, 2014). Thus, cities by their nature spatially concentrate the water demand of thousands or millions of people into a small area, which by itself would increase stress on finite supplies of available fresh water particularly near the city centre (McDonald et al. 2011, 2014). Cities also concentrate the sources of greenhouse gases caused by anthropogenic carbon emissions that impact the hydrological cycle and increase water stress. With increased concentration of greenhouse gases, cities are responsible for accelerated global warming impacting the hydrological cycle and increasing the water stress in cities. Increase in impervious surface and consequent depletion of groundwater reserves, increased water pollution from city waste and industrial affluent decrease the availability of finite water resources. Choked water bodies and urban drainage systems increase the vulnerability of floods. Two-thirds of the megacities are located in regions vulnerable to urban water scarcity.

Asia and the Pacific region continue to be affected by inadequate water security. Ten largest cities under water stress are located mostly in this region which includes Tokyo, Delhi, Shanghai, Beijing, Kolkata, Karachi and Moscow, in addition to Mexico City, Los Angeles and Rio de Janeiro. 60% of the households live without safe, piped water supply and improved sanitation (Asian Development Bank 2013, 2014). The Asia-Pacific Water Forum predicts that climate change impacts will impose additional threats to the vulnerable countries in Asia and the Pacific region, challenging the very concepts of sustainable development, poverty alleviation and improved water security. Poor water management is also responsible for the occurrence of natural disasters that include frequent and endemic floods and droughts, storm surges and landslides. The densely populated coastal cities in India, including Mumbai, Kolkata, Chennai, Surat and Thiruvananthapuram, are vulnerable to cyclones and associated hazards such as storm surges, high winds and heavy rainfall (Parikh et al. 2013, 2014). Although improved forecasting has reduced the number of deaths from water-related disasters, the cost of flood disasters in the region has increased over time (Asian Development Bank 2013).

Rapid urbanization is a cause of greenhouse gas emissions and global warming that is expected to present imminent challenges for urban areas, especially in the less developed countries of Asia, way beyond the physical risks such as sea level rise and increased frequency of extreme events. It will become increasingly difficult for cities to provide basic services like water supply and sanitation to the residents. Climate change would also increase the occurrence of extreme events like floods and droughts and alter the precipitation pattern by impacting the hydrological cycle with devastating consequences.

According to the population-weighted Regional Water Security Index developed by Asian Development Bank (2013) in countries of Western and Central Asia, East, Southeast and South Asia, as well as the Pacific region, it is evident that South Asia is more insecure than all other regions in terms of its household water security (including sanitation), urban water security, environmental water security and resilience to water-related disasters (Asian Development Bank 2013). West and Central Asia are least developed in economic water security (Asian Development Bank 2013). The lowest water security index of 1.6 is found in South Asia. Inequity in access is also the highest in South Asia (Asian Development Bank 2013). Impacts of climate change are likely to put these economies under additional water stress and affect the economies of these countries. The water security index value is 2.2 for West and Central Asia, 2.4 for Southeast Asia, 2.6 for East Asia and 2.5 for the Pacific region. The advanced economies including Hong Kong, Singapore, China and Japan have a water security index value of 3.3. A similar trend is also witnessed in the population-weighted disaster resilience and water security being lowest in South Asia.

However, the overall increase in total municipal water demand causes cities to search for new adequate, relatively clean water sources, leading to the creation of sometimes quite complex systems of urban water infrastructure (McDonald et al. 2014; Alcott et al. 2013; Brown et al. 2009; Chau 1993). Cities represent a concentration of economic and political power (Bettencourt et al. 2007), which cities use to build urban water infrastructure to satisfy their demand (McDonald et al. 2014). As this infrastructure can go out far from the city centre, cities exploit new sources of surface water and groundwater or adopt scientific knowledge and technology like desalination to escape water stress (McDonald et al. 2014). Cities can also reduce the emission of greenhouse gases and adapt to changing climate through policy initiatives. International Conventions are also emphasizing on such adaptations.

Local communities in urban areas are focusing on resilience in the urban water system and to future uncertainties in water supply because of climate change and population growth. It is now well accepted that the conventional urban water management approach is highly unsuited to addressing current and future sustainability issues (Wong and Brown 2009; Ashley et al. 2003, 2005; Newman 2001; Butler and Maksimovic 1999). Transforming urban water systems into more resilient and hence more sustainable systems would require innovative approaches. It is pertinent to not that there are a wide range of water sources that cities can have access to supplement the existing ones in the form of rain water harvesting, construction of storage devices, appropriate and sustainable extraction of groundwater, groundwater recharge, urban storm water use, treated wastewater and desalinisation options. Such sources may be found to already exist within urban areas. A strategy built around a diversity of water sources would allow cities the flexibility to access a portfolio of water sources (Wong and Brown 2008) and harvesting, treatment and delivery options.

This would include both centralized and decentralized water supply schemes, ranging from the simple rainwater tanks (Wong and Brown 2008) with secondary supply pipeline for non-potable use to city-scale indirect potable water reuse schemes and the pipeline grid, linking regional reservoirs (Wong and Brown 2009). The non-potable water from a variety of local sources can replace the use of potable water for such uses as toilet flushing, laundry, garden watering and open space irrigation (Wong and Brown 2008, 2009). Designing cities for climate change, particularly the sustainable management of water resources, requires a reversal in the conventional philosophical approach of urban communities drawing on their depleting ecosystems and natural environments (Wong and Brown 2008). The traditional value of open spaces and landscape features needs to be reinforced with an understanding of the ecological functioning of the urban landscapes that capture the essences of sustainable water management, micro-climate influences, facilitation of carbon sinks and use for food production (Wong and Brown 2008, 2009). Technological improvements and institutional capacity building is needed for sustainable urban water management. It is being argued that unless new technologies are socially embedded into the local institutional context, their development in isolation is insufficient to ensure their successful implementation in practice (Wong and Brown 2009; Brown 2008). There has been a significant boost in the involvement of local communities in defining the urban water problem, developing water-sensitive urban designs adapted to local conditions and participating in the processes of decision-making. Cities and countries of Asia have already adopted innovative measures in managing water stress and drought in urban areas. The densely populated country of Singapore, for example, with no source of freshwater supply, is highly water stressed with its demand exceeding the naturally occurring supply. Yet Singapore city is able to provide enough and sustained water for its industrial, agricultural, commercial and domestic use through investment in technology, international agreements and effective water management. Advanced rainwater detention systems account for 20% of the water supplied in Singapore. 40% is given by Malaysia. Use of grey contributes another 30% while desalination process provides the remaining 10% of the supply.

1.4 About the Book

It is already evident that urban areas are threatened by water insecurity under the impacts of rapid urbanization, increased water use and climate change-induced water stress particularly in the cities of Asia. This book attempts to comprehend the various aspects of urban water security and to analyse the threats to urban water security with case studies from Asia. Inefficient water use practices by households and industries, fragmented management of water resources between sectors and institutions, climate-induced water shortage, environmental degradation of water sources and inadequate use of alternate sources are issues of major concern. Despite recent advances in the literature, there exists a considerable gap in policy and practice in the integrated water resource management approach (Ibisch et al. 2016). The book thus intends to look into the indices to measure water security while evolving a water security index for urban areas and to explore the innovative approaches to sustainable urban water development and management. The book, constituting of 23 chapters, constitutes two segments. The introductory segment consists of overview articles on theoretical frameworks of water security, while the second part deals with specific examples from Asian cities. In this chapter, an attempt has been made to provide an insight into the concept of urban water security and its present-day relevance.

In Chap. 2, Ray and Shaw have highlighted the impact of rapid urbanization on urban water management along with the global concern for sustainable water systems. The chapter also indentifies water insecurity issues in selected cities across Asia and discusses the best practices adopted for urban water development in these Asian cities.

In Chap. 3, the authors Ray and Shaw highlight how climate change impacts urban drought, more so in urban areas disadvantaged by urban heat island effect. The chapter explores the linkage between urban areas and the changing climate. It also assesses the significance of global conventions and the modified local adaptations in finding suitable solutions.

Chapter 4 by the same authors explores the reasons for water insecurity, which is rarely one of physical scarcity. Based on a number of water stress and urban sustainability indicators, the chapter further explores the possibility of identifying various parameters for developing a water security index to assess sustainability of the urban water systems.

In Chap. 5, Mitra et al. discuss the water-energy-food nexus as an integrated tool to manage increasing water crisis in the context of rapid urban growth through multi-sectoral cooperation at in-boundary and transboundary level. The need for the water-energy-food nexus in urban areas arises from the twin impact of rapid urban growth on finite resources and the water footprint caused by intensive food and energy consumption. Highlighting some selected good practices of the nexus around the world, this chapter argues that through realizing synergies across urban water energy food system, cities deliver inclusive growth, even with increasing drought risk.

In Chap. 6, Marome focuses on the institutional factors including a lack of coordination among different agencies that hinder climate change adaptations in the city region of Udon Thani province in Thailand, aggravating its vulnerability to drought and flood. Udon Thani's capacity to cope with the risk of flood and drought was also assessed through the creation of three transient scenarios: business as usual, population increase according to the strategic plan of the Provincial Waterworks Authority and the scenario where all of the province's envisioned development projects of the strategic plan are materialized, leading to significant urban growth.

Chapter 7 by Iwasaki is a case of successful urban water management in Japan through the historical process of canal restoration in Yanagawa city. The case study highlights the traditional water management system linked to the concept of 'near water' and the continuing efforts to strengthen the water–human relationship in the Horiwari River system with active involvement of local people to bridge the gap between near water and far water.

In Chap. 8, Abedin and Kibria explore the vulnerabilities of water-related disasters like floods and groundwater depletion, their causes and impacts, as well as the causes of drinking water scarcity in Dhaka city and the role of the government in enhancing the resilience of the city.

Chapter 9 on Nagpur city by Deshkar discusses the role of urban infrastructure in building urban resilience and promotes an integrated approach involving local communities and the natural resources. It also highlights some of the challenges for planning urban water infrastructure in Nagpur particularly from the resilience perspectives.

Urban drought and water availability conditions are discussed in Chap. 10 for National Capital territory of Delhi by Singh and Singh. They also discuss in length the probable strategies that may be adopted for drought mitigation.

Chapter 11 on Kathmandu by Adhikari et al. is on the impact of urban growth and surface sealing in decreased infiltration and the inadequacy of the Kathmandu Upatyaka Khanepani Limited (KUKL) which supplies only around 30% of total demand. The rest comes from groundwater pumping, traditional water spouts, supply from private water vendors and bottled water companies. Hence, the need for water management strategies has been highlighted.

In Chap. 12, Porio et al. explore water security in the Philippines by examining the effect of drought on risk governance and its social impacts in the 1997–1998 and 2015–2016 El Niño episodes in Metro Manila, Iloilo City and Cebu City. During these periods, widespread dryness occurred in both urban and rural areas, as rainfall was reduced by more than 50%, affecting the urban poor and increasing the cost of provisioning of urban water services.

In Chap. 13, Ray and Shaw study the existing water supply system of Kolkata and highlight the inadequacies in the formal water system and the role of the parallel informal supply system in the light of Sustainable Development Goals for a resilient and sustainable water future.

In Chap. 14, Gupta and Nikam emphasize the use of various water conservation measures with the use of water conservation devices, leak detection and repair, water reuse, metering and incremental rates. Two applications of rainwater harvesting and wastewater reuse in Thane City in Mumbai Metropolitan Region have been described as a management strategy adapted against urban drought and water insecurity.

Chan and Ying-enHo discuss water supply in Hong Kong city in Chap. 15 which imports water from River Dongjiang of Guangdong province in China through a dedicated aqueduct. This chapter further examines the health risks associated with poor water resources management in the city based on three cases highlighting water pipeline pollution, climate variability and its impact and the potential threat of disasters in the urban context.

Chapter 16 by Ardalan et al. focuses on the water resource management and urban water issues in the megacity of Tehran. It highlights the impact of rapid urbanization, high growth of population, an arid environment and the resultant challenges in the urban water system. The authors also discuss short-term and medium-term solutions for sustainable urban water management.

Rahman et al. explore trend of population growth, soil sealing and its impact on water infiltration and loss of groundwater potential in Peshawar city district in

Chap. 17. The chapter has also highlighted the consequences impacts of soil sealing on human life and environment.

In Chap. 18, Aviruppola and Rekha Nianthi discuss water management issues in urban areas in Sri Lanka, aggravated by climate variability and improper practices. The paper further highlights the policy responses of Sri Lanka towards managing the issue of urban water crisis and examines the overall position of Sri Lanka in urban water crisis management. It provides profound understanding of the policy formulations on urban water crisis management in Sri Lanka.

Chapter 19 also focuses on urban water issues and possible solutions for sustainable urban water management in the city of Colombo where the author Wickramasinghe identifies the current threats in the existing urban water system including pollution by domestic sewage, municipal wastewater and industrial effluent, policy and governance issues, salinity intrusion in Kelani River and the threats posed by climate variability. The chapter also examines available options in enhancing water security for a sustainable water future.

Chapter 20 is a similar study on the water challenges in Ulaanbaatar by Dalai et al. and focuses on water scarcity and flash floods caused by climate change, urbanization and water consumption increase in the city.

Chapter 21 deals with the water crisis in hill towns based on the case of Darjeeling town with suggested measures to be adopted as well as planning interventions as proposed by Mondal and Roychowdhury.

In Chap. 22, Haque explores the present state of water governance in the slums of Dhaka metropolis and water pricing issues based on two survey reports. Slum dwellers often pay a higher price for illegal connections, while the middle-class households consume 7.5–10 times more water.

In Chap. 23, Rahman et al. discuss the urban water management issues and challenges in the post-2004 Indian Ocean Tsunami Recovery phase based on the learning experiences from Banda Aceh City, Indonesia.

In Chap. 24, the editors provide a summary of the book that covers all aspects of urban drought, the threats and challenges brought about by rapid urban growth, climate change, environmental degradation and increased frequency of extreme events across Asia. The case studies include cities from less developed nations of India and Bangladesh as well as from developed countries like Japan. This chapter also highlights the mitigation measures for increased urban resilience. It is now quite evident that the conventional systems of water management are inadequate to address water insecurity and water stress. Transforming urban water systems into more resilient and hence more sustainable systems would require innovative approaches. These approaches have been discussed under two broad categories—the infrastructure-based approaches and the system-based approaches.

1.5 Expected Readership

The book intends to look into water insecurity issues in urban areas while evolving water security index and explore the innovative approaches to water development and management with examples from Asian cities. Covering all aspects of urban drought and water insecurity, this book will be a valuable resource material for students, researchers, academia, policymakers and development practitioners. The main target groups would, however, be the students and researchers working in the fields of sustainable management of water resources and urban water systems, urban planning and disaster management, focused mostly on the less developed countries of Asia. The shared knowledge and experiences that may be gathered from the book will help the various stakeholders including local communities, service providers and policymakers to better understand the complexities of urban drought caused by rapid urbanization, unplanned urban growth and climate change and to evolve policies for sustainable water systems and resilient cities.

Acknowledgements The Editors would like to take this opportunity to humbly acknowledge the contribution and valuable suggestions provided for this book by all contributing authors.

References

Alcott E, Ashton M, Gentry B (2013) Natural and engineered solutions for drinking water supplies: lessons from the Northeastern United States and directions for global watershed management. CRC Press, Boca Raton, Florida

Ashley RM, Balmforth DJ, Saul AJ, Blanksby JR (2005) Flooding in the future. Water Sci Technol 52(5):265–274

Ashley RM, Blackwood DJ, Butler D, Davies JA, Jowitt P, Smith H (2003) Sustainable decision making for the UK water industry. Eng Sustain 156:41–49

Asian Development Bank (2014) Water, sanitation and hygiene for a healthy Asia and the Pacific. Asian Development Bank, Metro Manila, Philippines

Asian Development Bank (2013) Asian water development outlook: measuring water security in Asia and the Pacific. Asian Development Bank, Metro Manila, Philippines

Asian Development Bank (2016) Asian water development outlook: strengthening water security in Asia and the Pacific. Asian development Bank, Metro Manila, Philippines

Bartlett S (2003) Water, sanitation and urban children: the need to go beyond improved provision. Environ Urbanization 15(2):57–70

Bettencourt MA, Lobo J, Helbing D, Kuhnert C, West GB (2007) Growth, innovation, scaling, and the pace of life in cities. Proc Natl Acad Sci USA 104:7301–7306

Bhatia R, Falkenmark M (1993) Water resource policies and the urban poor: innovative approaches and policy imperatives, water and sanitation currents. World Bank, Washington, DC

Bradley RM, Weeraratne S, Mediwake TMM (2002) Water use projections in developing countries. J Am Water Works Assoc 94:52–63

Brown RR (2008) Social and institutional considerations. In: Fletcher TD, Deletic A (eds) Data requirements for integrated urban water management, 1st edn. Taylor & Francis, London

Brown RR, Keath N, Wong THF (2009) Urban water management in cities: historical, current and future regimes. Water Sci Technol 59:847–855

Butler D, Maksimovic C (1999) Urban water management—challenges for the third millennium. Prog Environ Sci 1(3):213–235

Chau K (1993) Management of limited water resources in Hong Kong. Water Resour Dev 9:65–73

Cheng J, Yang X, Wei C, Zhao W (2004) Discussing water security. China. Water Resour 1:21–23

Cook C, Bakker K (2012) Water security: debating an emerging paradigm. Glob Environ Change 22(2012):94–102

Falkenmark M, Lindh G (1974) How can we cope with the water resources situation by the year 2015? Ambio 3:114–122

Falkenmark M, Widstrand C (1992) Population and water resources: a delicate balance. Population Bull 47:1–36

Gerlak AK, Mukhtarov F (2015) Ways of knowing' water: integrated water resources management and water security as complementary discourse. Int Environ Agreements Polit Law Econ 15(3):257–272

Global Water Partnership (2009) GWP in action: annual report 2009. Global Water Partnership, Stockholm, Sweden

Global Water Partnership (2000) Towards water security: a framework for action. Global Water Partnership, Stockholm, Sweden. https://www.gwp.org/globalassets/global/toolbox/references/towards-water-security.-a-framework-for-action.-executive-summary-gwp-2000.pdf. Accessed on 22 Mar 2000

Grey D, Sadoff C (2007) Sink or swim? Water security for growth and development. Water Policy 9:545–571

Halmatov B, Lautze J, Manthrithilake H, Makin I (2017) Water security for productive economies: applying an assessment framework in southern Africa. Phys Chem Earth 100(2017):258–269

Hope R, Rouse M (2013) Risks and responses to universal drinking water security. Philos Trans Royal Soc A371:20120417. https://doi.org/10.1098/rsta.2012.0417

Howard G, Bartram J (2003) Domestic water quantity, service, level and health. World Health Organization, Geneva

Hussein D (2017) Urban mobility: challenges and opportunities In Hussein D (ed) Low carbon mobility for future cities: principles and applications, The Institution of Engineering and

Hutchinson CF, Herrmann SM (2008) The future of arid lands-revisited: a review of 50 years of drylands research. In: Advances in global change research series, vol 32. Springer, Berlin

Ibisch RB, Bogardi JJ, Borchardt D (2016) Integrated water resources management: concept, research and implementation. In: Borchardt D, Bocardi JJ, Ibisch RB (eds) Integrated water resources management: concept, research and implementation, 1st edn. Springer, Switzerland

McDonald RI, Weber K, Padowski J, Florke M, Schneider C, Green PA, Gleeson T, Eckman S, Lehner B, Balk D, Boucher T, Grill G, Montgomery M (2014) Water on an urban planet: urbanization and the reach of urban water infrastructure. Glob Environ Change 27(2014):96–105

Martin-Nagle R, Howard E, Wiltse A, Duncan D (2012) Bonn 2011 conference on the water, energy and food nexus: solutions for the green economy. 16–18 Nov, Federal Ministry for the Environment, Nature Conservation and Nuclear Safety (BMU), Federal Ministry for Economic Cooperation and Development (BMZ) and OOSKA news, Bonn, Germany

Mason N, Calow R (2012) Water security: from abstract concept to meaningful metrics an initial overview of options. Working Papers 357, Overseas Development Institute, London, UK

McDonald RI, Douglas I, Grimm NB, Hale R, Revenga C, Gronwall J, Fekete B (2011) Implications of fast urban growth for freshwater provision. Ambio 40:437–447

National Intelligence Community (2012) Global water security-intelligence community assessment. Office of the Director of National Intelligence, Washington, DC

Newman P (2001) Sustainable urban water systems in rich and poor countries: steps towards a new approach. Water Sci Technol 43(4):93–100

Oxford University Water Security Network (2012) Water security, risk and society: international water security conference, 16–18 Apr, St. Hugh's College, Oxford, United Kingdom

Parikh J, Sandal G, Jindal P (2014) Vulnerability profiling of cities—a framework for climate resilient urban development. Asian Cities Climate Resilience Working Paper Series 8 Integrated Research and Action for Development, New Delhi
Parikh J, Jindal P, Sandal G (2013) Vulnerability profiles of 20 Indian cities—a framework for climate resilient urban development. Integrated Research and Action for Development, New Delhi
Postel SL, Daily GC, Ehrlich PR (1996) Human appropriation of renewable fresh water. Science 271:785–788
Rouse M (2007) The institutional governance and regulation of water services: the essential elements. International Water Association, London, UK
Swaminathan M (2001) Ecology and equity: key determinants of sustainable water security. Water Sci Technol 43:35–44
Science Daily (2017) Water scarcity. Science Daily 2017. https://www.sciencedaily.com/terms/water_scarcity.htm
United Nations (2016) Water and jobs. The United Nations world water development report. United Nations Educational, Scientific and Cultural Organization, Paris, France
United Nations (2014) World's population increasingly urban with more than half living in urban areas. United Nations, New York
United Nations (2013) UN-Water defines water security, highlights threats and mitigation steps. United Nations, New York
United Nations Department of Economics and Social Affairs (2010) World urbanization prospects—the 2009 revision. United Nations, Department of Economic and Social Affairs, New York
United Nations Development Programme (2006) Human development report—beyond scarcity: power, poverty and global water crisis. United Nations Development Programme, New York
United Nations Population Division (2011) World urbanization prospects: the 2011 revision. United Nations Population Division, New York
Wong THF, Brown RR (2009) The water sensitive city: principles for practice. Water Sci Technol 60(3):673–682
Wong THF, Brown RR (2008) Transitioning to water sensitive cities: ensuring resilience through a new hydro-social contract. In: 11th international conference on urban drainage, 31 Aug–5 Sept 5, Edinburgh, Scotland, UK
World Economic Forum (2011) Water security: the water-food-energy-climate nexus. World Economic Forum Water Initiative, Washington DC

Chapter 2
Water Insecurity in Asian Cities

Bhaswati Ray and Rajib Shaw

Abstract Countries in Asia are likely to experience severe shortage of water by 2050 under the impact of increased economic activity, rapid growth of population, and climate variability. Already the world's driest continent in terms of per capita water availability, severe droughts are also common over vast regions of Asia extending from southern Vietnam to central India. The impacts of climate change would lead an additional one billion people of Asia to water-stressed conditions. Asian cities would be the worst hit. Expanding urban population and changing lifestyle along with climate variability are making water resource management a challenge particularly in the water-stressed areas. While more than 90% of urban growth is projected to be concentrated in less developed countries, 60% of that increase will be in Asia. Among the 27 Asian cities with population exceeding one million, the Indian cities of Chennai and Delhi are identified as the worst affected in terms of per day availability of water. Mumbai holds the second position while Kolkata is in the fourth. Most of the Asian cities are, however, fast adopting new improved approaches to water management. The state of Karnataka in India has successfully experimented with urban catchment models for small and medium towns and has initiated programs to restore and rejuvenate small lakes. Rainwater harvesting is promoted in Bangalore city for managing water crisis and heavy rains. Green infrastructure, permeable pavements, and subsurface detection systems are being considered under sponge city programme. In Thailand, Bangkok's stormwater management strategy incorporates the combined collection and treatment of stormwater in lakes and storage basins. This chapter intends to focus on the problems of water insecurity in the cities of Asia and to highlight some of the best practices adopted in urban water governance for a water-secure future.

Keywords Water insecurity · Urbanization · Water management · Best practices Urban governance

---

B. Ray (✉)
Sivanath Sastri College, University of Calcutta, Kolkata, India
e-mail: bhaswati173@gmail.com

R. Shaw
Graduate School of Media and Governance, Keio University, Fujisawa, Japan

B. Ray and R. Shaw (eds.), *Urban Drought*, Disaster Risk Reduction,
https://doi.org/10.1007/978-981-10-8947-3_2

2.1 Introduction

Urban areas, consisting of cities and their hinterlands, have often been defined as static built environments with high population concentrations and extensive impervious surface area (Canigilia et al. 2017). As centers of economic productivity, they influence economic growth rates. They are also the areas of demographic and social transformations particularly in the developing countries. Urbanization in developing countries has been analyzed in a wide range of disciplines and defined as an all-round transformation or urban transition, and hence, it is never a simple process of increased and concentrated population within a limited city area. Two processes seem to operate in urban areas, that of massive population concentration along with its impact on the socioeconomic and physical infrastructure and the inherent challenges of managing the stress in high-density urban areas and that of frequent conflicts among different population groups. These stresses are apparent in the inadequate provisioning of various infrastructures caused by continually rising demand without sufficient advancement in the infrastructure management systems.

Water management systems are particularly threatened by exponential growth of population, rapid and mass urbanization, higher consumption levels, changes in land utilization pattern, climate variability, and pollution of freshwater resources. Overuse and pollution of freshwater resources would result in increased water stress, reduced food production, impaired ecosystem functions, and stalled economic growth. Increased demand for water in almost every country across the globe has, on the other hand, tripled the consumption and withdrawal of water over the last 50 years (United Nations Educational Scientific Cultural Organisation 2012). The United Nations estimates place half of the world population in extreme water-stressed areas by 2030 (United Nations Environment Programme 2007). Pollution of the existing freshwater sources and consequent reduction in the availability of freshwater would render hundreds of millions vulnerable to the adverse impacts of food and energy shortage, hunger, and poverty (World Health Organization and The United Nations Children's Fund 2014). Climate change and the associated uncertainties in the availability and supply of water would also put additional pressure on the different stakeholders (Baker 2013). It may be noted that water scarcity issues are intrinsically linked with other challenges to sustainable development as exemplified by poverty, hunger, ill-health, low level of education, gender inequality, degraded ecosystems, climate variability, and frequently occurring disasters. Hence, the role of water for sustainable development is being increasingly recognized and addressed in all its dimensions in all international programs including the 2030 Agenda for Sustainable Development.

Countries in Asia are likely to suffer from acute shortage in the availability of water by 2050 because of rising demand resulting from escalating economic and commercial activities, a growing population and reduced water availability. Already identified as the world's driest continent in terms of per capita water availability, severe droughts are common over vast regions of Asia extending from southern Vietnam to central India. The average availability of freshwater in most of the Asian countries is half the global annual average of 6380 m^3 per person (Chellaney 2013), making Asia the driest continent in terms of per capita water availability. The per capita availability of freshwater in Asia is thus 3920 m^3 and is declining at the rate of 1.6% per year (Chellaney 2007). The decline is most rapid in countries located in the semiarid parts of central, southern, southwestern, and western Asia as well as in parts of northern China. The impacts of climate change are expected to lead an additional one billion people of Asia to water-stressed conditions (Massachusetts Institute of Technology 2016). Asian cities would be the worst hit. Expanding urban population and changing lifestyle along with climate variability are already making water resource management increasingly difficult particularly in the water-stressed urban areas. Hence, the Asian cities are all set to embrace innovative water management strategies in the urban areas to ensure a water-secure future. This chapter intends to focus on the problems of water insecurity in the cities of Asia and to highlight some of the best practices adopted in urban water governance for a water-secure future.

2.2 Urbanization and Water Security Issues in Asia

Asia is already identified as the world's driest continent in terms of per capita availability of freshwater. Inadequate water security appears to be one of the greatest challenges in Asian countries. 60% of the households in Asia live without access to safe water and improved sanitation (Asian Water Development Outlook 2013). 1.7 billion people lack access to basic sanitation facilities, and 80% of wastewater is discharged in water bodies without primary treatment (Asian Water Development Outlook 2016). Among the 27 Asian cities with population over one million, Chennai and Delhi in India are ranked as the worst affected in terms of per day water availability, while Mumbai ranks second and Kolkata ranks fourth. China has more than 400 cities with water shortage, 110 of which are facing severe scarcity (Li et al. 2016). The number of megacities in China is 32. As many as 30 megacities have various water shortage problems and challenges (Xu and Hou 2005). Key aquifers are declining at the rate of 3 m/10 feet per year, increasing the expenses for drilling and the total energy usage (Boteman 2014). It has also led to an increase in arsenic contamination. It is estimated that 20 million people in China may be affected by arsenic contamination in groundwater (Boteman 2014). Testing for contaminated wells, however, is difficult in China given its huge geographical size in terms of both resource and time usage. In Asia, conditions are also inadequate for wastewater management. In Indonesia, for example, only 14% of the urban wastewater undergoes treatment. In the Philippines, the percentage of water that is treated is 10%; in India, it is even less at 9% while in Vietnam, it is as low as 4% (Asian Development Bank 2012). Water availability in South Asia is likely to worsen in the future because of increased demand from rapid population growth and industrialization, climate change accelerating the melting of glaciers. The consequent sea-level rise increases salinity in aquifers while the withdrawal of groundwater occurs at an unsustainable rate. The region is also the most vulnerable to water-related disasters including floods, droughts, hurricane, storm surges, and landslides and continues to be inadequately prepared. Climate variability further exacerbates the situation. While parts of Asia suffer from severe floods worsened by climate change in recent times, other parts of Asia are threatened by water stress or severe drought. Asia is also frequented by disasters. It is estimated that 700 million people have died in Asia in the last 10 years and another 1.7 billion has been affected by severe storms, frequent floods, and heat waves (Asian Water Development Outlook 2016) resulting in an expenditure of about $1.4 trillion (Asian Water Development Outlook 2016). In 2015, Thailand suffered severe water scarcity in 8 of its 76 provinces while 31 others were under threat. It necessitated a fund allocation worth 6.8 billion baht ($208.65 million) to come out of the severe water crisis, an increase from a less significant amount of 430 million baht ($13.19 million) in the previous year (Reuters 2015). In 2011, Thailand's capital was hit by a severe flood that left the city with $45 billion in economic damages and losses, according to the World Bank. Cities in Asia are thus threatened by urban water insecurity following rapid urban growth, declining water resources, and climate change impacts (Table 2.1; Fig. 2.1). The Asia-Pacific Water Forum anticipates that climate change impacts will impose additional threats on the already vulnerable countries in Asia and the Pacific region, challenging the very concepts of sustainable development, poverty alleviation, and improved water security. Water-related natural disasters are also common in the Asia-Pacific region. The financial burden of repeated floods has increased over time in spite of a reduction in the number of deaths with improvement in the system of weather forecasting. The Asian cities would be worst hit under the impact of rapid and mass urbanization. Probable solutions include controlled urbanization, exploring alternate water sources like rainwater harvesting, desalinization, groundwater recharge, introducing sewage treatment plants, improving water infrastructure by replacing old worn-out pipelines and extending the water supply network up to the fringe areas, improved water governance and cost recovery through proper water metering, reduced greenhouse gas emissions, green infrastructure, rooftop gardens, permeable pavements, underground water detention systems, and disaster preparedness as already seen in many of the Asian cities.

The world has witnessed unprecedented urban growth since the beginning of the twenty-first century. The world urban population, which stood at just 10% of the global total at the start of the century, had by the first years of this century reached an unprecedented 50%, when the total urban population was 3.6 billion. It is now widely accepted that at global level, the number of people living in urban areas far exceeds that in rural areas. The global urban population was seen to outnumber the rural population, for the first time in human history, in 2007 and has remained predominantly urban ever since (United Nations Department of Economic and Social Affairs 2014b). In 1950, more than 70% of the world population resided in rural areas while less than 30% were living in urban settlements (United Nations Department of Economics and Social Affair 2014b). By 2015, the urban population stood at 2.38 billion, accounting for 60.1% of the world total (United Nations Human Settlement Programme, United Nations Economic and Social Commission for Asia and the Pacific 2015). The United Nations predicts that by 2050, the urban population of

Table 2.1 Key issues and the urban water challenges in Asia

Key issues	Urban water challenges
Rapid urbanization	Increased per capita demand and reduced per capita availability
	Unsustainable groundwater withdrawal
	Increase in impervious surfaces
	Increased runoff
	Reduced infiltration and recharge of groundwater
	Declining aquifers
	Increased salinity in aquifers
	Increased pollution of water bodies
Urban governance	Inadequate coverage of water supply in terms of area and population
	Intermittent supply
	Old dilapidated network
	High leakage loss
	Poor resource mobilization and cost recovery
	Inadequate stormwater drainage
Climate change	Erratic precipitation pattern
	Heat waves
	Severe storms
	Floods
	Droughts
	Sea-level rise
	Saline intrusion
	Urban heat islands
Water-related disasters	Chemical and microbiological contamination
	Storm surges
	Floods
	Droughts
	Landslides
Environmental	Changes in urban land use
	Reduced water bodies
	Degraded natural water systems like lakes and wetlands that store water
	Urban heat island
	Altered natural water cycles
	Increased wastewater generation
	Inadequate sewage treatment

Source by authors

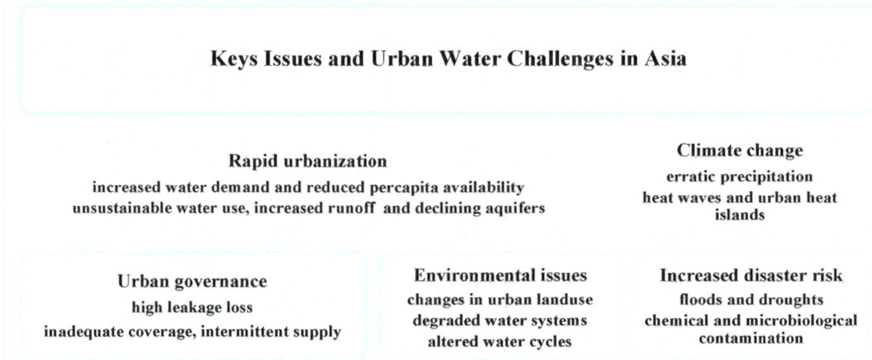

Fig. 2.1 Key issues and urban water challenges in Asian cities

the world will nearly double. The next few decades is expected to witness rapid urban growth, with an addition of 2.6 billion urban residents projected for 2050 (United Nations Population Division 2011). The global urban population would then reach 6.2 billion, an increase from the global population of 3.6 billion (United Nations Population Division 2011). The urban population would then comprise roughly 66% of the global population, roughly reversing the earlier global ratio of rural and urban populations (United Nations Department of Economics and Social Affair 2014b). 93% of the growth will be confined within the less developed countries with more than 60% in the cities of Asia. Urbanization in Asia is best described as pseudo-urbanization or mass urbanization with maximum concentration of urban population in the megacities of Asia. Of the 28 megacities in the world with a population of more than 10 million, as many as 17 are located in Asia, including the largest three cities of Tokyo, Delhi, and Shanghai.

The rapid increase in population, the fast pace of urbanization including mass urbanization, and climate change are making cities vulnerable to increasing water insecurity and water stress. As more people migrate to cities, associated lifestyle changes increase the consumption of water. With urbanization, per capita water use increases because of the increased use of water in toilet flushing, showers, and washing machines accompanied by increased consumption of meat and alcohol that require more water in washing and preparation. The two main challenges facing urban areas today include inadequate access to water and sanitation facilities and the increased frequency of water-related disasters like flood and drought with adverse consequences on economic growth, health, and well-being. The United Nations predicts an addition of 700 million people in the Asian cities between 2010 and 2025, (United Nations Department of Economic and Social Affairs 2014). Such phenomenal growth would require an investment worth US$59 billion in water supply and $71 billion for improved sanitation (Asian Development Bank 2013). Under the impact of climate variability and global warming, the availability of freshwater is further threatened. Sea-level rise and melting glaciers are likely to affect the large number of coastal cities of South Asia and Southeast Asia.

2.3 Water for Sustainable Development: A Global Concern

Appreciating the importance and contribution of water in sustainable development, at a time when global water systems are threatened by continued changes in climate, exponential growth of population and environmental degradation, efforts to address water security issues in all its dimensions are evident since the beginning of the century.

The importance of water in promoting human development and environmental sustainability has earned a prominent place in the United Nations Millennium Development Declaration at the Millennium Summit held in New York in September 2000 (Razzaque 2004). It urged global communities to adopt measures that would help combat desertification and lessen the adverse impact of floods and droughts, to formulate strategies for integrated water resources management, to prepare water efficiency plans by 2050, and to also support developing countries and countries with transition economies in their efforts to monitor water resources in terms of both quality and quantity (United Nations Department of Economics and Social Affairs 2015). The 189 member countries are committed to a global partnership for eradication of extreme poverty through eight quantitative and measurable Millennium Development Goals (MDGs) to be realized by 2015. Goal 7 on environmental sustainability aspires to reduce by 50%, the proportion of population without sustained access to safe water sources and basic sanitation by the same year. The Water for Life Decade introduced on March 22, 2005, on World Water Day, aims to realize international commitments on water and water-related matters by 2015, with special attention given to the involvement and participation of women.

In spite of the MDGs succeeding in achieving some progress in drinking water supply and sanitation, estimated 663 million people continue to use unimproved sources or surface water (United Nations Economic and Social Council 2016). Even for those who have access to water, service is often inadequate or unsustainable and the water is unsafe to drink. Water stress continues to affect more than two billion people across the globe, a figure that is projected to rise while 2.4 billion do not have access to improved sanitation facilities (United Nations Economic and Social Council 2016). The importance of water for sustainable development beyond 2015 was also recognized by the 192 Member States at United Nations Conference on Sustainable Development in 2012 (Rio+20). In the document prepared thereafter, 'The Future We Want', water was clearly identified as being at the center of the sustainable development concept and is often accountable for a large number of global challenges. Realizing the vital role of water in achieving sustainable development, the International Hydrological Programme (IHP) 2014–2021 of the United Nations Educational Scientific and Cultural Organization (UNESCO) 2014 recommends that Sustainable Development Goals must provide an assured framework to tackle water problems beyond mere access to water and sanitation as specified in the MDGs. It should focus on water security and responses to local, regional, and global challenges. It is highlighted that focus needs to be broadened to formulate an integrated approach to water resources management and include all critical water issues including water-use efficiency, water quality monitoring, wastewater and sewage management as well as managing water-related disasters. It proposed a separate goal in SDGs dedicated solely to water for an assured and holistic approach in managing global water problems and water security issues for sustainable development.

Built upon the United Nations Millennium Development Goals of 2000, the Sustainable Development Goals (SDGs) adopted in September 2015 aim to ensure sustainable development encompassing all dimensions of economic, social, and environmental development. The 17 goals and 169 indicators in the 2030 Agenda for Sustainable Development put adequate emphasis on poverty eradication, disease prevention, ecosystem conservation, infrastructure development, and sustainable cities. Goal 6 is devoted to universal access to safe water, sustainable management of water resources as well as right to basic sanitation. Goal 9 is about building resilient infrastructure through innovations while Goal 11 wants the cities to be safe, inclusive, resilient, and sustainable. Goal 15 promotes the sustainable use of ecosystems. The Asian cities, the future of the urban world, need to formulate strategies for sustainable urbanization, inclusive, and resilient cities with emphasis on innovative and resilient infrastructure, sustainable management of urban water systems, and proper use of urban ecosystems.

2.4 Water Insecurity in Selected Cities in Asia

Many cities in Asia suffer from acute water crisis under the impact of rapid urbanization, climate variability, and inadequate availability of water. Bangalore in Karnataka and one of the fastest growing cities in India faces severe water scarcity and an acute water shortage to meet the growing demand. In the absence of rivers as freshwater sources, Bangalore city has always been greatly dependent on local water bodies, lakes, and tanks for its water supply (Manasi and Umamani 2013). There were about 370 tanks in Bangalore (Manasi and Umamani 2013). However, the city has lost many of the lakes and tanks in an attempt to meet the demands for land for increased housing and business establishments and in the construction of stadiums, bus stops, and roads. The increase in built-up area has reduced the rainwater infiltration. Many of the lakes have also declined due to improper management. The present system of bringing water from River Cauvery about 300 km from the city against the gravitational forces (the city being located 500 m above the level of the river) and the requirement of electricity for pumping up water at various stages incur huge financial cost. Two lakhs borewells also supply water in Bangalore city resulting in the depletion of groundwater. The groundwater table has fallen by about 10 m between 1978 and 2003 (Manasi and Umamani 2013; Kumar 2005). Stormwater drains, and the sewage system is old and dilapidated with reduced capacities. Redesigning the drainage network even with the technical expertise and support of Bruhat Bangalore Mahanagara Palike (BBMP) is not only time-consuming but it would also involve huge costs. Improving the water services in the city of Bangalore with water conservation and revival of the tanks is urgently needed.

The coastal city of Chennai is another water-starved city in India under inadequate freshwater reserves and rain shadow location. The Chennai Metropolitan Water Supply and Sewage Board supplies water within the city and in the adjoining rural regions of Kanchipuram and Thiruvallur districts. The water demand in Chennai is 900 mL/day for the urban domestic sector (households), while the supply is only 700–730 mL/day (Sethuram 2014). Hence, there is a supply and demand gap. Although Chennai has 100% pipe connections, water supply is intermittent (Sethuram 2014).

Flooding is another problem common in most Asian cities. The increased pace of urbanization is accompanied by an increase in the impervious ground surface and the degradation of natural water systems like drainage channels and wetlands capable of storing water. Hence, it alters the normal water cycle by not allowing stormwater to infiltrate and replenish the groundwater storage. Most of the stormwater runoff is drained out of the city, making cities vulnerable to reduced availability of freshwater. Urban built-up areas also increase the chances of urban flooding. In 2010, out of the 351 cities in China, 214 cities suffered from inundation and flooding after a severe storm (Li et al. 2016). 137 cities suffered from urban floods thrice a year between 2008 and 2010, and 57 cities remained inundated for over 12 h during 2008–2010 (Li et al. 2016). During a severe storm in Beijing on July 21, 2012, flooding inundation with traffic dislocation was observed in 63 sites. It caused 79 deaths and affected another 1.19 million, while the estimated economic loss was about 11.84 billion yuan (Li et al. 2016).

Bangkok also suffers from inadequate stormwater drainage, resulting in localized flooding. The annual rainfall in Bangkok metropolitan area amounts to 1,500 mm, 88% of which is concentrated during the monsoon season extending from May to October (Phamornpol 2012). September and October are the wettest months. The high-intensity rainfall within a span of six months causes localized flooding in some low-lying areas and also along main roads because of poor drainage. Discharge from the Chao Phraya River, Bangkok Noi Canal, and Mahasawat Canal caused temporary dike damage and flooding in Bang Phlat, Bangkok Noi, and Thawi Watthana districts in 2011 following heavy rainfall (Phamornpol 2012).

Under the impact of rapid urbanization, cities in Asia also suffer from inadequate sewage treatment. In Yangon city, Myanmar, the existing sewerage system covers an area of only 4.33 km^2 and serves a population of 325,000 or 8.33% of the city population. The rest of the population is covered by the septic tank system. The sewerage system built in 1988 to improve urban sewerage and sanitation in the Yangon city area was originally meant for a population of 40,000 (Water World 2003). The only wastewater treatment plant in operation in Yangon is treating wastewater up to 12,300 m^3/day since 2005 from six townships in the downtown area. According

Table 2.2 Few examples of best practices adopted in Asian cities

Best practices	Examples from Asia
Rainwater harvesting and groundwater recharge	Bangalore, India
Rejuvenation of lakes and water bodies	Bangalore, India
Desalinization	Chennai, India
Sponge city programme	Lingang, China
Underground tunnels for stormwater drainage	Bangkok, Thailand
Sewage treatment facilities	Yangon, Myanmar
Deep tunnel sewage system	Singapore
Protection of wetlands	Kolkata, India
Water reclamation plants	Changi, Singapore

Source by authors

to the World Bank, with the estimated population growth of 2.6%, the total coverage capacity will reduce to 4.45% by 2040.

2.5 Best Practices in Urban Water Development in Asian Cities

Asian cities are seen to be affected by poor water management systems including inadequate drinking water availability and access, poor sewage treatment facilities, floods, and other disasters. It is hence imperative for the Asian cities to embrace new and innovative approaches to water management and undertake planning initiatives. Many of the cities have already adopted innovative measure to improve the water management system (Table 2.2). The Indian state of Karnataka, for instance, is experimenting with an urban catchment model for the small- and medium-sized towns and cities as well as rainwater harvesting to manage water stress. Program for the rehabilitation of lakes has also been undertaken. Desalinization plants in Chennai supply drinking water to the third largest city in India, an innovation in urban water management system. People's Republic of China is promoting green infrastructure through sponge cities with rooftop gardens and permeable pavements, to facilitate movement of water and infiltration and underground detention chambers to facilitate the collection and management of stormwater runoff. In Thailand, Bangkok's storm management is also promoting the use of detention tanks and drainage channels for improved stormwater management.

In Bangalore city, various institutions like Bangalore Water Supply and Sewerage Board, Karnataka State Council for Science and Technology, Bruhat Bangalore Mahanagara Palike (BBMP), Bangalore Development Authority (BDA) have taken major initiatives in promoting rainwater harvesting (RWH). It is expected to solve the dual problem of managing water stress as well as coping with heavy monsoon rains, thus ensuring adequate supply of water and best practices in water conservation. It is worth noting that Bangalore experiences nearly 70 days of rainfall, often heavy, round the year. The existing drainage network is designed to handle a rainfall amount of only 30 mm in one hour (Manasi and Umamani 2013). Against this backdrop, the civic authorities in Bangalore have made it mandatory for all houses covering an area of 2,400 ft^2 and above to have RWH facilities since the implementation of Bangalore Water Supply and Sewerage Board Amendment Act 2009 (Manasi and Umamani 2013). There has been a substantial increase in the number of buildings with RWH systems which currently stands at more than 25,000 households. Under the Rainwater Harvesting Regulations, 2009, it is mandatory for all existing buildings covering a ground area of 2,400 ft^2 and above and new constructions covering 1,200 ft^2 and above to have RWH facilities. The Water Supply and Sewerage Board has also initiated the installation of RWH facilities in 40 buildings under them. A theme park and resource center on RWH have been built in Jayanagar jointly by the Bangalore Water Supply and Sewerage Board and Karnataka State Council for Science and Technology (KSCTC) with 27 models of RWH on display (Manasi and Umamani 2013). KSCST is directly involved in the promotion of RWH with structures constructed in 20 prominent buildings and four exhibition plots to demonstrate the available techniques and the cost involved. The State Council for Science and Technology is also arranging training programs for various stakeholders including engineers, architects, planners, and masons. Gandhi Regional Institute for Rural Energy Development, Bangalore, functions as RWH training center under KSCST initiative along with 27 district nodal centers in 27 districts. Technical expertise is also being provided to enable the establishment of RWH facilities in 176 villages (at least one for each taluk) and 23,680 schools in rural Karnataka (Manasi and Umamani 2013). Bruhat Bangalore Mahanagara Palike (BBMP) has identified two different forms of rainwater harvesting process. The first is the storing of rainwater in surface water bodies including lakes, ponds, rivers, dams for direct use. The second type is groundwater recharge, by constructing recharge pits and reviving the defunct well and bore wells (Manasi and Umamani 2013). All apartments and properties with an area of 1 acre and above, palace grounds, parks and playgrounds, and houses above a specified plot size are required to implement RWH.

BBMP is also responsible for the rejuvenation of the dying lakes and tanks as well as their protection and fencing. The measures include desilting, wastewater diversion, improved stormwater inlet and outlet points, construction of toilets, sewage treatment plant, and wetland bunds for assimilation of sewage and urban waste. 132 lakes have been identified among which the project has been completed in 50 lakes. The Bangalore Development Authority is also responsible for the rejuvenation of lakes that fall within its jurisdiction, prevention of contamination, and to ensure groundwater recharge and improved sanitation. The Lake Development Authority (LDA) is responsible for the protection, conservation, reclamation, and restoration of both natural and man-made lakes. It is an autonomous planning body that was created for an integrated sustainable development of the lakes (Manasi and Umamani 2013) and implementation of government schemes, both state as well as central government. A database is also being prepared by them about the lakes using satellite imageries and Survey of India topographical sheets as well as through ground verification (Manasi and Umamani 2013). Karnataka Forest Department also promotes the revival of lakes and is involved in desilting and deepening of existing tanks and their fencing, constructing feeder channels and storage tanks, undertaking sewage water management, and total station surveys. Thus, the city of Bangalore is trying to revamp its old water supply system using the lakes and tanks and the rainwater harvesting system that was originally developed in the city in the 1860s.

Chennai suffers from inadequate freshwater resources and is located in the rain shadow area. Being a coastal city, it was appropriate for the city to bring in seawater for freshwater supply within the city. Currently, 20% of Chennai's population is living on desalinized water from the Minjur Seawater Desalinisation Plant and Nemmeli Seawater Desalinisation Plant with a capacity of 100 MLD each (The Times of India 2017). The Minjur Seawater Desalinisation Plant and Nemmeli Seawater Desalinisation Plant started operation in 2010 and 2013, respectively. The cost of production is also low at 6 paise a liter or Rs. 60 for 1,000 L compared to Rs. 15 for a liter in case of bottled water. The percentage of population covered is expected to increase to 60% soon with the initiation of new projects under consideration, namely the Nemmeli Seawater Desalination Plant with a capacity of 150 MLD, Perur Desalination Plant with the capacity of 400 MLD, and two projects at Kuthiraivaimozhi and Alalnthalai in Ramanathapuram district with a combined capacity of 120 MLD (Municipal Administration and Water Supply 2016). Here supply initiatives are the main focus, and hence, desalinization plants have been already initiated. The model may be adopted at least in the coastal cities of India.

To overcome the problems of urban flooding, China has evolved stormwater management practices that include, among others, low impact development and the sponge city programme (SCP). The SCP advocates sustainable development and construction practices to increase the resilience of a city to climate change impacts, reduce disaster risks of extreme rainfall, and increase the sustainability in ecological function (Wang et al. 2015). Launched on April 2, 2015 as pilot project in 16 cities of China, namely Qian'an (Hebei province), Baicheng (Jilin), Jiaxing (Zhejiang), Chizhou (Anhui), Xiamen (Fuzhou), Pingxiang (Jiangxi), Jinan (Shandong), Changde (Hunan), Nanning (Guangxi), Chongqing, Suining (Sichuan), Gui'an New District (Guizhou), Xixian New District (Shaanxi), He Bi (He Nan), Wu Han (Hu Bei) and Zheng Jiang (Jiang Su) (Li et al. 2016), the program attempts to reduce the intensity of surface runoff through the enhancement and even distribution of absorption capacities across spatial units. Conventional urban areas with a preponderance of impervious built-up surfaces in concrete roads, buildings, and walkways interfere with the normal water cycle. Sponge cities, on the other hand, adopt suitable measures like green roofs, constructed wetlands, and permeable pavements that capture stormwater for groundwater recharge and storage, thereby helping the normal water cycle to operate. Bioswales or ditches filled with native plants are common features in sponge cities to collect and filter rainwater (The Gaurdian 2015). The filtered rainwater is then allowed to infiltrate and replenish groundwater or gets collected in underground storage units. The resulting groundwater replenishment increases the water availability, reduces the chances of flooding, and enhances water supply security. Increased evaporation allows greater moderation of temperature. The sponge cities thus incorporate infiltration, detention, storage, cleaning, utilization, and discharge of stormwater for an integrated management of water resource. Green infrastructure is also more cost-effective than the gray infrastructure like concrete channels, piped drainage system, and treatment plants. The city of Lingang in Shanghai's Pudong district exemplifies a typical sponge city with rooftop gardens, wetlands, and permeable pavements to store excess runoff. The concept is similar to other water management practices like low impact development in the USA, water sensitive urban design programs of Australia, and the sustainable drainage systems of the UK.

Following frequent floods in Bangkok city efforts are directed toward effective drainage in Bangkok metropolitan area (BMA), and the capacity of the drainage system has been increased to manage rainfall intensity up to 60 mm/h (Phamornpol 2012). A total of 1682 canals have been constructed running for a length of 2,600 km and an elaborate pipe system running for 4,760 km (Phamornpol 2012). Pumping system has been improved with pumping stations installed at 409 places with a total capacity 1,638 cm (Phamornpol 2012). Twenty-five stormwater retention ponds are also located within the city with a capacity of 12.88 million cm (Phamornpol 2012). Drainage tunnels have also been built at a depth of 15–22 m below the ground by BMA to drain out the excess stormwater by pumping the water through rivers. It has been particularly helpful in areas where the drainage system is insufficient. At present, there are seven drainage tunnels 19 km in length with a pumping capacity of 155 cm (Phamornpol 2012).

The inadequate sewage treatment facilities in Yangon have initiated the process of constructing another sewage treatment plant in the eastern part of the city on a six-acre plot identified by Yangon City Development Corporation. The construction of the new treatment plant at Than Lyet Soon is expected to improve the existing sewage disposal system that continues to be in operation since 1888 (Water World 2003). The site being located at the confluence of Pazundaung Creek with River Bago enables easy disposal of treated urban sewage into natural waterways (Water World 2003). It is based on the Deep Tunnel Sewerage System (DTSS) in Singapore worth $7.3 billion construction cost. The system compromises of smaller link sewers that connect to two major sewer lines crisscrossing Singapore. Three water reclamation plants located at Kranji, Changi, and Tuas at the northern, eastern, and western margins of the city also form an integral part of the sewerage management system at Singapore. The same idea can be implemented in Yangon covering the whole region.

2.6 Observations and Conclusion

It is already evident that world population is getting overwhelmingly concentrated in urban areas. More than half of the world's population lives in towns and cities and is expected to increase to 60% by 2050. Though urbanization is a global phenomenon, the developing countries of Asia and Africa are witnessing the fastest pace of urbanization with adverse environmental consequences and increasing infrastructural inadequacies. Water management systems are particularly threatened by exponential growth of population, rapid and mass urbanization, higher consumption levels, land use changes, increase in built-up areas, and climate variability. Reduced infiltration and increased pollution level in freshwater threaten the sustainability of urban water systems. The resultant decrease in the availability of freshwater resources is expected to further reduce food production, hinder proper functioning of ecosystems, and limit economic growth. The region is also vulnerable to disasters resulting from floods, droughts, hurricanes, storm surges, and landslides and continues to be inadequately prepared. The Asian cities are particularly threatened by rapid increase of population, changing lifestyles, and increased water use, old, dilapidated infrastructure, poor urban governance, and climate variability, making the cities vulnerable to increasing water insecurity and water stress. Effective and efficient management of water resources is hence essential for sustainable urban development. Lack of rainfall is only a part of the water scarcity; it is more an issue of water governance as exemplified by Aswan in Egypt, which in spite of being one of the driest cities is not water stressed because of proper water management. While water resources are fast depleting with over-exploitation, pollution, and climate change, water management strategies are already being implemented to ensure sustainable urban development. The innovative best practices in vogue in many of the Asian cities encompass all aspects of urban water management in compliance with the Sustainable Development Goals and include adequate drinking water supply in terms of accessibility and affordability, green infrastructure, stormwater drainage, wastewater treatment, and reuse. Examples are already aplenty in Asian cities. However, more innovations are needed to make cities resilient and urbanization sustainable with financial and technical support of the developed world in infrastructure development. Community-based and community-managed initiatives, like the community-based rainwater harvesting projects, may be undertaken wherever possible as such initiatives would be suitable for local conditions and are expected to be environment friendly. Solutions that are generated locally involving stakeholder participation are expected to be more flexible and adaptive to local needs. They would thus strengthen the capacity and resilience of both the local governments and local communities. With a significant increase in the slum population, urban water management policies need to have a strong social impact since access to safe water is often related to poverty. The poor and the underprivileged sections are often excluded from the urban water systems. Institutional mapping to identify the water institutions and their role in urban water management and for promoting public–private partnership is recommended. It is expected to ease the resource crunch in the cities of the less developed Asian countries.

Community-based urban water management alone would not be able to ensure sustainable urban water systems. The community-based initiatives must be supported by techno-centric approaches. Translating scientific knowledge and improved technology like green infrastructure, desalinization techniques for increased water availability, permeable pavements must be made available to civic bodies and local communities. A new Web-based information sharing platform could be considered as a communications tool for data sharing and for strategic planning in the urban water sector. Technocrats need to be available to facilitate learning and scaling up of improved water management practices and innovations in the urban water systems. Building stakeholder platforms for idea sharing, co-developing, and monitoring strategic plans with the different stakeholders including the technocrats would bridge the gap between local communities and scientific knowledge for effective resilience and sustainable water management. Enhanced stakeholder understanding, improved urban water assessment, increased access to modern water management techniques and disaster warning systems are needed to reduce vulnerabilities and make cities resilient to urban water insecurity.

References

Asian Development Bank (2016) Asian water development outlook 2016: strengthening water security in Asia and the Pacific. Asian Development Bank, Metro Manila, Philippines

Asian Development Bank (2013) Asian water development outlook: measuring water security in Asia and the Pacific. Asian Development Bank, Metro Manila, Philippines

Asian Development Bank (2012) Wastewater management and sanitation in Asia. 19 November. http://www.adb.org/features/wastewater-management-and-sanitation-numbers. Accessed 30 May 2016

Baker C (2013) Water and international security: talk for CSCAP 2nd meeting for study group on water. Siem Reap, Cambodia

Boteman J (2014) China's looming water crisis. Ecologist Setting the Environmental Agenda, 25 Feb 2014

Canigilia B, Vallee M, Frank B (2017) Resilience, environmental justice and the city. Routledge, United Kingdom, Abingdon

Chellaney B (2013) Water: Asia's new battleground. Georgetown University Press, Washington D.C.

Chellaney B (2007) Climate change and security in Southern Asia: understanding the national security implications. R U Serv Inst J 152(2):59–68

Kumar AR (2005) Amruthavarshini—a guide for rainwater harvesting. Karnataka State Council for Science and Technology, Bangalore

Li X, Li J, Fang X, Gong Y, Wang, W (2016) Case studies of the sponge city program in China. https://doi.org/10.1061/9780784479858.031

Manasi S, Umamani KS (2013) Water conservation in urban areas: a case study of rain water harvesting initiative in Bangalore city. In: Nautiyal S, Rao KS, Kaechele H, Raju KV, Schaldach R (eds) Knowledge systems of societies for adaptation and mitigation of impacts of climate change, 1st edn. Sringer, Berlin

Massachusetts Institute of Technology (2016) Severe water stress likely in Asia by 2050-Water problems in Asia's future? https://www.sciencedaily.com/releases/2016/03/160330152053.htm. Accessed 28 Sept 2016

Municipal Administration and Water Supply Department (2016) Policy Note 2015–16 of Government of Tamil Nadu, Chennai

Phamornpol K (2012) Flood mitigation and management in Bangkok Metropolitan Area. Department of Drainage and Sewerage, Bangkok Metropolitan Administration, Bangkok

Razzaque J (2004) Trading water: the human factor. Rev Eur Commun Int Environ Law 13(1):15–26

Reuters (2015) Thai crops to suffer worst drought in 15 years. Environment, 5 Feb 2015

Sethuram S (2014) Case study: water management issues in Chennai, India. Univ J Environ Res Technol 4(6):299–306

The Guardian (2015) China's sponge cities: soaking up water to reduce flood risks, 1 Oct 2015

The Times of India (2017) How Chennai depends on the sea for its water supply, 7 June 2017
United Nations Department of Economics and Social Affairs (2015) Sustainable development knowledge platform-water and sanitation. https://sustainabledevelopment.un.org/topics/waterandsanitation
United Nations Department of Economic and Social Affairs (2014) International decade for action: water for life 2005–2015. United Nations Department of Economic and Social Affairs, New York
United Nations Department of Economics and Social Affair (2014) World urbanization prospects the 2014 revision. United Nations Department of Economic and Social Affairs, New York
United Nations Population Division (2011) World Urbanization Prospects: The 2011 Revision. United Nations Population Division, New York
United Nations Educational Scientific Cultural Organisation (2012) World Water Assessment Programme-The United Nations world water development report 4: managing water under uncertainty and risk. United Nations Educational Scientific Cultural Organisation, Paris, France
United Nations Economic and Social Council (2016) Progress towards the sustainable development goals: 2016 session, 24–27 July 2016 United Nations Economic and Social Council, New York
United Nations Environment Programme (2007) State-and-trends of the environment: 1987-2007, Global environmental outlook 4. United Nations Environment Programme, New York
United Nations Human Settlement Programme, United Nations Economic and Social Commission for Asia and the Pacific (2015) The State of Asian and Pacific Cities 2015 – Urban transformations. Shifting from quantity to quality, United Nations, New York
Wang W, Li J, Wang E, Zhang L, Cao Y, Xu H (2015) Introduction of sponge city construction. Construction Science and Technology 01:19–21
Water World (2003) Myanmar plans new sewage treatment plant. Water World 18(2). https://www.waterworld.com/articles/wwi/print/volume-18/issue-2/regulars/worldwide-news/myanmar-plans-new-sewage-treatment-plant.html
World Health Organisation and United Nations Children's Fund (2014) Progress on water and Sanitation-2014 Updates. Switzerland, Geneva
Xu H, Hou L (2005) Zhai: Speed up water saving work and improve water resource utilization efficiency. Finance and economics. http://finance.people.com.cn/GB/1037/3496676.html. Accessed 28 Jan 2016

Chapter 3
Climate Change: Implication on Urban Drought

Bhaswati Ray and Rajib Shaw

Abstract There is clear evidence that the earth is becoming warmer. With the continued emission of greenhouse gases, the increase in global average temperature by 2100 is expected to range between 1.1 and 6.6 °C higher than the 1990 levels. The evidences include rise in mean sea level during the last century, a significant increase in global temperature, shrinking ice sheets, glacial retreat and an increased frequency of extreme events. Many regions have experienced changes in the pattern and intensity of rainfall, resulting in more frequent and intense floods and droughts, as well as severe heat waves. The recent El Nino activities were responsible for widespread droughts in Asia, and many cities have suffered instances of water crisis and massive flooding since 2000. Urban areas, characterized by increased concentration of population, economic activities and infrastructural facilities are likely to bear the most severe impacts of climate change. Matters are expected to be worse for the urban areas in the less developed countries of Asia, experiencing the fastest rate of urbanization and the growth of megacities. According to recent estimates, two-thirds of the megacities of the world are located in regions most vulnerable to the impacts of climate change that would affect water systems and urban water utilities. Cities are also responsible for the creation of micro-climates with effects on temperature and wind conditions. The projected shifts in climate variables for the twenty-first century along with the observed impacts of extreme weather events are likely to make adaptations to climate change extremely difficult and a major challenge for urban areas over the coming decades. International Conventions are thus emphasizing on the adaptation of climate change responses. This chapter aims to look into the recent trend of climate change and its implications on urban areas particularly in the less developed countries of Asia. The chapter would explore the linkage between urban areas and climate change and assess the role of various international conventions and local adaptations in finding solutions.

B. Ray (✉)
Sivanath Sastri College, University of Calcutta, Kolkata, India
e-mail: bhaswati173@gmail.com

R. Shaw
Graduate School of Media and Governance, Keio University, Fujisawa, Japan

B. Ray and R. Shaw (eds.), *Urban Drought*, Disaster Risk Reduction,
https://doi.org/10.1007/978-981-10-8947-3_3

Keywords Water stress · Vulnerability · Climate change · Urbanization
Heat island

3.1 Introduction

Climate change refers to any change in the state of climatic conditions that are easily identified by changes in the mean values and variability of its properties that persists for an extended period, covering decades (Intergovernmental Panel on Climate Change 2013). The term climate change has been explained by United Nations Framework Convention on Climate Change (1992) as changes in the climatic conditions attributed directly or indirectly to human activities that alters the composition of the global atmosphere in addition to natural climate variability observed over comparable time periods. Although climate change did occur in the past, in response to natural drivers, recent climate change has been more rapid than previous episodes. It has been unambiguously linked to increasing concentrations of greenhouse gases in earth's lower atmosphere and the short time span within which these gases have accumulated (Intergovernmental Panel on Climate Change 2013). The major causes of this rapid uploading of the greenhouse gases into the atmosphere include the use of fossil fuels since the beginning of the industrial revolution, incorrect agricultural practices, increases in livestock grazing and deforestation (Gilbert 2012). According to the Intergovernmental Panel on Climate Change (IPCC), warming of the climate system is evident from observations of global increase in atmospheric and oceanic temperatures, widespread melting of snow and ice and rise in average sea level (Intergovernmental Panel on Climate Change 2013). The international conventions and institutions like the United Nations Framework Convention on Climate Change, the Intergovernmental Panel on Climate Change and the Kyoto Protocol have served to establish climate change as a legitimate global concern (Lankao 2008; Bulkeley and Betsill 2003).

It is already evident that the earth is becoming warmer. Since 1990 global average temperature has shown an increase of about 0.74 °C. IPCC 1996 opines that global temperature increase by 2100 would vary over a wide range, with an average global mean temperature expected to be 2 °C higher in 2100 than in 1990 (uncertainty range 1–3.5 °C), at the baseline scenario for global emissions. One of the climate models given by Hadley Centre for Climate Prediction and Research, United Kingdom and used in the IPCC 1996 assessment predicted an increase of 3 °C by 2100. Global annual surface temperature increase has been at the rate of 0.13 °C per decade during 1950–2000. This rate is double the rate observed during 1900–1950. Besides, 11 out of 12 years during 1995–2006 are identified as the warmest years in the instrumental records of global surface temperature since 1850 (Intergovernmental Panel on Climate Change 2013). Climate models further indicate that average surface temperature will increase globally at a rate of 0.2 °C per decade for the period 2000–2020 and will increase by at least the same amount per decade during 2020–2080. Based on a number of emission scenarios, the IPCC projected an average increase in surface temperatures of 1.7–4 °C by 2100 compared to 1980 through 1999 levels, within a range of 1.1–2.2 °C to account for the uncertainties in climate science (Intergovernmental Panel on Climate Change 2013). Approximately half of this warming is the result of past greenhouse gas emissions and will occur even if the emissions were halted at 2000 levels. This trend is expected to accelerate as humans add more greenhouse gases to the atmosphere. Projections by IPCC in 2007 revealed that greenhouse gas emissions would rise by 25–30% by 2030, which would increase the temperature of the earth by 3 °C. By 2100, the additional rise in global average temperature is projected to be in the range of 1.1–6.6 °C above 1990 levels (Association of Metropolitan Water Agencies 2007).

Global climate change has resulted in an increase in the occurrence and intensity of climate-related disasters like storms, floods, droughts, extreme temperature events and forest fires, aggravated by growth of population and consequent greenhouse gas emission, land-use changes, industrialization and urbanization.

While rapid urbanization is often a cause of greenhouse gas emissions, climate change and its impacts are bound to present major challenges for urban areas and their growing population, especially in the less developed countries of Asia. These challenges are predicted to extend beyond the apparent risks posed by climate change such as rise in sea level and increased frequency of extreme weather events. The greatest challenge for the cities would be to provide basic services to the urban residents including water supply. Climate change would also disrupt the proper functioning of ecosystems, interrupt local economies and livelihoods. Two-thirds of the world's megacities are located in regions vulnerable to climate change impacts that are bound to affect urban water utilities. Climate change can also disturb the socio-economic fabric of cities by reinforcing existing inequalities and enhancing poverty.

However, the cities could also be the source of reduced emissions of greenhouse gases and improved adaptations to climate change through policy initiatives. International Conventions are emphasizing on the adaptation of climate change responses. Apart from highlighting the trends of climate change, this chapter would take into account the implications of climate change on urban areas and urban water systems for the less developed countries of Asia. The chapter would also explore the linkage between urban areas, climate change and water scarcity and assess the role of various international conventions and the major challenges in mitigating the adverse impacts.

Urban water scarcity or urban drought implies a lack of enough potable unpolluted fresh water because of growing freshwater use and increased water demand in the domestic and industrial sectors, for power generation and for the proper functioning of ecosystems. All events that hamper any of these activities would be considered as triggering water scarcity. Hence, urban drought or water scarcity may be caused by global warming resulting in physical decline in water resource availability and socio-economic scarcity brought about by inadequate access to water resources and inequitable power distribution. Apart from rising demand for water due to rapid increase of population and greater water use in different sectors of the economy, environmental consequences of climate change like increased frequency of extreme events like drought and flood due to erratic precipitation pattern, increased chances of water contamination during such events, sea-level rise, saltwater intrusions, coastal erosion, overexploitation of groundwater, unplanned urban growth in vulnerable areas, impaired ecosystem services and altered hydrological cycle are believed to be the cause of urban drought, impacting availability of fresh water either directly or indirectly. Hence all these aspects of global warming have been taken into account while analysing urban drought.

3.2 Linkage Between Urban Areas and Climate Change

The linkage between urbanization and global climate change are complex (Sánchez-Rodríguez et al. 2005). The global environmental change brought about by greenhouse-gas-induced warming and deforestation is triggered by rapid population growth. The greenhouse effect is actually a natural phenomenon in which water vapour and carbon dioxide present in the atmosphere trap infrared radiation. However, the world has witnessed a phenomenal increase in the concentration of greenhouse gases, carbon, methane, nitrous oxides and halogenated compounds like chlorofluorocarbons due to anthropogenic interference causing an enhanced greenhouse effect and global warming. Urban areas concentrate people, economic activities and built environment over a relatively small area and use a disproportionate share of resources making the cities key drivers of global environmental change. Urban areas also contribute much of the anthropogenic carbon dioxide emissions from the burning of fossil fuels for heating and cooling, in industrial units and in the transport sector. Cities thus concentrate the sources of greenhouse gases. Svirejeva-Hopkins et al. (2004) suggest that more than 90% of anthropogenic carbon emissions are generated in cities (Grimmond 2007). In addition to that, clearing of forest land for the growth and expansion of cities and for development of road network as well as the increased demand for goods and resources by urban dwellers since historic times has resulted in massive regional land-use change, deforestation and reduction in the area of natural sinks like wetlands. Such changes have resulted in the reduction of the magnitude of global carbon sinks. One of the major effects of such development is urban warming; globally, cities are almost always warmer than the surrounding rural area (Grimmond 2007; Oke 1973). On an average, the temperature of urban areas is expected to be 1–3 °C warmer than the surroundings and under-worsened meteorological conditions like calm, cloudless nights in winter and air temperatures can be more than 10 °C warmer than surrounding rural environments (Grimmond 2007; Oke 1981). Urban areas thus act as urban heat islands.

The morphology of the city, characterized by increasing height and density of buildings, increases the absorption of short-wave solar radiation during the daytime while decreasing the long-wave terrestrial radiation loss after sunset. It also results in reduced wind speed and increased surface heat storage. Urban areas also alter the hydrological cycle because of increased surface run-off and reduced infiltration. It results in groundwater depletion and increased water stress. In combination with anthropogenic emissions of heat, carbon dioxide and greenhouse gases, these result in distinct urban climates (Grimmond 2007; Landsberg 1981; Oke 1997).

The construction materials used also have profound effect on increased heat absorption. Given that the rate of radiation and cooling is influenced by the sky view factor, higher building density, narrower streets with lesser sky view factors result in reduced long-wave radiation loss and remain warmer than more open areas with higher sky view factor (Grimmond 2007). Cities have higher building densities in the centre, so warmer temperatures tend to be found in these locations (Grimmond 2007). Most of the urban areas in the less developed countries of South and Southeast Asia are experiencing rapid and mass urbanization associated with increased population and building density, increased number of high-rise buildings and reduced green spaces. All these changes alter the local thermal environment and micro-climate in cities. Warmer conditions in cities and extreme weather conditions increase the use of air conditioners that have significant adverse impact on the local external climate. Significant increase in the use of air-conditioning systems in the countries of North America, Europe and Asia has already been documented.

3.3 Implications of Climate Change on Urban Areas

Global warming indicates an increase in the global average temperature caused mostly by a greater concentration of carbon dioxide and other greenhouse gases in the atmosphere due to anthropogenic interference allowing more heat to be trapped and resulting in global warming. According to the Intergovernmental Panel on Climate Change (2007), emissions of greenhouse gases and aerosols continue to modify the atmosphere such that the effects on climate are evident. Estimates indicate that the presence of aerosols has compensated for nearly 50% of the total warming till date, caused by the greenhouse gases (Intergovernmental Panel on Climate Change 1996). They, however, have a shorter lifespan in the atmosphere and cannot get distributed far. Hence the effects are short-lived and mostly regional. Global concentrations of the greenhouse gases are projected to increase further from 1990 levels by 2050, CO_2 by 45% (from 354 to 512 ppmv), CH_4 by 80% (from 1.60 to 2.84 ppmv) and N_2O by 22% (from 310 to 377 ppbv) (Intergovernmental Panel on Climate Change 1996). The source of greenhouse gases can be traced to land-use changes, rapid urbanization, greater industrialization, development of transportation facilities, increased use of fossil fuels, air conditioning and refrigeration that distort the carbon cycle. It results in higher surface temperatures, thermal expansion of ocean water, melting of glaciers, rising sea level, changes in the hydrological cycle, frequent heavy rainfall, intense tropical storms and El Nino conditions.

Thermal expansion of ocean waters and melting of the ice sheets due to global warming are responsible for global sea-level rise. The Intergovernmental Panel on Climate Change (IPCC) predicts that sea-level rise will continue within a range of 0.18–0.59 m above 1990 levels by the end of the twenty-first century (Intergovernmental Panel on Climate Change 2007). The rise in sea level would be accompanied by a decline in the total area under Arctic Ocean ice sheets by an average of 2.7% per decade since the 1980s (Intergovernmental Panel on Climate Change 2007). It is already evident that the maximum area of seasonally frozen ground has reduced by nearly 7% in the northern hemisphere compared to 1900 while the reduction in spring season reaches nearly 15% (Intergovernmental Panel on Climate Change 2007). Satellite surveys over Western Antarctica indicate that melting of Antarctica glacier is consistent with a 0.2 mm rise in sea level per year (Thomas et al. 2004). The survey further indicates that melting of the Antarctic glaciers has accelerated during the early 2000s compared to the late 1990s (Thomas et al. 2004). Estimates of sea-level rise due to ice loss from Antarctica and Greenland from 1993 to 2003 are about 0.21 mm per year for both; but loss of these sheets in the future, even partially, could greatly alter the projections of sea-level rise (Intergovernmental Panel on Climate Change 2007). It would result in increased urban flooding, prolonged inundation, rising coastal water tables and a higher rate of coastal erosion, increased salinity in coastal aquifers as well as changes in the functions of coastal ecosystems, wetlands, mangroves, swamps and coral reefs. These mangroves and swamps form natural barriers to the impacts of sea-level rise for coastal cities. Hence their degradation due to global warming would make cities more vulnerable to the impacts of sea-level rise. It is already being predicted that sea-level rise and its associated impacts will, by the 2080s, affect five times as many coastal population as they did in 1990 (Nicholls et al. 1999). A 1–2 °C increase in temperature will severely affect the Egyptian cities in the Nile Delta, including Port Said, Alexandria, Rosetta and Damietta (Bigio 2009). Inundation of highways and erosion of road bases and bridge supports are common in coastal cities affected by intense rain, inundation and sea-level rise, thereby increasing the maintenance cost in the transport sector. In India, for example, landslides caused by heavy rain resulted in 14 days of disrupted train service and an estimated loss of US$2.2 million (Shukla et al. 2003) during July 2000. 20% of major repairs in the Konkan railways between Mumbai and Mangalore are due to climatic factors.

Storms would also increase in frequency and intensity. Although the relationship between temperature and formation of storm systems is not completely understood, increased temperature does correlate with increased occurrence of tropical cyclones and extra-tropical storms (United Nations Human Settlements Programme 2011; Intergovernmental Panel on Climate Change 2007). Rising sea surface temperatures change the earth's water cycle, disrupting ocean currents and altering precipitation patterns, which may lead, in part, to the increases in storm intensity observed over the past several decades (United Nations Human Settlement Programme 2011; Donnelly and Woodruff 2007). Storms and consequent flooding due to inadequate stormwater drainage would result in water stress, increased risk of water contamination, health hazards, injuries and deaths. Flooding introduces contaminants into water supply systems leading to increased incidence of diarrhoeal illnesses in urban areas.

By altering the hydrological cycle, urban areas will contribute to an overall increase in rainfall intensity. Precipitation increases have been documented in the tropical latitudes while precipitation has decreased in the mid-latitude regions. More frequent occurrences of heavy precipitation would have far-reaching implications in the urban areas through flooding and landslides. Recent flooding in Manila (the Philippines) and surrounding areas, following typhoon Ketsana, affected an estimated 1.9 million people and killed at least 41 (United Nations Human Settlement

Programme 2011; Ruth and Ibarrarán 2009). Floods following heavy rains in the cities and towns located in north-eastern Brazil caused large-scale damage to life and property in 2010. Many of the world's largest cities are already located in areas vulnerable to climate change impacts, storm surges and flooding in the low-lying coastal areas that concentrate 13% of the world's urban population while occupying only 2% of the total land area of the world. Industries and facilities located in the coastal cities are also prone to damage and relocation of such activities incurs substantial cost.

A ranking of cities based on vulnerability to flooding found that the top ten cities in terms of exposed population were Mumbai (India), Guangzhou (China), Shanghai (China), Miami (US), Ho Chi Minh City (Vietnam), Kolkata (India), Greater New York (US), Osaka-Kobe (Japan), Alexandria (Egypt) and New Orleans (US) (United Nations Human Settlement Programme 2011; Nicholls et al. 2008). It is predicted that by 2070 almost all cities in the top ten exposure risk category will be located in developing countries (particularly in China, India and Thailand) because of the rapid population growth occurring in these areas (United Nations Human Settlement Programme 2011). On a national scale, the study predicts that the concentration of future exposure to sea-level rise and storm surges will be in the rapidly growing cities of developing countries in Asia, Africa and, to a lesser extent, Latin America (United Nations Human Settlement Programme 2011). It is anticipated that the majority of high-exposure coastal land area (90%) will be located in only eight countries: China, US, India, Japan, The Netherlands, Thailand, Vietnam and Bangladesh (United Nations Human Settlement Programme 2011). Less developed structural defence and negligence to follow building codes increases the vulnerability of cities particularly in high-risk coastal areas. Structural defences across cities in Japan, for example, would result in negligible cyclone damages as compared to cities in the Philippines, though cities in Japan are at a higher risk compared to the cities in the Philippines. Unplanned population growth, development and expansion into hazardous areas further increase the vulnerability of cities to such hazards. In cities such as Dhaka (Bangladesh), residents of informal settlements are found to inhabit slopes surrounding the urban core, putting themselves at increased risk from flash floods and landslides (United Nations Human Settlement Programme 2011).

Consequent to climate change, coastal erosion and saltwater intrusions is likely to affect cities across the world particularly in the mega deltas of South, East and Southeast Asia (Intergovernmental Panel on Climate Change 2007) and may even render some areas uninhabitable. The city of Kochi in India is located at an altitude of 2 m above mean sea level and compromises a network of rivers and canals (United Nations Human Settlement Programme 2011). Saltwater intrusion into these rivers particularly during dry pre-monsoon months induces salinity in the freshwater system resulting in drinking water scarcity and affecting agriculture and economy. Coastal erosion is also often a major problem for cities with a coastal location particularly those that depend on tourism as their source of sustenance and livelihood. Mombasa (Kenya), for instance, could lose approximately 17% of its land from a 0.3-m rise in sea level, causing the loss of hotels, cultural monuments and beaches that draw tourists (United Nations Human Settlement Programme 2011; Awuor et al. 2008).

Extreme heat events are also predicted to increase in intensity and frequency, due to increasing concentrations of atmospheric greenhouse gases. With the urban heat island effect, the impact would be more in the urban areas. Cities retain more heat and hence have a higher air and surface temperature. For the average developed country, the annual mean air temperature of a city with a population of 1 million can be 1–3 °C higher than the surrounding areas (Oke 1997a). Compact cities with higher density of population and built-up area, common in the developing countries of South and Southeast Asia, would have stronger heat island effect and increased water demand. Cities would lead to greater greenhouse gases emissions as they struggle to cope with the increasing temperature through artificial cooling systems. Greater use of air-conditioning systems because of rising temperatures has worsened the urban heat island effect and has resulted in increased demand for cooling systems (Hunt and Watkiss 2007) leaving room for further increase in global warming. Extended periods of heat waves increase the incidence of heat stress and have adverse health impacts, increased morbidity and mortality. Urban residents are in fact at a much higher risk of heat stress because of the urban heat island effect. Given their low adaptive capacities, the urban poor are particularly vulnerable to the adverse impacts of extreme heat events. Urban areas suffering from extreme drought conditions are expected to further increase in future as a result of changes in precipitation patterns (Bates et al. 2008). Currently, as much as 1% of all land area is considered as being under extreme drought conditions (United Nations Human Settlement Programme 2011; Bates et al. 2008). By 2100, this could increase to as much as 30% impacting a greater number of cities.

Stress in the urban water system results in excessive withdrawal of groundwater making urban areas vulnerable to subsidence. The rate of subsidence can be as rapid as 1m per decade, resulting in significant damage to pipelines, building foundations and other infrastructure (United Nations human Settlement Programme 2011; Klein et al. 2003). Subsidence has been noted in several megacities throughout the world, including Tokyo (Japan), Dhaka (Bangladesh), Jakarta (Indonesia), Kolkata (India), Metro Manila (the Philippines), Shanghai (China), Los Angeles (US), Osaka (Japan) and Bangkok (Thailand) (United Nations human Settlement Programme 2011; Klein et al. 2003). During the late 1980s, Tianjin (China) experienced as much as 11 cm of subsidence per year (United Nations human Settlement Programme 2011; Klein et al. 2003).

Extreme events and global warming would also affect the tourism industry particularly the winter sports-based tourism activities because of global-warming-induced declines in natural snowfall and fewer days of snow cover. Coastal cities like Rio de Janeiro (Brazil) would also have reduced tourist activities with its beaches becoming increasingly vulnerable to sea-level rise and increased coastal erosion. Frequent coral bleaching would not only damage coral reefs but would also affect the tourism industry. A 2 °C rise in temperatures might even result in irreparable annual bleaching of coral reefs along many coastal areas. Destruction of wetlands and coral reefs would in turn impair ecosystem functions with immense environmental implications including reduced oxygen production, loss of carbon sinks, hampered natural filtering of toxins and increased vulnerability of the coastal cities to flooding and storm surges. Extreme climate events may also destroy livelihood assets with adverse impact on economic development, unsustainable livelihoods and increased poverty. The economic loss due to climate change impacts is estimated at 19% of gross domestic product by the UNFCCC (2007). Sea-level rise, extreme events, flood and heat waves would affect large areas and result in enormous loss to property and assets making the insurance industry in turn vulnerable to climate change impacts.

3.4 Climate Change and Urban Water Issues

The complex relationship between water resources, climate change impacts and urban development (Fig. 3.1) is getting increasingly recognized globally with urban areas becoming home to an increased proportion of global population. The urban water sector is recognized as one of the most vulnerable areas in the rapidly urbanizing world and the impact of climate change may entail additional cause–effect relationship. While there is much evidence documenting climate change, impacts of climate change are still uncertain. Climate change would result in drier summers and hence to increased water demands. These water shortages are themselves being further exacerbated by increases in extreme climate events, greater monsoon variability, endemic drought, flooding and resource conflict (Gosain et al. 2006; Mall et al. 2006).

Extreme heat events would increase the demand for water when water availability would decline, making the population vulnerable to water contamination and water stress. Climate change would also impact the availability of water in the long run by altering the hydrological cycle. It would also result in groundwater depletion and increased water stress. Changes in the climate variables including temperature, rainfall, surface run-off, recharge of groundwater and surface water bodies and vegetation cover would also produce major baseline changes in the existing urban water systems, frequency and intensity of urban drought and flooding, and in future-predicted renewable water supplies. Increased frequency and intensity of rainfall is one of the most immediate effects of global warming that is already apparent in streamflow records from the last several decades (Tolkou and Zouboulis 2016) and may result in severe flooding and additional water pollution. Urban water suppliers are also faced with the challenge of increased influent, side effects of which are manifest in health threats. Such threats to safe water programmes caused by climate change would result in dire consequences for urban water utilities and capital resources across Asia. Climate change impacts are accompanied by lifestyle changes that demand increasing water use, like the introduction of washing machines and dishwashers in urban households, thereby complicating the water issues in urban areas.

The existing institutional relationships of utilities with other agencies responsible for water management would come under stress due to climate change impacts. The environmental and social problems associated with water scarcity point to a crisis in urban water resource management, and one that threatens the security and livelihood of the population and the environment over the coming decades (Mukherjee et al. 2009). While empirical evidence is limited, some scholars suggest that climate change may force the pace of rural-to-urban migration to increase over the next few decades, due in part to an ongoing water shortage in the agrarian communities (Baier 2011; Sainath 2002), contributing to increased urban water scarcity. Behavioural changes in urban areas because of climate change such as increased demand for heating and cooling, also impact urban water use.

Urban water utilities are already facing an increasing need to improve the management of water resources and associated infrastructure. Diversifying sources of water supply will become increasingly important whether through the construction of new storage facilities, the appropriate and sustainable extraction of groundwater, water trading or conservation, the use of recycled or desalinated water (Danilenko et al. 2010a, b).

3.5 Role of Various International Conventions and Local Adaptations

Various international conventions have addressed the issue of climate change though not necessarily restricted to urban areas. The United Nations Conference on the Human Environment (Stockholm Conference) held in Stockholm, Sweden from 5 to

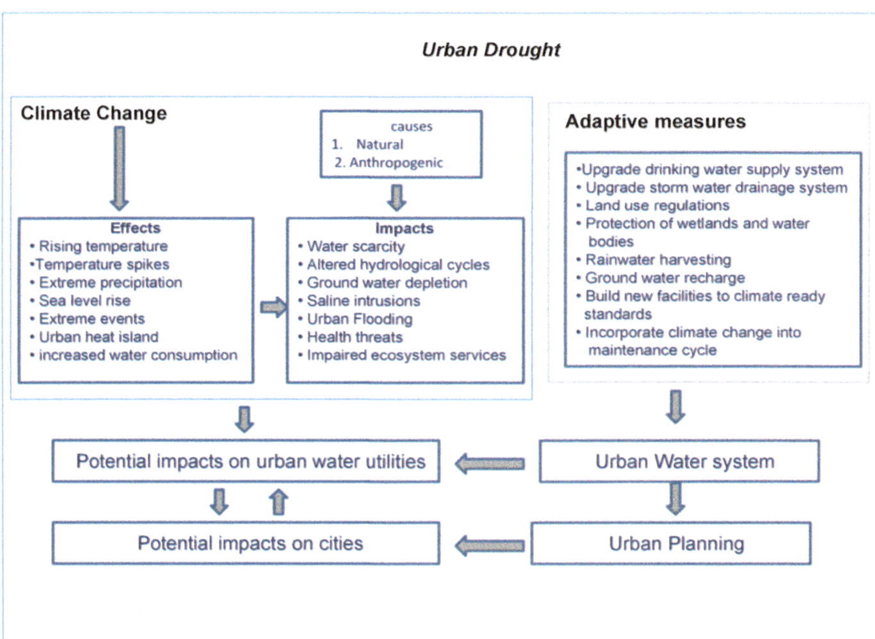

Fig. 3.1 Climate change and urban drought

16 June 1972 was the first major global conference to focus on environmental issues. The major areas of concern included urbanization and chemical pollution. The decision to establish a United Nations Environment Programme was also adopted to enable nations to work together on environmental issues. By the mid-1980s, scientists and researchers were convinced that global warming beyond natural climate variability was occurring and that irrational human behaviour and increased anthropogenic emissions of greenhouse gases were largely responsible for such warming. In the year 1985, the International Conference on Assessment of the Role of Carbon Dioxide and of Other Greenhouse Gases in Climate Variations and Associated Impacts was held in Villach, Austria (Law Teacher 2013). Scientists imposed upon the politicians to collaborate on the formulation of effective policies to mitigate the adverse impacts of human-induced climate change. By the late 1980s, after a series of incidents and discoveries, including the leakage of poisonous gases from a chemical plant at Bhopal, India, the explosion and radioactive release from Chernobyl, Ukraine and the discovery of the ozone hole in the Antarctic ozone enhanced the urgency for global action (Law Teacher 2013). The conference was closely trailed by the World Commission on Environment and Development held in Nairobi in June 1987.

The Commission's document or the Brundtland Report was an overview of the state of environmental degradation, economic inequality and poverty. The conference also addressed the future challenges with the concept of global sustainable development as the major outcome. The recommendations for sustainable development covered, among others, issues related to population and human resource, food production, ecosystem preservation, sustainable energy and consumption patterns, industrial production and rapid urbanization. The Brundtland Report paved the way for the Earth Summit to be held in Rio de Janeiro in June 1992. The summit was a major landmark with agreements on climate change, forests and biodiversity, the United Nations Convention on Biological Diversity, the United Nations Framework Convention on Climate Change (UNFCCC), United Nations Convention to Combat Desertification (UNCCD), Principles of Forest Management. Agenda 21 at the Rio Declaration on Environment and Development required participating countries to prepare the blueprint of a national strategy to promote sustainable development. The summit also prompted the establishment of United Nations Commission on Sustainable Development.

The main objective of UNFCCC was to stabilize greenhouse gas concentrations at a level that would thwart dangerous anthropogenic interference with the climate system. It was further stated that such a level should be achieved within the time frame that would allow ecosystems to adapt naturally to climate change, ensure food production is not threatened and substantiate sustainable economic development (Batuwitatge 2012). Special attention needs to be paid to the less developed countries which lack resources.

With a population of 14.35 million, the megacity of Kolkata is ranked third amongst the cities in India and is the eighth largest city in the world. Located on the Ganga delta, 145 km from the Bay of Bengal, the average elevation of the city ranges from 1.5–9.0 m above mean sea level. The climate is tropical, hot and humid, with the wet season lasting nearly four months. The summer temperature can reach up to 45 °C. The mean annual temperature is 27 °C, while the mean monthly temperature ranges from 19–30 °C. The annual precipitation is 1582 mm.

Like most of the megacities in India, Kolkata also suffers from drinking water scarcity. No part of Kolkata receives water supply for 24 h. While the daily per capita supply is 130 L, 35% of the total urban population remains uncovered by municipal water supply (Ray and Shaw 2016) and is forced to depend on groundwater sources that supply 30 million gallons of water per day. The consequent fall in the groundwater level not only increases future threat to freshwater availability, it also enhances the chances of subsidence and may be responsible for increased saltwater intrusions into the urban water system. These regions also suffer from arsenic contamination making the population vulnerable to diseases (Ray and Shaw 2016) which often turn fatal.

The city is also susceptible to climate-related disasters. In order to accommodate the rapidly growing city population, Kolkata has reclaimed a significant area under the East Calcutta Wetlands, a declared Ramsar site since 2002, resulting in a reduction in the number of wetlands. Fewer wetlands and increase in impermeable surfaces has reduced the water retention capacity of the city making it more vulnerable to urban floods. The old dilapidated drinking water, sewerage and drainage systems increase the chances of flooding. Flooding is quite frequent in the monsoon months of August and September resulting in enormous economic losses.

Under the impact of climate change, Kolkata has been experiencing a decrease in annual precipitation and warmer climatic conditions. The average rise in temperature during the last century is estimated to be 16 °C. The city is thus vulnerable to sea-level rise and suffers from land subsidence. The estimated mean land subsidence rate is 13.53 mm/year and for a 1 m drop in the piezometric head, the mean subsidence is 3.28 cm (Sahu and Sikdar 2011). Affected by sea-level rise and increased rate of subsidence, the city of Kolkata is likely to experience saltwater intrusions, aided further by overexploitation of groundwater. Droughts have increased in frequency and intensity in the last few decades. The IPCC predicts that thermal stress and water scarcity in the countries of Asia would significantly affect agricultural productivity. For example, a 0.5 °C increase in winter temperature will result in reduction in wheat yield in India by 0.45 tons per ha (Sivakumar and Stefanski 2011). It is also indicated that wheat and maize yield in India will reduce by 2–5% for a temperature increase of 0.5–1.5 °C (Sivakumar and Stefanski 2011). Climate change impacts to crop yields will impact food availability within cities, reducing the ability of people to meet their dietary needs.

Regional population projections indicate that Kolkata's population will continue to grow and water shortages will increase significantly. Due to the threat of increased frequency and intensity of tropical storms, storm surges and saltwater intrusion, increased protection, enforcement and restoration of mangroves and wetlands are already being done. The coastal city of Kolkata has a low elevation and needs additional protection. These ecosystems provide protection against storms and storm surges in addition to providing food and livelihoods for millions of Indians. Revamping the booster pumping stations and the drinking water supply system is expected to improve the availability of drinking water. Attempts have also been undertaken to ensure reduced dependence on groundwater sources to supplement the municipal water supply while rainwater is being promoted for groundwater recharge and also for human consumption.

The World Summit on Sustainable Development 2002 also referred as the Earth Summit 2002, held in Johannesburg, builds on the earlier United Nations Conference on the Human Environment in Stockholm in 1972 and the Earth Summit in Rio de Janeiro in 1992. In addition to increasing governmental commitments, it helped in extending the concept's reach amongst various stakeholders including business and administration, civil society and the private sector.

Meanwhile, the Vienna Convention was held in 1985 at Vienna with an emphasis on the Protection of the Ozone Layer. The multinational environmental agreement was ratified by 197 nation states. It was designed as a framework to decide on international efforts to protect the stratospheric ozone layer. The ozone layer prevents global warming and reduces the prevalence of skin cancer and cataracts by filtering out the harmful ultraviolet radiation. However, the reduction goals were not made legally binding and mandatory for the release of chloroflurocarbons, the main agent responsible for ozone depletion. These were formulated and spelt out in the Montreal Protocol, finalized in 1987. A global agreement for the protection of stratospheric ozone layer, the protocol emphasized on the phasing out of the production of ozone-depleting substances such as chlorofluorocarbons and halons.

The Montreal Protocol was followed by the Kyoto Protocol adopted on 11 December 1997 at Kyoto Japan. It came into force in February 2005. There are 192 nations for the Protocol. It extends and implements the basic objectives of 1992 United Nations Framework Convention on Climate Change to fight global warming and reduce anthropogenic contributions to greenhouse gas emissions. It is based on the consensus that anthropogenic interference is the cause of global CO_2 emissions and of global warming. The major responsibility to reduce current emissions rested on the developed countries being historically responsible for contributing to the current levels of greenhouse gases in the atmosphere. The Protocol's first commitment period commenced in 2008 and lasted till 2012, followed by a second commitment period agreed upon in the Doha Amendment 2012 with binding targets set for 37 countries. Negotiations to continue with the initiative even after the second commitment period ending in 2020 resulted in the adoption of the Paris Agreement in 2015.

The Paris Climate Agreement adopted in December 2015, within the United Nations Framework Convention on Climate Change (UNFCCC), is concerned with reduction and mitigation of greenhouse gas emissions where each concerned nation needs to regularly report its own contribution to reduce global warming. By October 2017, 195 UNFCCC member countries signed the agreement, 168 of them have ratified it.

The Cancun Conference or the United Nations Climate Change Conference held in Cancun Mexico was the sixteenth conference of the 194 parties to UNFCCC and the sixth conference of the 192 parties to the Kyoto Protocol. The conference is committed to restrict global temperature increase within 2 °C above pre-industrial levels, to establish Green Climate Fund in developing countries to finance new climate friendly technology and to set up an adaptation committee under the Cancun Adaptation Framework to promote the implementation of stronger, effective and cohesive adaptation measures. Emission reduction targets were agreed upon by developed countries. They are also committed to develop and strengthen low carbon national plans and strategies as well as to increase reporting frequency. Developing countries were also encouraged to develop low carbon national plans and strategies to be adopted with technical support and financial assistance from the developed nations.

Following such international agreements on reducing human interference in the global climate system and accepting that the cities are one of the main reasons for global production of greenhouse gases, cities are increasingly involved in transnational and national networks (Lankao 2008) of environmental governance or multi-governance, happening across multiple scales like the Cities for Climate Protection (CCP), the C40 Cities and Climate Alliance. Local governments participating in the International Council for Local Environmental Initiatives Programme, Cities for Climate Protection (CCP), commit to undertake and complete five milestones that include conducting an energy/emissions' inventory list and forecast, establishing an emissions' target (Lankao 2008), developing and obtaining approvals for local action plan, implementing policies and measures as well as monitoring and verifying the results. The Large Cities Climate Leadership Group, also known as the C40 Cities, represent a group of diverse cities like Chicago, Cairo, Mumbai and Sydney committed to reducing urban carbon emissions and adapting to climate change urgently as cities contain around 50% of the world's population, and consume a high share of the world's energy (Lankao 2008). The Climate Alliance is an alliance of European cities and municipalities that have developed a partnership with indigenous rainforest communities (Lankao 2008). Its aim is to preserve the global climate through a reduction in greenhouse gas emissions by high-income countries and conservation of forests in middle and low-income countries (Lankao 2008). Some municipalities have developed a systematic approach to climate policy, through the stages of undertaking inventories of greenhouse gas emissions, determining emission reduction targets, climate change action plans and various implementation plans (Bulkeley et al. 2009). For example, in Sweden approximately half of all municipalities have adopted climate mitigation goals in accordance with the national objective of reduced climate impact as formulated in the Swedish climate strategy (Bulkeley et al. 2009; Granberg and Elander 2007), while in Japan about one-third of the local governments had adopted the national reduction target of the Kyoto Protocol (a 6% reduction of 1990 CO_2 levels by the period 2008–2012) (Bulkeley et al. 2009; Sugiyama and Takeuchi 2008). Most cities who adopted the reduction targets for the greenhouse gases, however, failed to pursue an efficient and structured approach.

Realizing the implications of climate change and its linkage with other sustainable development challenges like reduction of poverty and hunger, improvement in infrastructural facilities, health and hygiene, most of the Sustainable Development Goals adopted in September 2015 address the challenges faced by urban areas. The 17 goals include no poverty, zero hunger, health and well-being, quality education, gender equality, clean water and sanitation, affordable and clean energy, decent work and economic development, industry, innovation and infrastructure, reduced inequalities, sustainable cities and communities, responsible consumption and production plan, life below water, life on land, peace, justice and strong institutions and partnerships for the goals (Gazattee 2018). It was preceded by the Millennium Development Goals 2005 which also emphasized targeted reduction of poverty and hunger, improved access to potable drinking water and improvement in the life of the slum dwellers.

Climate change impacts particularly in the cities located in less developed countries are obvious and may range from increased frequency of extreme weather events and flooding to higher air temperatures and adverse impact on public health. Climate change affects human well-being as well as economic development, posing threats to assets and livelihoods and making infrastructure inadequate and unaffordable. This is particularly true for the Asian cities accounting for 46% of Asia's urban population. This share is expected to increase to 52% with a total population of 2.2 billion by 2020. Some of the obvious impacts of climate change include the increased variability in rainfall patterns for both monsoon and post-monsoon seasons, increase in average air temperature, increased intensity and frequency of extreme weather events, sea-level rise and increased salinity in coastal areas, reduced stream discharge and disrupted ecosystems. Adaptation efforts in South Asia have been far from adequate and fragmented, without a strong link between national climate change adaptation strategies and existing measures for disaster risk reduction.

While there is a consensus about the extensive impact of climate change in South Asia, understanding of the precise impacts of climate change at a more local level and community adaptations are still inadequate. At a national level, the less developed countries of South Asia like Bangladesh and Nepal have developed National Adaptation Programmes of Action. Bangladesh had it developed in 2005 and Nepal in 2010.

Moreover, several countries have developed national strategies and action plans addressing the concerns of climate change. China's National Action Plan on Climate Change in 2007, the Bangladesh Climate Change Strategy and Action Plan (BCCSAP) and the National Action Plan on Climate Change in India in 2008, the National Climate Change Adaptation for Sri Lanka 2011–2016 and the National

Climate Change Policy in Pakistan in 2012 are a few such examples. These plans and policies have not yet been successful in ensuring significant adaptation efforts at the national level and only a limited resource has so far been committed. The climate change adaptation (CCA) policies developed so far continue to be fragmented, and the climate change strategies and plans are not properly linked with existing disaster risk reduction, agricultural and other relevant policies, partly due to a lack of conceptual understanding, lack of meteorological data and inaccurate modelling. Various international and national NGOs have already started to develop programmes addressing climate change adaptations in the countries of South Asia, many of which are stand-alone projects to suit the local conditions. If climate change impacts are left unaddressed, cities would become vulnerable to the existing challenges interfering with sustainable development goals.

3.6 Conclusion

Impacts of climate change and global temperature increase under anthropogenic interference are increasingly evident in the urban water sector particularly in the Asian cities challenged by rapid growth of urban population. Urban water infrastructure is plagued by outdated infrastructure, competing water use, climate-change-induced water shortage and physical water scarcity. Pollution adds to the growing threats to water resources, increasing the treatment requirements for providing safe water to the city population (Danilenko et al. 2010a). Groundwater supplies are under stress due to decreasing and erratic precipitation rates and increasing extraction (Danilenko et al. 2010a). Extreme weather conditions including flood and drought, sea-level rise and saline intrusions into the water system are impacting the availability of freshwater resources and municipal water supply. The environmental degradation and destruction of natural habitats such as swamps and forests, associated with urban expansion in Asia, also have a strong impact on water availability (Cities Development Initiative for Asia 2011). Climate-change-related natural disasters threaten urban habitats and aggravate inequality and urban poverty (Cities Development Initiative for Asia 2011). Large cities in Asia including Bangkok in Thailand, Chennai in India, Dhaka in Bangladesh, Jakarta in Indonesia, Manila in the Philippines, by virtue of being located along coastal zones and floodplains, become more vulnerable to climate-induced water scarcity and natural disasters. Urban water utilities in Asia are hence facing an increasing need to improve the management of water resources and associated infrastructure (Danilenko et al. 2010a) to meet the demands of expanding needs and diminishing water resources.

Urban water systems are typical examples of a complex socio-ecological system (Kirono et al. 2014), consisting of a geophysical unit managed by multiple institutions. Climate impact studies and urban water provisioning must consider both the biophysical and social dimensions of the urban water system when addressing the problem of urban water scarcity. Climate adaptation policies need to adopt a multidisciplinary-integrated approach involving all the various stakeholders, adjusting to changes in regional climatic and socio-economic conditions like population increase, economic development and urbanization. Urban communities in the less developed countries of Asia are crippled by severe financial and institutional constraints and poor water governance that necessitates better understanding and management of water security issues including adaptation options.

Water management strategies should thus aim to augment supply through diversification of water sources including rainwater harvesting and recycled water, construction of improved storage facilities, sustainable levels of groundwater extraction, water trading and conservation and setting up of desalinization plants wherever feasible. Climate monitoring, assessment of water stress, demand management, land-use planning, community-based water schemes and resource mobilization would also ensure better urban water systems. Urban areas must also explore strategies to mitigate urban warming involving the changing of material properties of individual buildings to reduce energy demand, introduction of water detention ponds and preservation of wetlands to reduce run-off in urban areas and provision of enough open spaces to increase the sky view factor. It may be made mandatory for new residential development to go for water-sensitive urban designs and effective water management strategies like the use of grey water for nonpotable consumption. Proper development and design of wetlands would also play an important role in enhancing the quality of the stormwater. Building of sea walls and dykes together with introduction of water pumps can reduce the chances and intensity of flooding and saltwater intrusions. Institutional capacity building, promoting scientific research and awareness, evolving new and innovative approaches of urban water governance would ensure increased urban water resilience to climate-change-related disaster risks. Cities being at the hub of all economic activities must promote innovative approaches for sustainable urban development.

References

Association of Metropolitan Water Agencies (2007) Implications of climate change for urban water utilities. Association of Metropolitan Water Agencies, Washington D.C., USA

Awuor BC, Ayo Orindi V, Ochieng Adwera A (2008) Climate change and coastal cities: the case of Mombasa, Kenya. Environ Urban 20:231–242

Bagio AG (2009) Adapting to climate change and preparing for natural disasters in the coastal cities of North Africa. The World Bank, Washington D.C., USA

Bates BC, Kundzewicz ZW, Wu S,Palutikof JP (2008) Climate change and water. Intergovernmental Panel on Climate Change, Cambridge University Press, Cambridge

Batuwitatge LP (2012) National climate change policy and role of the citizens in the post-Durban era. Climate Change Policy Network, Institute of Policy Studies of Sri Lanka, Friday 20 Apr 2012

Bulkeley H, Betsill M (2003) Cities and climate change: urban sustainability and global environmental governance. Routledge, London

Bulkeley H, Schroeder H, Janda K, Zhao J, Armstrong A, Chu SY, Ghosh S (2009) Cities and climate change: the role of institutions, governance and urban planning. Report for the World

Bank Urban Research Symposium on Cities and Climate Change, Durham University, United Kingdom
Cities Development Initiative for Asia (2011) Asia-Pacific in the era of climate change—building urban resilience and green cities. Cities Development Initiative for Asia, Singapore
Danilenko A, Dickson E, Jacobsen M (2010a) Climate change and urban water utilities: challenges and opportunities. Working Water Note 24. Water Sector Board of the World Bank Sustainable Development Network, Washington D.C., USA
Danilenko A, Dickson E, Jacobsen M (2010b) Climate change and urban water utilities: challenges and opportunities. Water P-Notes, No. 50. World Bank, Washington, D.C.
Donnelly JP, Woodruff JD (2007) Intense hurricane activity over the past 5000 years controlled by El Nino and the West African monsoon. Nature 447:465–468
Gilbert N (2012) One-third of our greenhouse gas emissions come from agriculture, farmers advised to abandon vulnerable crops in face of climate change. Nature News, 31 October
Gosain A, Rao S, Basuray D (2006) Climate change impact assessment on hydrology of Indian river basins. Current Science 90(3):346–353
Granberg M, Elander I (2007) Local governance and climate change: reflections on the Swedish experience. Local Environ 12(5):537–548
Grimmond S (2007) Urbanization and global environmental change: local effects of urban warming. Geogr J 173(1):83–88
Hunt A, Watkiss P (2007) Literature review on climate change impacts on urban city centres: initial findings. Organ Econ Co-oper Dev, Paris France
Intergovernmental Panel on Climate Change (1996) Climate change 1995: the science of climate change. Cambridge University Press, Cambridge UK
Intergovernmental Panel on Climate Change (2007) Climate change 2007: fourth assessment report. Cambridge University Press, Cambridge, UK
Intergovernmental Panel on Climate Change (2013) Climate change 2013: the physical science basis. In: Stocker T, Qin D, Plattner GK, Tignor M, Allen S, Boschung J, Nauels A, Xia Y, Bex V, Midgley P (eds) Working Group I contribution to the fifth assessment report of the intergovernmental panel on climate change, Ist edn. Cambridge University Press, New York, USA
Kirono DGC, Larson S, Tjandraatmadja G, Leitch A, Neumann L, Maheepala S, Barkey R, Achmad A, Seintung M (2014) Adapting to climate change through urban water management: a participatory case study in Indonesia. Reg Environ Change 14(1):355–367
Klein RJT, Nicholls RJ, Thomalla F (2003) The resilience of coastal megacities to weather-related hazards. In: Kreimer A, Arnold M, Carlin A (eds) Building safer cities: the future of disaster risk, 1st edn. The World Bank, Washington D.C., USA
Landsberg HE (1981) The urban climate. Academic Press, New York and London
Lankao PR (2008) Urban areas and climate change: review of current issues and trends, issues paper for the 2011 global report on human settlements, National centre for Atmospheric Research, Boulder, USA
Law Teacher (2013) The United Nations framework convention on climate change (UNFCCC). https://www.lawteacher.net/free-law-essays/international-law/the-united-nations-framework-convention-on-climate-change-international-law-essay.php?vref=1. Accessed 11 May 2018
Mall RK, Gupta A, Singh R (2006) Water resources and climate change: an Indian perspective. Curr Sci 90(12):1610–1626
Mukherjee S, Shah Z, Kumar MD (2009) Sustaining urban water supplies in India: increasing role of large reservoirs. Water Res Manag. https://doi.org/10.1007/s11269-009-9537-8
Nicholls RJ, Hoozemans FMJ, Marchand M (1999) Increasing flood risk and wetland losses due to global sea-level rise: regional and global analysis. Glob Environ Change 9:S69–S87
Nicholls R, Hanson S, Herweijer C, Patmore N, Hallegate S, Corfee-Morlot J, Chateau J, Muir-Wood R (2008) Ranking port cities with high exposure and vulnerability to climate extremes: exposure estimates. OECD Environment Working Papers, No. 1, Organization for Economic Co-operation and Development, Paris, France

Oke TR (1973) City size and the urban heat island. Atmos Environ 7:769–779
Oke TR (1981) Canyon geometry and the nocturnal urban heat island: comparison of scale model and field observations. J Climatol 1:237–254
Oke TR (1997a) Urban environments. In Bailey WG, Oke TR, Rouse WR (eds) The surface climates of Canada, 1st edn. Queens University Press, Montreal
Oke TR (1997b) Urban climates and global environmental change. In: Thompson RD, Perry A (eds) Applied climatology: principles and practices. Routledge, NewYork
Ray B, Shaw R (2016) Water stress in the megacity of Kolkata, India and its implications for urban resilience. In: Shaw R, Rahman A, Surjan A, Parvin GA (eds) Urban disasters and resilience in Asia, 1st edn. Elsevier, United Kingdom
Ruth M, Ibarraran M (2009) Distributional effects of climate change: social and economic implications. In: Ruth M, Ibarraran M (eds) Distributional impacts of climate change: social and economic implications, 1st edn. Edward Elgar Publishers, Cheltenham, England
Sahu P, Sikdar PK (2011) Threat of land subsidence in and around Kolkata City and East Kolkata Wetlands, West Bengal, India. J Earth Syst Sci 120(3):435–446
Sainath S (2002) Everybody loves a good drought: stories from India's poorest districts. Penguin, New Delhi, India
Sánchez-Rodríguez R, Seto K, Simon D, Solecki W, Kraas F, Laumann G (2005) Science plan: urbanization and global environmental change. IHDP Report 15 International Human Dimensions Programme on Global Environmental Change, Bonn, Germany
Saudi Gazattee (2018) 3 innovative solutions to support UN's SDGs. 11 May, 2018 World Government Summit, Dubai
Shukla PR, Sharma SK, Garg A, Bhattacharya S, Ravindranath NH (2003) Vulnerability and adaptation: challenges ahead. In: Shukla PR, Sharma SK, Ravindranath NH, Garg A, Bhattacharya S (eds) Climate change and India: vulnerability assessment and adaptation, 1st edn. Universities Press, Hyderabad, India
Sivakumar MVK, Stefanski R (2011) Climate change in South Asia. In: Lal R, Sivakumar MVK, Faiz MA, Mustafizur Rahman AHM, Islam KR (eds) Climate change and food security in South Asia, 1st edn. Springer, Singapore
Sugiyama N, Takeuchi T (2008) Local policies for climate change in Japan. J Environ Dev 17(4):424–441
Svirejeva-Hopkins A, Schellnhuber HJ, Pomaz VL (2004) Urbanised territories as a specific component of the global carbon cycle. Ecol Model 173:295–312
Thomas R, Rignot E, Casassa G, Kanagaratnam P, Acuna C, Akins T, Brecher H, Frederick E, Gogineni P, Krabill W (2004) Accelerated sea-level rise from West Antarctica. Science 306:255–258
Tolkou AK, ZouboulisA I (2016) Effect of climate change in WWTPs with a focus on MBR infrastructure. Desalin Water Treat 57(5):2344–2354
United Nations Framework Convention on Climate Change (1992) United nations framework convention on climate change. United Nations. Available from https://unfccc.int/files/essential_background/background_publications_htmlpdf/application/pdf/conveng.pdf. Accessed 5 Aug 2015
United Nations Framework Convention on Climate Change (2007) Climate change: impacts, vulnerabilities and adaptation in developing countries. United Nations Framework Convention on Climate Change, Bonn, Germany
United Nations Human Settlements Programme (2011) Cities and climate change: global report on human settlements. Earthscan, London, United Kingdom

Chapter 4
Developing Water Security Index for Urban Areas

Bhaswati Ray and Rajib Shaw

Abstract The rapid growth of urban population is a major demographic trend along with the physical expansion of cities. It is estimated that by 2050, nearly 70% of the world population would be concentrated in urban areas, particularly in the less developed countries where a population of 5 million is added every month. It is expected to be accompanied by the accelerated growth of megacities and metacities. There is also significant increase in the number of slum dwellers who do not have adequate access to proper infrastructure, potable drinking water and basic sanitation. The urban poor are thus forced to pay more to procure water than their richer counterpart who enjoy municipal supply at negligible or no cost. Inadequate provisioning of drainage and sewerage makes the urban areas susceptible to increased risk of urban flooding. The proper functioning of the ecosystems is also hampered because of increased pollution levels. The chapter explores the reasons for water insecurity, which, as is widely understood, is rarely one of physical scarcity. Based on number of water stress and urban sustainability indicators, the chapter further explores the possibility of identifying various parameters for developing a water security index based on weighted aggregate score to assess water security in urban areas.

Keywords Urbanization · Water security · Indicators · Index · Weighted score

4.1 Urban Water Security: A Challenge

The Dublin Conference held in 1991 concluded that because water sustains all life form, effective management of water resources demands a holistic approach, linking social and economic development issues with protection of natural ecosystems (International Conference on Water and Environment 1992). Anticipated anthro-

B. Ray (✉)
Sivanath Sastri College, University of Calcutta, Kolkata, India
e-mail: bhaswati173@gmail.com

R. Shaw
Graduate School of Media and Governance, Keio University, Fujisawa, Japan

B. Ray and R. Shaw (eds.), *Urban Drought*, Disaster Risk Reduction,
https://doi.org/10.1007/978-981-10-8947-3_4

pogenic climate change impacts of higher temperatures, drought, erratic precipitation patterns and intense storms are expected to intensify water demands (Intergovernmental Panel on Climate Change 2007). The Millennium Development Goals 2000 and the Sustainable Development Goals 2015 emphasized the importance of water resource management and innovations in infrastructure, for sustainable development and resilient cities, in an era of rapid urban growth. Many of the cities worldwide have outgrown the capacity of their localities to provide adequate sustainable water supplies.

It is estimated that by the year 2050, almost 70% of the total world population would be concentrated in urban areas. World population will show an increase from 6 billion to more than 10 billion in this century, and it is estimated that nearly 8.5 billion will live in urban areas, an increase from 3 billion in the year 2000 (Organisation for Economic Co-operation and Development 2013). Associated with rapid increases in urbanization is the growth of urban water uses, combined industrial and domestic sectors (Jenerette and Larsen 2006; Falkenmark 1998; Jury and Vaux 2005; Gleick 2003). The additional 5.5 billion urban dwellers would thus increase the stress on limited water resources. The growth of population would be particularly high in the developing countries where cities gain an average of 5 million residents every month (United Nations Environment Programme 2011) along with the growth of megacities and metacities. Megacities and metacities, with more than 10 million and 20 million inhabitants, respectively, as defined by UN-Habitat, are steadily increasing in the Asia-Pacific region under the impacts of population growth and economic development. In 1970, Tokyo was the only megacity of the region. By 1990, there were five megacities; Tokyo and Osaka-Kobe in Japan, Mumbai and Kolkata in India, and Seoul in the Republic of Korea. By 2014, out of the 28 megacities of the world, 17 were located in the Asia-Pacific region (United Nations Human Settlement Programme, United Nations Economic and Social Commission for Asia and the Pacific 2015).

Megacities dominate the urban population as exemplified by Tokyo and Osaka-Kobe. Together these cities account for 49% of the total urban population of Japan (United Nations Human Settlement Programme, United Nations Economic and Social Commission for Asia and the Pacific 2015). The megacities attract inhabitants in great numbers from the rural areas because of their obvious employment opportunities and better living conditions. However, such migration in overwhelming numbers has its obvious adverse impacts, in the proliferation of slums and inadequate infrastructure provisioning. There is significant increase in the slum population that lack adequate access to proper infrastructure including safe drinking water and proper sanitation. Hence, the urban poor are forced to depend on private vendors for their daily water supply and end up paying 50 times more for a litre of water than their wealthier counterpart (United Nations Environment Programme 2011). Lack of reliable and safe drinking water supply and sanitation further impacts health and well-being and increases poverty and economic backwardness. Urban areas are also becoming increasingly vulnerable to water scarcity and to increased risks of flooding and drought. According to the Fifth Assessment Report (2013) of the Intergovernmental Panel on Climate Change, water scarcity will be the greatest challenge facing most of the region in future. Climate change will increase manifold with multiple stresses that include rapid urbanization, industrialization, polluted water resources, degraded ecosystems and poor water governance.

Ensuring sustainable urban water systems is predicted to be one of the major challenges of this century and a daunting task for the scientific community and policymakers. Cities cannot be sustainable unless they are able to guarantee reliable access to safe drinking water and adequate sanitation. Crippled with massive urban growth, increased number of slum dwellers without access to proper infrastructure, water supply and sanitation facilities, increased water pollution and degraded ecosystems, cities in the Asia and Pacific region would benefit from sustainable management of water resources and well-developed equitable water supply systems. The major challenges in the water sector include water insecurity, inadequate access to safe drinking water and sanitation, poor water quality, flood risks and transboundary conflicts. It is important to assess the degree of water scarcity with the help of suitable indicators and undertake effective water management measures, including both supply management and demand management in a proactive system-based approach.

4.2 Water Security Index: Dimensions and Parameters

There appears to be no universal definition of water security. The concept differs amongst various disciplines and areas. These diverse notions, however, include certain common factors such as availability, accessibility, affordability, quality, safety and stability (Gain et al. 2016; Bizikova et al. 2013; Hoff 2011). According to Grey and Sadoff (2007), water security refers in particular to the availability of an acceptable quantity and quality of water for health, livelihoods, ecosystems and production, coupled with an acceptable level of water-related risks to people, environment and economies (Gain et al. 2016). Water security thus emphasizes that sufficient quantity of water resources is available for meeting the needs of the human population and for the survival of ecosystems, the water is accessible and affordable, the environment and society are free from water-related disasters, and there is good water governance and water management. Consideration of governance and management aspect is central to implementing a sustainable approach to water security (Gain et al. 2016; Pahl-Wostl et al. 2013).

It is being widely recognized that human water systems are traditionally viewed through the lens of physical water scarcity (Gain et al. 2016; Gunda et al. 2015), either demand-driven water scarcity or water stress and population-driven scarcity or water shortage (Gain et al. 2016). The demand-driven scarcity is measured by calculating the ratio of estimated annual freshwater demand to availability, with a threshold value exceeding 0.4 (Gain et al. 2016; Vörösmarty et al. 2005). The supply-driven scarcity is instead measured by calculating per capita availability of renewable freshwater resources. Water is scarce when the availability goes below 1000 m^3 per person per year (Gain et al. 2016; Falkenmark 1989). But such assessment of water scarcity excludes the socio-economic and institutional dimensions of scarcity. Water stress or water scarcity due to population growth and climate variability is intensified by economic disparity, poor water management and water governance issues as well as institutional inefficiency. Hence, water security parameters must include physical, socio-economic and institutional water scarcity as well as water-related disasters and their impacts. Vörösmarty et al. (2000) developed an integrated approach to water scarcity parameters, bringing together the physical and socio-economic dimensions. The International Water Management Institute (IWMI) used a similar assessment of water scarcity including both physical water scarcity and economic water scarcity. Physical water scarcity is implied by a withdrawal of more than 75% of river flows for agricultural and industrial use and for domestic consumption. Indicators of physical water scarcity include acute environmental degradation, diminishing groundwater and water allocations that support some sectors over others (Brown and Matlock 2011). Countries having adequate renewable resources with less than 25% of water from rivers withdrawn for human purposes, but needing to make significant improvements in existing water infrastructure to make such resources available for use, are considered as economically water scarce (Brown and Matlock 2011). The IWMI assessed the status of global freshwater resources and mapped it to indicate regions with little or no physical and economic water scarcity. Kummu et al. (2010) mentioned that research on water scarcity should continue to extend towards the inclusion and scrutiny of the concepts of water governance, water management, water policy, environmental integrity and water's role in societal and economic development (Gain et al. 2016).

The Falkenmark indicator measures water scarcity and water stress in terms of the fraction of total annual run-off that is available for human consumption. This indicator relates to the fixed amounts of renewable freshwater resources with the world population, using a per capita estimate of water requirement. Based on the per capita water usage, the water conditions may be categorized as one of no stress, stress, scarcity or absolute scarcity based on the water usage per person in the multiple countries surveyed for the purpose. The availability of water is used as thresholds values, with the per capita availability of water at 1,700 and 1,000 m^3 per year used as the thresholds between water-stressed and water-scarce areas, respectively (Falkenmark 1989). While large parts of South Asia is under water stress and west Asia is water scarce, major parts of central Asia are vulnerable to water stress and water scarcity (Fig. 4.1). No stress conditions prevail when the annual per capita availability of water exceeds 1,700 m^3. Water stress conditions are indicated by an annual per capita availability of water between 1,000 and 1,700 m^3. Water scarcity implies an annual per capita availability between 500 and 1,000 m^3. If the average annual availability of water per capita is less than 500 m^3, water conditions are one of absolute scarcity. This index is commonly used for country-level assessment of water stress and water scarcity where the data is readily available. However, water quality parameters and seasonal fluctuations in water availability were not taken into consideration while developing it.

In 1987, Shiklomanov and Markova from the State Hydrological Institute in St. Petersburg published an estimated current and predicted water resources use by region and sector (Brown and Matlock 2011; Shiklomanov 1993). Using population

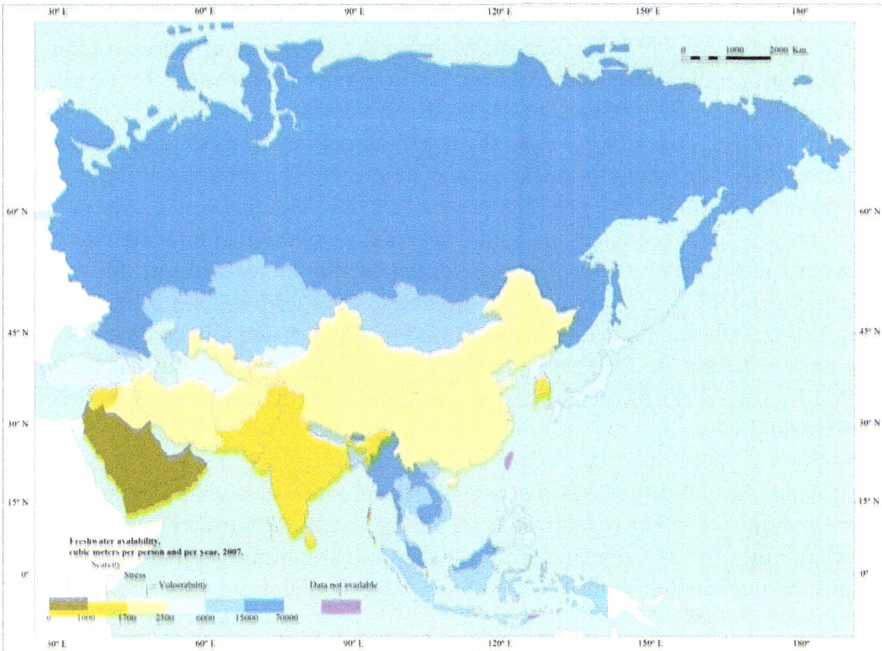

Fig. 4.1 Water scarcity and water stress in Asia

and economic variables as major parameters, water use was assessed in terms of water use in the industrial, agricultural and domestic sectors. Water lost from reservoir evaporation was also considered (Brown and Matlock 2011). Raskin et al. (1997) modified the approach proposed by Shiklomanov while using the water resource data, substituting water withdrawals for water demand. Since water demand varies between societies, cultures and regions, the term is subjective (Brown and Matlock 2011; Rijsberman 2006) and its use as a universal variable may lead to inaccurate assessments. The Water Resources Vulnerability Index, sometimes referred to as the WTA ratio, was then developed by Raskin et al. in 1997 as the ratio of total annual withdrawals to available water resources (Brown and Matlock 2011). A country is considered water scarce if annual withdrawals are between 20 and 40% of annual supply and severely water scarce if withdrawals exceed 40% (Brown and Matlock 2011; Raskin et al. 1997). This method and 40% threshold are commonly used in water resources analyses and have been termed the criticality ratio—the ratio of water withdrawal for human use to total renewable water resources (Brown and Matlock 2011; Alcamo et al. 2000).

Gleick (1996) developed a water scarcity index known as the Basic Human Needs Index as a measurement of the ability to meet all water requirements for basic human needs; drinking water for survival, water for human hygiene, water for sanitation and household needs for preparing food (Brown and Matlock 2011). The accepted minimum amount of water required by a person for drinking is 5 L per day in addition to 20 L per person per day for sanitation. Water needed for personal hygiene is 15 L per capita per day. Studies have suggested that the minimum amount of water needed for adequate bathing is 15 L per person per day (Brown and Matlock 2011; Gleick 1993). An additional 10 L of water is required per person per day to meet the demands for food preparation according to most regional standards. The total water requirement as proposed thus stands at 50 L per person per day, independent of climate and culture. International organizations and water providers are recommended to adopt this overall basic water requirement as a new threshold for meeting these basic needs, independent of climate, technology and culture (Brown and Matlock 2011; Gleick 1996; Kalbermatten et al. 1982). Both Falkenmark and Gleick developed the benchmark indicator of 1,000 m^3 per capita per year as a standard that has been accepted by the World Bank (Brown and Matlock 2011; Gleick 1993; Falkenmark and Widstrand 1992).

Meigh et al. developed the Water Availability Index (WAI) in 1999 based on Global Water Availability Assessment (GWAVA) model that takes into account temporal variability of water resources, both surface and groundwater, and compares total availability to total demand from all sectors (Singh and Kumar 2014). Total availability is calculated in terms of surface water availability as 90% reliable run-off and groundwater availability (Singh and Kumar 2014) as the potential recharge. The index is calculated based on monthly surface water balance or the potential aquifer yield, and it is the lower value that is usually considered. The index is normalized within the range -1 to $+1$. When the index is zero, availability and demands are equal (Singh and Kumar 2014).

$$\mathrm{WAI} = R + G - D/R + G + D$$

where R is the surface run-off, G is the groundwater resources, and D is the sum of demands of all sectors (Singh and Kumar 2014).

Ohlsson (2000) integrated the adaptive capacity of a society to consider how economic, technological and other factors affect the overall freshwater availability in a region (Brown and Matlock 2011). He was of the opinion that capability of a society to adapt to difficult scenarios is a function of the distribution of wealth, education opportunities and political participation (Brown and Matlock 2011). The widely accepted set of indicators to assess the social variables is the UNDP Human Development Index (HDI). The HDI functions as a weighted measure of the Falkenmark indicator in order to account for the ability to adapt to water stress and is termed the Social Water Stress Index (Brown and Matlock 2011).

The Water Poverty Index (WPI) (Sullivan 2002; Lawrence et al. 2002) was developed by the Centre for Ecology and Hydrology (CEH), Wallingford. This water poverty index is an attempt to assess the connection between water scarcity and socio-economic conditions. Under this index, countries are ranked according to the provisioning of water, combining five different components including resource availability, access to resources, their use, capacity building and environmental considerations. Each of these components is arrived at from two to five indicators, normalized on a scale from 0 to 1. For resource parameter, total availability of water per unit area is considered and includes both internal and peripheral water resources. Access to safe drinking water, proper sanitation and irrigation water is considered under access. The third category or use includes per capita per day domestic water consumption, water used in industrial sector and also the agricultural water use. Industrial and agricultural water use is calculated as the ratio between GDP derived from the industry or agriculture sector and the proportion of water used in industry or agriculture. Under environment, there are water quality indicators like dissolved oxygen, phosphorous, suspended solids and conductivity, regulations and policies on water management, informational capacity, sustainable and biodiversity parameters measured in terms of percentage of threatened mammals and percentage of threatened birds. The proposed water poverty index incorporates ecosystem productivity, community and human health, and economic welfare (Brown and Matlock 2011; Vörösmarty et al. 2005). Sullivan (2002) noted that depleted freshwater resources are linked to ecosystem degradation, and therefore, any index of water poverty should include the condition of ecosystems that maintain sustainable levels of water availability (Brown and Matlock 2011). For giving equal weightage, the sub-index and component values would be calculated as an average of the corresponding indicator values, to be multiplied by 20 that would give a summation value between 0 and 100. A value of 100 is achieved only if a country ranks best in all the five components.

Vörösmarty et al. (2010) assessed global threats to human water security and biodiversity based on 23 indicators (Gain et al. 2016). The selected indicators were mostly bio-physical in origin and there was little consideration for human dimensions like governance issues (Gain et al. 2016). Later, Lautze and Manthrithilake (2012) assessed water security considering the five critical dimensions of basic needs, agricultural production, environment, risk management and independence, for 46 countries in the Asia-Pacific region (Gain et al. 2016). Recently, Dickson et al. (2016) consolidated a comprehensive and flexible list of indicators for assessing community water security (Gain et al. 2016).

The SIPE approach to water scarcity was developed by Abedin and Shaw (2014) covering the socio-economical, institutional, physicochemical and environmental dimensions of water scarcity. Each dimension consisted of five primary indicators (Abedin and Shaw 2014). Each primary indicator was made up of five secondary indicators amounting to a total of 20 primary and 100 secondary indicators. It is further demonstrated that the inclusion of five dimensions (physical, social, economic, institutional and natural) (Abedin and Shaw 2014; Joerin and Shaw 2011) is essential to measure resilience including urban water resilience.

The Arcadis Sustainable Cities Water Index was prepared in association with the Centre for Economics and Business Research. It assessed 50 global cities on issues impacting their resiliency, efficiency and quality and their impact on long-term sustainability in the water sector. Resiliency in the water index meant proper management of water resources and water-related risks and vulnerabilities.

The resiliency index is measured in terms of water stress, availability of green space, water-related disaster risk, water balance and water reserves. The indicators include freshwater withdrawn as percentage of total water availability, number of water-related disasters like flood and drought, percentage of city area covered by green spaces, the monthly deficit or excess of rainfall and also the reservoir capacity within 100 km of the city as compared to the total water supply.

The efficiency sub-index gauges the present status and efficiency of water systems in urban areas and includes parameters like leakage loss, water charges, adequate and sustainable supply of safe drinking water, metered water supply, wastewater reuse and sanitation. The indicators for the efficiency sub-index are proportion of water lost in transit, percentage of households with metered supply and the tariff involved, average cost of per cubic metre of water use, total wastewater produced and the proportion of wastewater reused, continuity of service, hours of supply, percentage of households with access to safe drinking water and sanitation and the incidence of water-related diseases.

The quality sub-index assesses drinking water quality, water pollution and water-related diseases, sanitation, wastewater treatment and reuse. Amongst the 50 cities studied, Tokyo occupies the highest rank and is the most efficient city in terms of water resource management despite the lack of wastewater reuse. Singapore ranks 22nd despite the vulnerability to flood hazards. The last four cities in terms of sustainable water index are Delhi, Mumbai, Manila and Jakarta ranked 50th, 49th, 48th and 47th, respectively.

The water security dimensions proposed by the Asian Development Bank 2016 comprised of water security of household and urban areas, water security for economic development, urban water use, river health and resilience to disaster. The water security status is determined from five dimensions of water security index (WSI): WSI1 or basic water (renewable supply and sanitation), WSI2 or sufficient water (for water supply, consumption, use in agriculture), WSI3 or development water (irrigation water, industrial water use, water used in the energy sector and water for aquaculture), WSI4 or water disaster (floods and drought) and WSI5 or water for future (population growth, population growth in urban areas, water footprint). The index status thus analysed was correlated with water productivity (US $ per cubic metre of water) with countries categorized into four groups in terms of per capita GDP. The indicators available from various sources were normalized and the index for each country is determined with ranking from average and standard deviation values. Based on the index, the distribution of water security status of 146 countries was calculated.

The global water security index (GWSI) as developed by Gain et al. in 2016 takes into account physical, socio-economic and governance dimensions of water management. It aggregates the indicators using MCA method. Security is conceptualized as a function of four main criteria, namely availability, accessibility, safety and quality and management (Gain et al. 2016), and the assessed indicators are chosen accordingly (Table 4.1) and included such measures as water security index, access to safe water drought index, flood frequency index.

For assessing availability, WSI alone is inadequate as it does not include considerations of aridity, drought and groundwater depletion. Therefore, in addition to WSI, two more indicators, groundwater depletion (Gain et al. 2016; Wada et al. 2012) and the drought index (Gain et al. 2016; Wada et al. 2013), were added to obtain

Table 4.1 Indicators for assessing water security

Water security criteria	Indicators	Definition
Availability	Water scarcity index (WSI)	Ratio of total water withdrawal to water availability including environmental requirements. Higher WSI indicates lower water security (Gain et al. 2016)
	Drought index (DI)	Calculated using PCR-GLOBWB. Higher DI leads to decrease in water security. (Gain et al. 2016; Wada et al. 2013)
	Groundwater depletion	Depletion rate (million m^3 year^{-1}) calculated using PCR-GLOBWB. The values with higher DI lead to decrease in water security (Gain et al. 2016)
Accessibility	Access to drinking water	Percentage of population with access to improved drinking water source. Higher values indicate increase in water security (Giupponi and Gain 2016; Gain et al. 2016; Hsu et al. 2014)
	Access to sanitation	Proportion of population with improved sanitation as percentage of total. Higher values indicate increase water security (Gain et al. 2016; Hsu et al. 2014)
Safety	Water quality index	Higher index value leads to increase in water security (Gain et al. 2016; Srebotnjak et al. 2012)
	Flood frequency index	Frequency of flood events
Management	World governance index	The values with higher index value lead to increase in water security (Gain et al. 2016; Kaufmann et al. 2010)
	Safe water adaptability index	5-point rating scale of primary and secondary indicators
	Transboundary legal framework	Effective transboundary legal agreements
	Transboundary political tension	Risk of potential hydro-political tensions

Source By authors

a more comprehensive notion of water availability. Drought frequency is derived by counting the occurrences of drought events, i.e. when stream flow falls below the threshold and drought index was calculated by dividing drought frequency by the average frequency over the period 1960–2010 (Gain et al. 2016; Wada et al. 2013). Groundwater depletion is the persistent removal of groundwater from aquifer storages due to abstraction (Gain 2016).

For quality and safety, water quality index (WATQI) developed by Yale Centre for Environmental Law and Policy and the Centre for International Earth Science Information Network at Columbia University has been used. Flood frequency index has also been considered because floods also make the water unsafe for drinking.

For management issues, the World Governance Index developed by World Bank (Gain et al. 2016; Kaufmann et al. 2010) was considered. The index is calculated through aggregation of six governance dimensions, voice and accountability, political stability and absence of violence, government effectiveness, regulatory quality, rule of law and control of corruption (Gain et al. 2016; Kaufmann et al. 2010). Trans-boundary issues and hydro-political tension were also considered. The indicators were suitably modified based on the availability of data and their relative importance in assessing global water security index. The indicators were then normalized with a value function approach.

4.3 Developing a Water Security Index to Assess Sustainability of the Urban Water System

Various concepts and parameters define water security and the indicators used to measure global water security. In this chapter, the authors attempt to identify the parameters and indicators to assess water security in urban areas and resilience in the urban water system (Ray and Shaw 2016) for sustainable development. Indicators have been used as a tool to describe the physical, socio-economic and institutional dimensions of urban water security. They summarize complicated measurements into simple ones easy to comprehend and highlight the main characteristics of the system. The importance of indicators lies in their ability to communicate relevant information and facilitate decision-making. Apart from physical water scarcity, cities in Asia also suffer from water insecurity, often magnified by improper water management. Hence in the listing of indicators, greater emphasis has been given to institutional dimensions of urban water security.

Indicators are regarded as pieces of information, which have a wider implication than their immediate meaning (Bakkes et al. 1994). An indicator is useful as one of the fundamental inputs in decision-making that simplifies or summarizes important properties, visualizes different phenomena, quantifies and communicates relevant information (Gallopin 1997). In addition to these essential quantifications, indicators also have an important role to play in assessing conditions and trends for spatial comparisons, providing early warning and anticipating future conditions.

An urban society needs a number of infrastructural facilities to function properly. The provision of adequate and safe drinking water, storm water drainage and wastewater treatment facilities are prerequisites for urban systems to function properly. Urban water systems are primarily designed to meet these objectives until recent times when, because of increased awareness about environmental degradation, disaster risk reduction is also considered for efficient and sustainable urban water systems.

50 indicators have been chosen under the physical, socio-economic and institutional dimensions of urban water security (Table 4.2). These indicators were normalized to a scale from 1 to 5, ranging from very poor or not available score of 1 to a best score of 5. Simple arithmetic functions such as weighted mean index and aggregate weighted mean index would be used to calculate the scores for the indicators under the different dimensions (Abedin and Shaw 2014). In addition, all

Table 4.2 Indicators for urban water security

Dimension	Criteria	Parameters/indicators
Physical	Water quantity	Surface water reserve
		Groundwater reserve
		Source of freshwater supply to city
		Annual rainfall in mm
		Water availability per capita per year (>1700 m^3)
	Water quality	Total coliform and faecal coliform
		Total dissolved solids
		Hardness
		Contaminants like iron, fluoride and arsenic
		Industrial pollutants
	Provisioning	Total withdrawal for municipal supply in MGD
		Hours of supply
		% of area covered by municipal supply
		% of population covered by municipal supply
		% of population dependent on alternate water supply
Socio-economic	Affordability	% of households with individual tap connection
		% of metered connection
		% of population in slums
		Convenient source of water in slums
		Household expenditure on water services
	Livelihood	Population with access to safe water sources
		Daily per capita water supply
		Water vendors
		Manpower employed in the water sector
		Household expenditure on waterborne diseases
	Education and knowledge	% of population with knowledge on safe water
		% of students as dropouts to collect water
		% of population affected by waterborne diseases
		% of population using purification techniques
		Community-based capacity building
Institutional	Water management	Loss due to leakage

(continued)

Table 4.2 (continued)

Dimension	Criteria	Parameters/indicators
		Groundwater in city water supply in MGD
		Water treatment
		Wastewater treatment and reuse as % of total generated
		Booster pumping stations
	Disaster risk	Frequency of disasters like flood and drought
		% of population affected by floods and drought
		Fall in groundwater level in metres
		Storm water drainage
		Sewage generated and capacity of sewage treatment plant
	Resilience	Cost recovery
		Rainwater harvesting
		Desalinization options and capacity
		Protection of natural water bodies
		Revamping the existing municipal water supply system
	Governance	Budget allocation for the water sector
		External fund sourcing
		Private–public partnership
		Legislation related to water supply
		Water audit

Source By authors

five variables within a parameter would have to be ranked (W_1, W_2, W_3, W_4, W_5) between each other in the range of not important (1) to very important (5) in order to give a particular variable a higher or lower weightage in the calculation of the aggregate scores (Joerin and Shaw 2011). The constant use of five choices, ranks and weights allows the adoption of a formula named weighted mean (Joerin and Shaw 2011) (Eq. 4.1, Formula for weighted) to calculate the urban water security for each indicator, parameter and dimension (Abedin and Shaw 2014).

$$\frac{\sum_{i=1}^{n} W_i X_i}{\sum_{i=1}^{n} W_i} = \frac{W_1 X_1 + W_2 X_2 + W_3 X_3 + W_4 X_4 + W_5 X_5}{W_1 + W_2 + W_3 + W_4 + W_5} \quad (4.1)$$

Cities may then be ranked according to the aggregate weighted mean index combining all the components. It allows a transparent adoption of the formula of weighted mean (Abedin and Shaw 2014). The dimensions and indicators used may be modified to suit local conditions to ensure their global applicability.

The index may thus be applied to different urban areas to assess water security issues. A questionnaire survey would be conducted at institutional level within the study area. The questionnaire would contain both scientific as well as perception-based questions as some indicators may be best assessed through perception study. Officials of all such institutions that are responsible for ensuring urban water security would be interviewed on all selected parameters and indicators. Weighted parameters would then be applied to each indicator. It is best to have a micro-level study. Hence, the unit for data collection would be micro-level administrative and functional units like zones in each city. The survey unit may be even smaller consisting of wards (many wards make up a zone), the smallest urban unit in most cities of Asia. Apart from questionnaire survey, relevant data may have to be collected from the micro-level institutions responsible for urban water security. Focus group discussions may also be necessary with officials from various water institutions in the study area. After having collected all data, analysis and formulation of aggregate score may be done using Excel software.

The scoring would help highlight the issues with respect to each specific urban unit and to identify gaps for the provision of effective water security. The weighted index and the information available during questionnaire survey would be used to depict the scenario of various intensities of urban water insecurity and to assist in framing policy initiatives for micro-level planning for universal water security in urban areas. By identifying the available water resources and the gaps in urban water security, the water security index would be an essential part of policy formulations for urban water security at macro-level as well.

4.4 Conclusion

A global water crisis is apprehended as the capacity of the hydrological cycle is being outstripped and threatened by increasing human demand, pollution of water resources and poor water management (Ray and Shaw 2016). While global and regional climate change threatens water availability, increasing population and enhanced per capita water use with changing lifestyles are already putting stress on the existing water resources. Cities are also becoming increasingly at risk of water shortage as population and economic activities concentrate in cities. Urban areas also demand increasing amounts of water to support the intense pace of activities. The challenge is to make cities resilient to future urban growth and ecological degradation (Ray and Shaw 2016). Cities should be resilient to environmental conditions that affect water use, the stresses in the existing water supply infrastructure and to lifestyle changes. However, micro-level units in each city and the whole city in general differ considerably in terms of vulnerability and water stress. Hence, separate action plans may be formulated for cities to ensure sustainable water systems and resilient cities. An assessment and identification of the vulnerabilities and challenges in the existing urban water systems is a prerequisite to formulate water action plans in the urban areas. A listing of the indicators has thus been prepared based on the various water security indices and variables that are in use to identify water stress and potential threats to human and environmental water security in urban areas. These indicators may be modified and newer dimensions added to improve the assessment of water security and global adaptability.

References

Abedin MA, Shaw R (2014) Safe water adaptability for salinity, arsenic and drought risks in southwest of Bangladesh. Risk, Hazards Public Policy 4:62–82. https://doi.org/10.1002/rhc3.12033
Alcamo J, Thomas H, Thomas R (2000) World water in 2025: global modeling and scenario analysis for the world commission on water for the 21st century. Kassel World Water Series Report No. 2, Center for Environmental Systems Research, Germany
Asian Development Bank (2016) Asian water development outlook 2016: strengthening water security in Asia and the Pacific. Asian Development Bank, Philippines
Bakkes JA, Vanderborn GJ, Helder JC, Swart RJ, Hope CW, Parker JDE (1994) An overview of environmental indicators: state of the art and perspectives. Environmental Assessment Technical Reports, United Nations Environmental Programme, New York, USA
Bizikova L, Roy D, Swanson D, Venema HD, McCandless M (2013) The water-energy-food security nexus: towards a practical planning and decision-support framework for landscape investment and risk management. Canada, The International Institute for Sustainable Development, Manitoba
Brown A, Matlock MD (2011) A review of water scarcity indices and methodology. White Paper 106, The Sustainability Consortium, University of Arkansas
Dickson SE, Schuster-Wallace CJ, Newton JJ (2016) Water security assessment indicators: the rural context. Water Resour Manage 30:1567–1604
Falkenmark M (1998) Preparing for the future: water for a growing population. J Water Serv Res Technol-Aqua 47(4):161–166
Falkenmark M (1989) The massive water scarcity threatening Africa-why isn't it being addressed. Ambio 18(2):112–118
Falkenmark M, Widstrand C (1992) Population and water resources. Popul Bull 47(3):1–36
Gain AK, Giupponi C, Wada Y (2016) Measuring global water security towards sustainable development goals. Environ Res Lett 11(124015):1–13. https://doi.org/10.1088/1748-9326/11/12/124015
Gallopin GC (1997) Indicators and their use: information for decision-making. In: Moldan B, Billharz S, Matravers R (eds) Sustainable indicators: a report on the project on indicators of sustainable development, 1st edn. Wiley, Chichester, England
Giupponi C, Gain AK (2016) Integrated spatial assessment of the water, energy and food dimensions of the sustainable development goals. Reg Environ Change 1:1–13
Gleick PH (2003) Water use. Annu Rev Environ Resour 28:275–314
Gleick PH (1996) Basic water requirements for human activities: meeting basic needs. Water International 21:83–92
Gleick PH (1993) Water and conflict: fresh water resources and international security. Pacific Institute, Oakland
Grey D, Sadoff CW (2007) Sink or swim? water security for growth and development. Water Policy 9:545–571
Gunda T, Benneyworth L, Burchfield E (2015) Exploring water indices and associated parameters: a case study approach. Water Policy 17:98–111
Hoff H (2011) Understanding the nexus: background paper for the Bonn 2011 conference on the water, energy and food security nexus. Stockholm Environment Institute, Stockholm
Hsu A, Emerson J, Levy M, de Sherbinin A, Johnson L, Malik O, Schwartz J, Jaiteh M (2014) The 2014 environmental performance index. Yale University, New Haven, Connecticut, Yale Center for Environmental Law & Policy

Intergovernmental Panel on Climate Change (2013) Climate change 2013: the physical science basis, Fifth Assessment Report. United Nations Environment Programme, Cambridge University Press, Cambridge
Intergovernmental Panel on Climate Change (2007) Asia. In: Parry M, Canziani O, Palutikof J, van der Linden P, Hanson C (eds) Climate change 2007: impacts, adaptation and vulnerability. Contribution of working group II to the fourth assessment report of the intergovernmental panel on climate change, 1st edn. Cambridge University Press, Cambridge
International Conference on Water and Environment (1992) The Dublin statement of water and sustainable development. International conference on water and environment, Dublin, Ireland
Jenerette GD, Larsen L (2006) A global perspective on changing sustainable urban water supplies. Glob Planet Change 50(3–4):202–211
Joerin J, Shaw R (2011) Mapping climate and disaster resilience of cities. In: Shaw R, Sharma A (eds) Climate and disaster resilience of cities, 1st edn. Emerald Publisher, UK
Jury WA, Vaux H (2005) The role of science in solving the world's emerging water problems. Proc Natl Acad Sci USA 102(44):15715–15720
Kalbermatten JM, Julius DS, Gunnerson CG, Mara DD (1982) Appropriate sanitation alternatives: a planning and design manual. World Bank Studies in Water Supply and Sanitation 2. The Sustainability Consortium, University of Arkansas
Kaufmann D, Kraay A, Mastruzzi M (2010) The worldwide governance indicators: methodology and analytical issues. World Bank Policy Research Working Paper No. 5430, World Bank
Kummu M, Ward PJ, de Moel H, Varis O (2010) Is physical water scarcity a new phenomenon? Glob Assess Water Shortage Over Last Two Millennia Environ Res Lett 5(034006):1–10
Lautze J, Manthrithilake H (2012) Water security: old concepts, new package, what value? Nat Res Forum 36:76–87
Lawrence P, Meigh J, Sullivan, C (2002) The water poverty index: an international comparison. Keele Economics Research Papers 1–17. http://ideas.repec.org/p/kee/kerpuk/2002–19.html. Accessed 24 Aug 2008
Ohlsson L (2000) Water conflicts and social resource scarcity. Phys Chem Earth 25(3):213–220
Organization of Economic Co-operation and Development (2013) Green growth in cities, OECD Green Growth Studies, OECD Publishing. http://dx.doi.org/10.1787/9789264195325-en. Accessed 13 July 2015
Pahl-Wostl C, Palmer M, Richards K (2013) Enhancing water security for the benefits of humans and nature—the role of governance. Curr Opin Environmental Sustain 5(6):676–684
Raskin P, Gleick P, Kirshen P, Pontius G, Strzepek K (1997) Water futures: assessment of long-range patterns and prospects. Stockholm Environment Institute, Stockholm, Sweden
Ray B, Shaw R (2016) Water stress in the megacity of Kolkata, India and its implications for urban resilience. In: Shaw R, Rahman A, Surjan A, Parvin GA (eds) Urban disasters and resilience in Asia, 1st edn. Elsevier, United Kingdom
Rijsberman FR (2006) Water scarcity: fact or fiction? Agric Water Manag 80:5–22
Shiklomanov IA (1993) World fresh water resources. In: Gleick P (ed) Water in crisis: a guide to the world's fresh water resources, 1st edn. Oxford University Press, New York
Singh RB, Kumar D (2014) Water scarcity—handbook of engineering hydrology. Environmental Hydrology and Water Management, Taylor & Francis Group, New York
Srebotnjak T, Carr G, de Sherbinin A, Rickwood C (2012) A global water quality index and hot-deck imputation of missing data. Ecol Ind 17:108–119
Sullivan C (2002) Calculating a water poverty index. World Dev 30(7):1195–1210
United Nations Environment Programme (2011) Emerging issues in our global environment. United Nations Environment Programme, Nairobi
United Nations Human Settlement Programme, United Nations Economic and Social Commission for Asia and the Pacific (2015) The state of Asian and Pacific cities 2015—urban transformations, shifting from quantity to quality. United Nations, New York

Vörösmarty CJ, McIntyre PB, Gessner MO, Dudgeon D, Prusevich A, Green P, Glidden S, Bunn SE, Sullivan CA, Reidy Liermann CR, Davies PM (2010) Global threats to human water security and river biodiversity. Nature 467:555–561

Vörösmarty CJ, Douglas EM, Green PA, Revenga C (2005) Geospatial indicators of emerging water stress: an application to Africa. Ambio J Human Environ 34:230–236

Vorosmarty CJ, Green P, Salisbury J, Lammers RB (2000) Global water resources: vulnerability from climate change and population. Growth Science 289:284–288

Wada Y, van Beek LPH, Wanders N, Beirkens MFP (2013) Human water consumption intensifies hydrological drought worldwide. Environment Research Letters 8(034036):1–14

Wada Y, van Beek LPH, Weiland FCS, Chao B F, Wu Y, Beirkens MFP (2012) Past and future contribution of global groundwater depletion to sea-level rise. Geophys Res Lett 39(9). https://doi.org/10.1029/2012gl051230

Chapter 5
Water-Energy-Food Nexus: A Provision to Tackle Urban Drought

Bijon Kumer Mitra, Rajib Shaw, Wanglin Yan and Tomoko Takeda

Abstract Unprecedented urban growth adds pressure on the finite water resources to meet the growing demands. Urban does not only account large direct consumption of water, but also account large water footprint because of consumption of water-intensive food and high-energy consumption. Consequently, urban areas have experienced competition among various water users, particularly in the dry season. Failure of conventional approach to manage drought risk has translated realization urgency of integrated approach. Water-energy-food nexus (WEFN) approach emerges as an integrated tool to manage increasing water crisis in the context of rapid urban growth through facilitating multi-sectoral cooperation at both in-boundary and trans-boundary scales. Taking selected good practices of nexus around the world, this chapter argued that through realizing synergies across the urban water-energy-food system, cities could deliver inclusive growth, even with increasing drought risk. Scale-up of WEFN approach needs to be supported by enabling conditions. Mainstreaming of synergistic interactions across the WEFN is determined by inter-sectoral coordination, common vision, harmonizing policies, and the alignment of strategies and regulations, and economic incentives.

Keywords Urban drought · Water-energy-food nexus · Synergies
Inter-sectoral coordination · City community

5.1 Introduction

Water is a critical natural resource for growing urban areas, the home to more than half of the world's population. The rapid population growth rate in many cities is much faster than the rate of increasing services by the city authorities. Along with the rise in

B. K. Mitra (✉) · T. Takeda
Institute for Global Environmental Strategies, Hayama, Japan
e-mail: b-mitra@iges.or.jp

R. Shaw · W. Yan
Keio University, Fujisawa, Japan

B. Ray and R. Shaw (eds.), *Urban Drought*, Disaster Risk Reduction,
https://doi.org/10.1007/978-981-10-8947-3_5

rapid and largely unplanned urbanization, the demand for water in cities is projected to increase by 50–70% within the next three decades; reduced water availability and water quality degradation will lead to seasonal water shortage for 1.9 billion urban dwellers in 2050 (Water Scarce Cities Initiative[1]). Erratic rainfall is not the only cause of urban drought; conventional poor water management also contributes to increase water scarcity in urban. Growing cities often destroy their own water sources through over-abstraction of natural water resources and increasing pollution loads to the natural water bodies, which leads to worsen urban drought. For example, in Shanghai, the largest megacity of China, consumption of goods and services led to 11.6 billion m^3 of freshwater consumption, 796 thousand tons of chemical oxygen demand (COD), and 16.2 thousand tons of NH_3–N in discharged wastewater in 2007 (Zhao et al. 2016).

Current inadequate approach of drought management that attempts to manage different aspects of urban water cycle in isolation failed to provide solutions of major challenges including growing competition, conflicts, shortages, waste and degradation of water resources make. One of the main reasons for this failure is not to recognize the interlinkage relationships of water and other sectors. Therefore, it is imperative to shift from a sectoral approach to an integrated approach supported by all sectors.

In recent years, water-energy-food nexus (WEFN) has been discussed and proposed as a useful concept to promote cross-sectoral integration for sustainable management of the interconnected resource systems and to deal with challenges to access to these resources. Water, energy, and food are basic elements for survival of human beings, economic growth, and sustainable development. Water, energy, and food securities are inherently inter-linked and inter-dependent. For example, about 4% of total world's energy consumption is accounted for water delivery (IEA 2016). Energy sector needs water for fuel extraction, cooling, and hydropower generation. Based on IGES (2013) estimate, Indian energy sector's water demand will reach to 90 billion cubic meter (BCM) if thermal power plants are cooled down with conventional cooling system, which is about 8% of total utilizable water. Food security also relies on supply of water and energy. Indian agriculture sector consumes about 90% of total water withdrawals and 18% of total electricity consumption (WWAP 2012; Ministry of power 2015). It is therefore important to bring water, energy, and food securities together to address them in a tandem.

As homes of more than half of the world's population cities have appeared as demand center of food, water, and energy, it is projected that urban population will be increased by 2.6 billion between 2011 and 2050 (United Nations 2011). In addition to large direct consumption of water, urban resident also accounted large water footprint because of consumption of water-intensive food and high-energy consumption. Growing urban population will require large supply of water, energy, and food. As a result, cities will face increased risk of water, food, and energy that would affect the entire city community including homes, businesses, and industries (DVRPC 2011; Denver's Climate Resiliency Committee 2014). In fact, these risks

[1]http://pubdocs.worldbank.org/en/588881494274482854/Water-Scarce-Cities-Initiative.pdf.

became as reality in various parts of the world. In India, water shortage-related thermal power plants shutdown cost roughly 14 terawatt hours (TWh) of electricity generation (WRI 2017). Cities in developed countries also experienced competition among various water users in the dry season. For example, the governor of California announced mandatory regulations that prohibited the watering of ornamental grass, required new homes to use drip irrigation, and directed water agencies to set up new pricing structures to maximize conservation to deal with two hottest years between 2014 and 2015 in its history (The Guardian 2016).[2]

Urban sustainable water management requires multi-sectoral, inclusive, and comprehensive strategies, and an integrated approach. Efficient and equitable water use requires integrated planning that reflects interdependencies and trade-offs. Urban water-energy-food nexus can be harnessed as a holistic policy tool to build resilience in the face of global change. This chapter aims to explore synergistic cross-sectoral links across the water-energy-food nexus to improve drought management in both the urban boundary layers and urban–rural regional layer.

5.2 Brief History of Water-Energy-Food Nexus (WEFN) Thinking

Until recently, in the absence of nexus thinking in planning and policy making for water, energy and food resources, interactions between the systems have been overlooked. Such uni-sectoral approaches have resulted in incoherent policy making, contradictory strategies, and the inefficient use of natural resources (Foran 2015). The insecurity of each of the individual sectors is also aggravated when they are considered together. The United Nations University (UNU) initiated the first nexus program to acknowledge inter-dependent nature of food and energy. In the following years, WEFN concept has gathered momentum in various international and regional forums (Fig. 5.1). WEFN concept was first proposed by the World Economic Forum (2011) and presented the WEFN framework to promote the inseparable links between the use of resources to provide basic and universal rights to food, water, and energy security. A subsequent version has taken on water resources as a central component, which is known as Bonn 2011 Nexus Framework. WEFN thinking is concerned with addressing externalities across multiple sectors, with a focus on system efficiency, rather than on the productivity of isolated sectors (Hoff 2011). The Bonn 2011 Nexus Conference was a significant catalyst for increased nexus attention from international organizations, the private sector, and other major global players.

The nexus is becoming increasingly prominent on policy makers' agendas, notably in relation to the post-2015 development agenda (Weitz et al. 2014). Research institutes such as the German Development Institute and the Stockholm Environment Institute have proposed a nexus approach to support cross-sectoral integration for

[2]https://www.theguardian.com/cities/2016/jul/29/where-world-most-water-stressed-cities-drought.

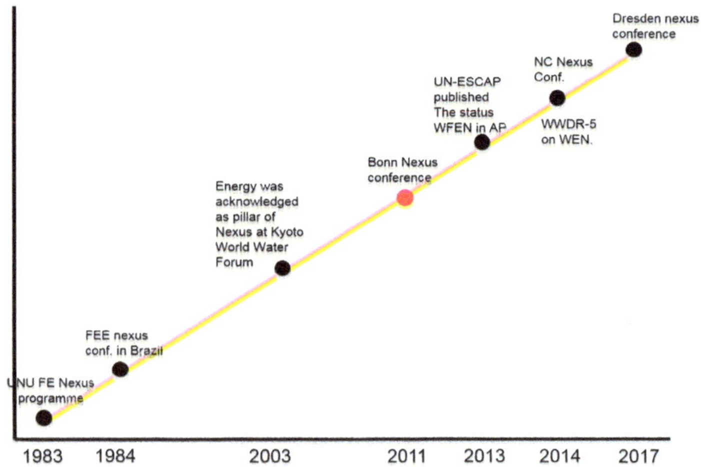

Fig. 5.1 Chronology of major WEFN events (compiled from Endo et al. 2015 and other sources)

the SDGs (Brandi et al. 2013; Weitz et al. 2014). The Colombian Government has shown support for a nexus approach to the SDGs to foster dialogue on broad development issues rather than sectoral challenges, thereby enabling interactions to emerge as central foci (Weitz et al. 2014).

5.3 Urban WEF Nexus Framework

The citywide water, energy, and food (WEF) demand is met through both local in-boundary WEF production and trans-boundary production. Therefore, cities also have to recognize that urban demand for WEF has far-reaching environmental impacts from in-boundary to outside of city boundaries. Production and supply of WEF to cities is a key aspect of the framework that focuses on WEF interactions in both city in-boundary process and trans-boundary boundary process which addresses synergies and trade-offs between individual environmental impact categories. The development of water footprint and energy footprint would be helpful to evaluate trade-offs and co-benefits among the different environmental impact categories. The water foot print informs volume of water removed from the watershed, operational risk of industries or power plants to water scarcity, production risk of crops due to low water availability. Evaluation of trade-offs and synergies will help to take adequate action or set of actions at appropriate level among four key categories to deal with water scarcity such as (a) changes in community-wide urban WEF demand; (b) reposition between in-boundary versus trans-boundary WEF supply; (c) inter-

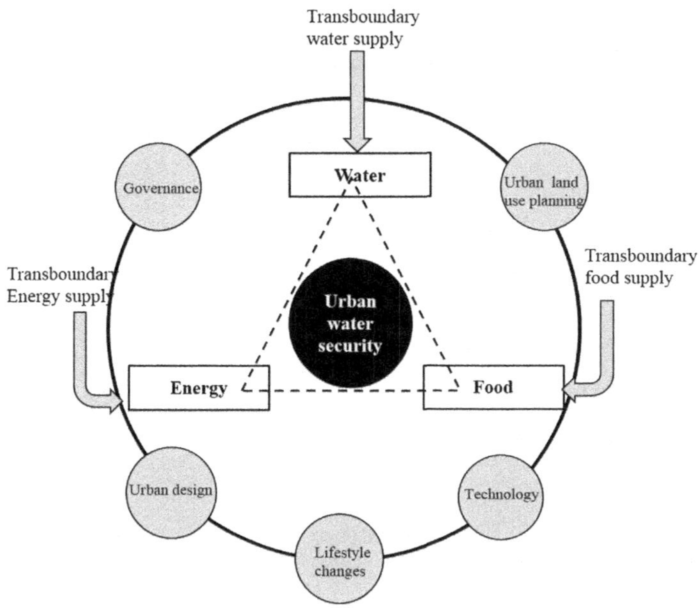

Fig. 5.2 Illustration of urban WEFN framework. *Source* Modified from Ramaswami et al. (2017)

vention in trans-boundary production systems; and (d) intervention of in-boundary production and cross-sectoral WEF interactions. Implementation of this action will require innervations in supply demand governance, land use planning, urban design, citizen's lifestyle, and technological innovations (Fig. 5.2).

5.3.1 Water-Energy Nexus at the Urban Water System Interface

Urban water system heavily dependents on energy supply in every step of water supply use—disposal cycle. Water abstraction, water conveyance, water treatment, wastewater treatment are generally energy-intensive activities in urban. Table 5.1 shows energy use intensity of each component of the urban water cycle. It shows that energy use intensity differs considerably among the component. Water source for the energy intensity of urban water supplies on a whole water cycle basis can range from a low of 0.27 kWh/m^3 to a hypothetical high upward of 9.50 kWh/m^3. Even energy intensity of a component also differs depending on source of water and technologies. In selecting water source, for instance, energy intensity of gravity-fed surface water is almost zero, whereas for use of groundwater, brackish groundwater, desalinated

Table 5.1 Energy intensity of urban water supply systems

Water use cycle segment	Electricity intensity (kWh/m^3)	
	Low	High
Water supply and conveyance	0.00	3.70
Water treatment	0.03	4.20
Water distribution	0.07	0.31
Wastewater collection and treatment	0.18	1.20
Wastewater discharge	0.00	0.10
Total	0.27	9.50

Source Modified by authors from Griffiths-Sattenspiel and Wilson (2009)

water, and recycled water, the energy cost rise to 0.50, 0.85, 3.60, and 0.30 kWh/m^3.[3] It implies that energy availability is a determining factor to choose source of water to mitigate stress situation. Desalinated water may not be a sustainable solution if supply of energy is not sufficient.

5.3.2 *Water-Energy Nexus at the Urban Energy Interface*

Energy sector uses water for extraction of fossil fuels, cooling of thermal power generation, hydropower generation, and biofuel production. Global aggregates show that 400 billion cubic meter water was abstracted from the sources for primary fuel production and electricity generation (IEA 2016). Cities account nearly two-thirds of global energy supply, which means they account largest portion of global water footprint for energy production and generation. Therefore, urban energy interface is also vulnerable to drought, which ultimately affects energy-dependent production and service system including urban water system. Water footprint of urban energy system enables cities to understand potential impact of drought not only on energy supply but also its consequence of increasing water competition among users. Since urban economy and life are energy intensive, its energy demand meets from in-boundary power station and trans-boundary power supply sources. Water footprint of energy supply chains is used by power plant designers, operators, and policy makers to identify water-related risk factors to critical power generation. In addition, it also communicates to urban energy users their collective impact on local and regional water resources in terms of billions of liters of freshwater withdrawals. In the context of a city where many residents routinely face seasonal shortages and daily rationing of both water and electricity supply, the trade-off between water and energy is all the more apparent. For example, annual water footprint for Delhi's energy system is nearly 560 million cubic meter, which is roughly half of the water volume used by residential and commercial sectors together (Cohen 2014). The water use intensity

[3]http://www.pacinst.org/resources/water_to_air_models/index.htm.

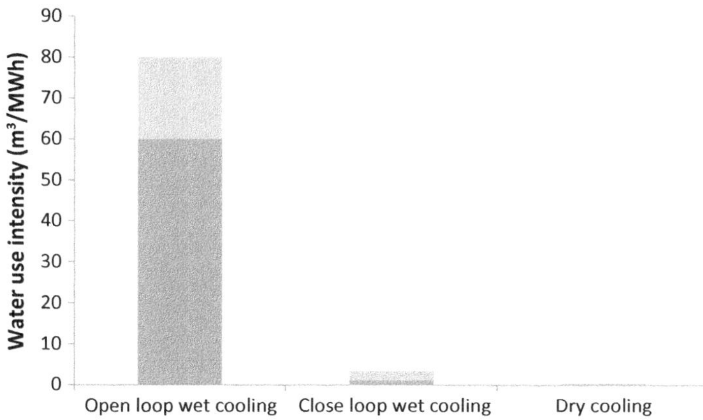

Fig. 5.3 Water use intensity of different cooling technology of thermal power plant. *Source* IGES power plant survey in 2012–2015

varies significantly, depending on power generation technologies, fuels, and cooling technologies. For instance, the use of closed-loop cooling system requires significantly less amount of water supply than open-loop cooling system for cooling down thermal power plants. Dry cooling system requires negligible amount of water, but it may reduce efficiency of power plants, particularly during hot season. Figure 5.3 represents water use intensity for different cooling system in India. Therefore, selection of appropriate cooling technologies of power generation system can contribute to drought management by reducing water footprint of power generation both within urban boundary and outside of boundary. Estimates show that phasing out of all open-loop cooling systems of thermal power plant by 2030 could reduce cooling water demand by 48% in India (Fig. 5.4).

5.3.3 WEFN at the Urban Agriculture Interface

As urbanization increases globally and the natural environment becomes increasingly fragmented, urban agriculture including home gardens has a vital role in improving food self-sufficiency index of cities, which heavily rely on trans-boundary food supply. Urban agriculture requires water for growing crops that is met via a combination of direct rainfall recharging soils and application of supplemental irrigation, competing with the other urban water users. The competition over water can be exacerbated especially during the summer months, and the situation will be worsened in the future by climate change, increasing urbanization, and population growth. However, good strategies of urban agriculture create symbiotic relation with water infrastructure, which can bring positive effect on the whole urban water cycle including freshwater supply and wastewater management. Cities are made up mainly by extended sealed

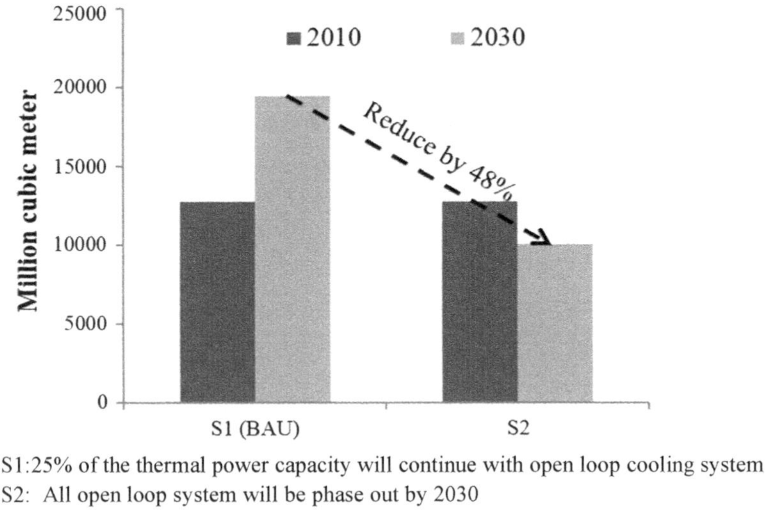

Fig. 5.4 Change in cooling water demand by phaseout of open-loop cooling system in Indian thermal power plants

surfaces (e.g., streets, roofs, and car parks) and are more and more frequently affected by risks of floods and landslides due to difficulties in the stormwater management. Urban agriculture can allow rainwater and runoff to drain through the soil, and the need for costly stormwater sewers and drainage can be minimized. To minimize competition with other water users, urban agriculture can improve water self-sufficiency by harvesting alternative water sources including rainwater, treated wastewater, and gray water. On the supply side, generally, urban agriculture is comparatively water resource because adoption of intensive growing methods, selection of low water-consuming crops, use water-efficient irrigation system, and use of alternative water sources. Urban agriculture also would also allow its residences to make choice of food products based on the water footprint, production method, and water availability at the location of origin.

5.4 Practical Responses in Practice to Realize Synergies Across Urban WEFN

Cities in various parts of the world are demonstrating its possibilities to continue growth, even with water scarcity by realizing synergies of WEF nexus in the water, energy, and food sectors. This section discusses selected such cases of practical responses to realize synergies across the urban WEFN.

5.4.1 Using Municipal Wastewater in Power Plant to Mitigate Water Crisis in Delhi, India[4]

Delhi is located in the Yamuna River basin, which is classified as water-stressed river basin. It is likely that rapid population growth and economic will lead to greater water demand. Consequently, competition among major users over water will be increased in coming days. The power supply in Indian capital region, Delhi, has not kept pace with the increasing population and the growth of economic activities. Installed capacity for power generation in Delhi has increased fourfold in the last two decades. By 2021–22, the installed capacity requirement will be 11,731 MW, compared to current installed capacity of 2,590 MW. As per Central Electricity Authority (CEA) guideline, water use intensity for power generation is 3.6 m^3/MWh. Based on the CEA guideline, the estimation shows that 2,100 million cubic meter (MCM) water will be required for power generation in 2021–22.

Management of wastewater in the capital city is also a big challenge. With rapid expansion of the city and domestic water supply, quantity of wastewater is increasing in Delhi. Currently, the city generates 1,650 MCM wastewater. But City's current wastewater treatment capacity is about 850 MCM per year. The growth of sewage treatment capacity has not kept pace with the increased wastewater generation. In last four decades, treatment capacity increased only sevenfold compared to a 12-fold increase in wastewater generation in that same period. Untreated wastewater discharge is the main source of pollution in surface water bodies such as Yamuna River. While water availability is an issue, reuse of wastewater can mitigate water crisis and minimize risk of pollution of natural water bodies.

If Delhi's wastewater can be properly collected and treated, and reused in power plants, freshwater demand of power sector would reduce up to 75% of its water use. Indraprastha Power Generation Co Ltd (IPGCL), is the electricity generation company of the Government of Delhi state, is moving toward a cleaner and more water-smart power generation by using treated municipal wastewater in operation of Pragati Combined Cycle Gas Turbine (CCGT) power plant instead of freshwater withdrawals.

The use of treated wastewater in Pragati CCGT is a win–win solution for water crisis mitigation and wastewater management. One of the salient features of this power plant is, the raw water requirement for operation of the power plant is met through Delhi Municipal sewage water treatment plants instead of using fresh surface water. As per this agreement, in exchange for free 20 mld of daily intake of treated wastewater from the sewage treatment plants, Pragati Power will operate and maintain the plants as well as cover the cost of electricity at the plants.

The model provides a greater opportunity to plan integrated water and energy infrastructure for the future. This model has many co-benefits because it:

i. Ensures the cost of operation, and maintenance of the STP ($594,964 USD) is borne by the power company, not the community,

[4]Observation from IGES power plant survey in 2015.

ii. Ensures reliable and continuous water supply, particularly for dry season,
iii. Avoids conflict with other users over water especially the agriculture sector who receive the excess treated water from the STP,
iv. Provides long-term economic benefit,
v. Reduces pressure on freshwater,
vi. Contributes to improved water quality in Yamuna River because the plant effluent is discharged to the river after naturalizing, and thus, the effluent discharge is better than sewage water.

5.4.2 *Water-Energy-Food Nexus Approach for Wastewater Management in Da Nang, Vietnam*[5]

Da Nang is the third largest city in Vietnam, which has the highest urbanization ratio among the municipalities. As a result, demand of resources including water food and energy is increasing. However, the Government of Da Nang city has committed to develop it as a green city by 2025. To achieve this goal, Da Nang city adopted nexus approach to manage wastewater and sanitation challenges. The city established a nexus task force with representative from different key agencies and stakeholders. The Department of Planning and Investment (DPI) and Department of Natural Resources and Environment (DoNRE) of Da Nang will be implementing a pilot project to implement integrated resource management approach to address wastewater and sanitation challenges in An Hai Bac Ward, benefitting 110 households, with funding from World Bank and technical assistance from GIZ. The pilot project is based on the concept of separate sewerage systems, to increase the organic load in the wastewater. The kitchen waste from households will be added to increase organic load in the wastewater for energy production; treated wastewater will be used for irrigation and the agricultural residue will be used for urban farming.

5.4.3 *Capturing Solar Energy to Operate Desalination Plants in Freshwater Water-Scarce Environment, Saudi Arabia*[6]

Desalination is becoming a strategic choice to meet growing demand for water in regions with limited freshwater sources. Saudi Arabia produces 5 million m^3 of drinking water from its 35 desalination plants. These desalination plants consume valuable oil resources. Given the country's abundant solar resources, utilizing solar energy for desalination can reduce the cost of energy inputs of water production.

---

[5]http://www.unescap.org/sites/default/files/The%20Urban%20Nexus_Policy%20Brief.pdf.

[6]http://www.irena.org/DocumentDownloads/Publications/IRENA_Water_Energy_Food_Nexus_2015.pdf.

The King Abdullah Initiative for Solar Water Desalination is a key initiative of the National Plan for Science, Technology, and Innovation led by King Abdulaziz City for Science and Technology.

The initiative is to be executed in four phases:

i. Building a desalination plant powered by solar energy with a capacity of 30,000 m^3 per day of water to meet the drinking water needs of Al Khadji city. The project will use 10 MW solar systems as well as domestically developed membranes.
ii. Building a desalination plant capable of producing 300,000 m^3 of drinking water per day, enough to meet the needs of 1 million inhabitants.
iii. Building a series of solar-powered desalination plants in various parts of the country.
iv. Implementing the initiative in the agricultural sector.

5.4.4 Urban–Rural Collaboration Water Environment Conservation Case of Kanagawa Prefecture, Japan

Although Kanagawa Prefecture houses two of the two most populated cities in Japan (Yokohama and Kawasaki, respective populations 3.7 million and 1.5 million), the western area is rural, with a declining and aging population. The water supply of Kanagawa Prefecture is sourced in the rural area as well as outside prefectural boundaries and then transported to the urban areas. 90% of the water supply comes from Sagami River and Sakawa River, of which 80% of the water catchment area is located in neighboring Yamanashi and Shizuoka Prefectures (Kanagawa Prefecture 2015) (Fig. 5.5).

In the water catchment area, privately owned plantation forests, which account for approximately 70% of the total catchment forest, have deteriorated over the decades since forestry became unprofitable. As a result, the quality of prefectural water sources unfortunately does not satisfy the expectations of all prefectural citizens. To resolve this issue, Kanagawa Prefecture adopted the Basic Policy for Kanagawa Water Source Environment Conservation and Restoration. This Basic Policy laid out a long-term plan to conserve and restore the water sources from 2005 to 2025 (Kanagawa Prefecture 2005). In line with this Basic Policy, the Kanagawa Prefecture formulated the five-year action plan for conservation and restoration of water source environment, in its third phase as of 2018. To sustainably finance the programs under these frameworks, it became necessary to secure stable financial sources. To address this, Kanagawa prefectural inhabitant tax is increased by adding an annual burden of average 890 JPY per taxpayer, comprised of fixed rates (300 JPY, regardless of income) and those based on income levels (4% of total tax per capita). The tax rate is determined every 5 years based on the required total expenses to implement the programs stipulated in the action plan, which is currently 20 billion JPY over the five-year period. This tax revenue is used not only to address forest conservation

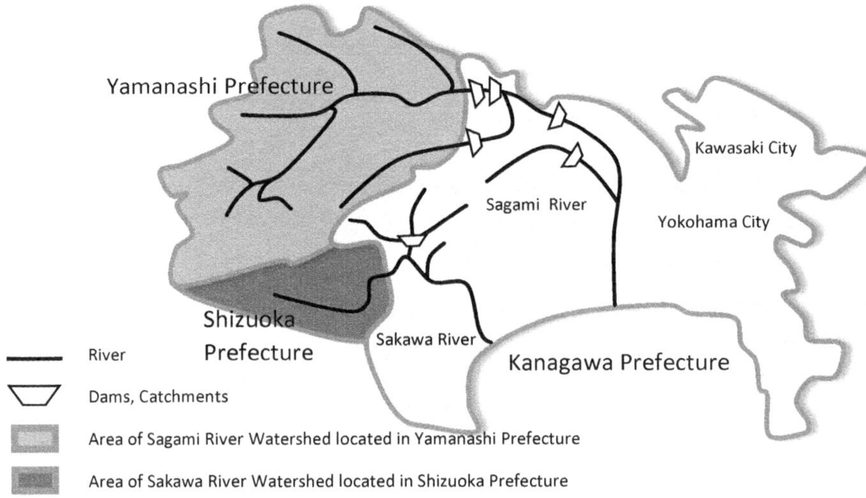

Fig. 5.5 Watersheds of the Kanagawa Prefecture water source (Compiled from Kanagawa Prefecture 2015)

but also includes sewage collection and treatment, water environment monitoring, awareness raising, river and groundwater conservation. It is noteworthy to mention that in surveys conducted to the residents of Kanagawa Prefecture, 77% responded in favor of spending the prefectural tax revenue outside the prefectural borders as needed, and 61% responded in favor of the taxation system for water source conservation programs. The evaluation of the action plan is carried out according to each stage of the outcomes, as well as collectively (Fig.5.6).

5.4.5 Economic Incentive Mechanism for Urban Agriculture Project with Stormwater Management Functions, New York, USA

It is important to note that urban agriculture brings multilayers of benefit to cities, neighbors, and residents. Urban agriculture does not only help to improve access to fresh and nutritious foods but also provide socioeconomic benefit by improving food self-sufficiency, job creation and offer recreation and relaxation. It also helps to improve urban water cycle by managing stormwater.

The New York Department of Environmental Protection launched a grant scheme to support urban agriculture project with stormwater management functions. Table 5.2 shows three awarded urban agriculture project.

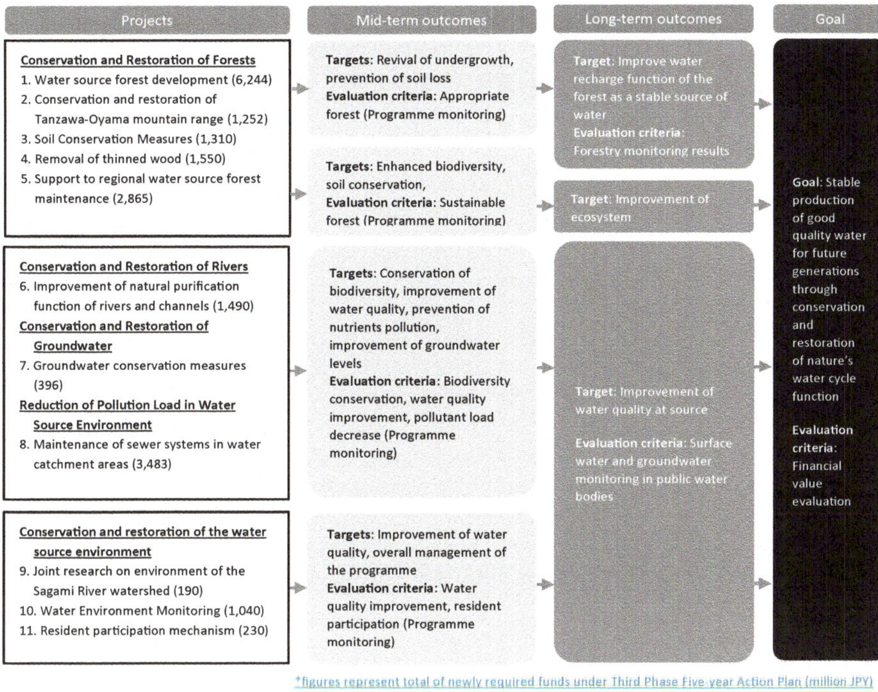

Fig. 5.6 Evaluation framework of Kanagawa Prefectural five-year action plan for conservation and restoration of water source environment (compiled from Kanagawa Prefecture 2016)

5.5 Enabling Environment for Translating Water-Energy-Food Concept to Actions

Recognizing the importance of water-energy-food nexus for sustainable societies, efforts to generate evidence-based conceptual analysis have increased through various forums involving academia, think tanks, non-profit organizations, and international agencies, etc. However, these efforts only have value when the analytical results are translated into actions. How WEFN concept can be translated to actions in a continuous cycle of urban development is a major governance issue. The following measures and interventions are critical to mainstream WEFN concept in urban drought management (Table 5.3).

5.5.1 Institutionalize Multi-Stakeholder Collaboration

Water-energy-food nexus concept encourages governments to shift from conventional approach to integrated approach for more efficient resource governance. Estab-

Table 5.2 Three awarded rooftop urban garden with stormwater management functions, New York

Project type	Socioeconomic benefit	Stormwater manage capacity (m^3)	Awarded amount (USD)
1. A green roof on a building that houses 61 local, a local bar, and restaurant	i. Varieties of drought-tolerant herbs on the green roof, which is used in the food and drinks served at the bar	227	41,975
2. Construct a 40,000-square foot commercial rooftop farm	i. Fresh local foods ii. Job creation	3,785	592,730
3. Two rooftop gardens	i. Fresh local foods	240	40,000

Source Compiled from Mississippi water shed management organizations, 2013 (http://www.arboretum.umn.edu/UserFiles/File/2012%20Clean%20Water%20Summit/Freshwater%20Urban%20Ag%20White%20Paper%20Final.pdf)

Table 5.3 Summary of selected good practices of water-energy-food nexus approach for urban drought management

Good practices	Areas of integration	Benefit for drought management
1. Municipal wastewater use for cooling of thermal power plant, Delhi	Integration at system	Reduce completion over freshwater between power generation and other users
2. Capturing energy in the process of wastewater treatment, Da Nang	Integration at system	Reduce water footprint for urban electricity use and reuse of treated wastewater for irrigation
3. Use of solar energy is a solution of reducing cost of delinization, King Abdulaziz City	Integration at system	Reliable and affordable supply of desalinated water in dry areas
4. Urban–rural collaboration for water environment conservation, Kanagawa	Integration of scale	Ensure reliable amount and good quality water in all season
5. Economic incentive for agriculture with stormwater retention capacity, New York	Integration in policy	Improve water retention capacity of urban agriculture

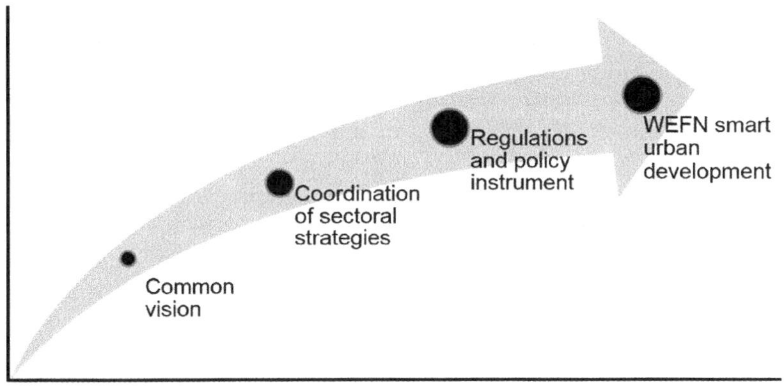

Fig. 5.7 Multi-stakeholder collaboration for operationalized WEFN in urban development

lishment and scale-up WEFN solutions would require institutionalization of multi-stakeholder collaboration in both horizontally and vertically. Establishing an inter-sectoral new institute would be the first step to formalize multi-stakeholder collaboration to integrate water, energy, and food systems for more efficient resource use in urban areas. Nexus Task Force of Da Nang City can be a model case for other cities. The city mayor may lead the new inter-sectoral institute. This institute will identify and guide WEFN solutions in the urban development policies, regulations, and projects to meet local needs. It will work to establish a common vision with maintaining a proper balance among the sectoral interests (Fig. 5.7).

5.5.2 *Incentive Mechanism for Design-Led WEFN Management in Urban*

Water-energy-food nexus approach offers solutions for water balance management and improves or maintains water environment, conserve water resources. WEFN smart urban design emerges as options for urban drought management, and these are often more promising for individual building or household. This design often consists of decentralized small technologies rooftop gardening, rainwater harvesting, installation of efficient technologies. Adoption of decentralized design-led WEFN management can generate large public benefits including reliable resource supply, financial, and environmental. But direct benefit of adopting resource smart technologies and practices may not be sufficing to cover installation cost and maintenance cost by individual households that often hinder promotion of design-led nexus management. Therefore, governments should introduce economic incentive tools such as subsidies, tax exemption that make visible direct benefit for individual households. For example, major cities of Australia provide subsidies for installation rainwater tank (Zhang et al. 2015).

5.5.3 *Enhancing WEFN Knowledge in Urban Communities*

Public involvement is critical for designing and scaling-up of WEFN approach, because they are the major implementer of this new solution. It is also important to address local needs in the design. However, WEFN is a new concept, which would require the widespread education of the urban communities, who are the main victims of drought. Educational institutions including universities, research institutes, and training institutes can play a key role to build WEFN thinking and behavior in urban communities through integrating and promoting the WEFN into the curricula.

5.6 Conclusions

As the home to more than half of the world's population, urban areas are vulnerable to growing water scarcity. Conventional silo approach of drought management fails to provide solutions to major challenges including growing competition, conflicts, shortages, waste, and degradation of water resources. One of the main reasons is not recognizing the interlinkages in relationships between water and other sectors, particularly energy and food sectors. Sufficient access to any one of these resources affects the others. Therefore, it is imperative to shift from a sectoral approach to an integrated approach supported by all sectors to enhance the resilience of urban systems to drought. Urban WEFN as a policy framework can facilitate more integrated, sustainable, and resilient planning for urban drought management.

Since the WEF supply to cities depend on both local in-boundary WEF production and trans-boundary production, cities must recognize that demand for WEF has far-reaching environmental impacts both within and outside city boundaries WEFN system. Production and supply of WEF to cities is a key aspect of the framework of WEF interactions in both urban in-boundary process and trans-boundary process, which addresses synergies and trade-offs between individual environmental impact categories. It helps to identify actions or set of actions to deal with water scarcity in the following four key categories: (a) changes in community-wide urban WEF demand; (b) shifts between in-boundary versus trans-boundary WEF supply; (c) changes in trans-boundary production systems; (d) changes of in-boundary production and cross-sectoral WEF interactions.

Cities in various parts of the world are demonstrating the possibility to continue growth, even with water scarcity, by realizing synergies of WEFN in the water, energy, and food sectors. The practical responses by cities highlighted above demonstrate that WEFN approach can build resilience in the urban water system by capturing multiple benefits. Governments should take proactive steps in promoting WEFN approach in managing urban water system. Establishment of inter-sectoral institute would be the first step to promote WEFN, which will facilitate the process of developing common vision, coordination of sectoral strategies, enabling policies, etc. Economic incentive mechanisms are also crucial to motivate individual household to adopt

WEFN management. Moreover, it is important to educate city community about the synergistic relationship of WEF and benefits of this approach for themselves and resilience of urban water system.

References

Brandi C, Richerzhagen C, Stepping K (2013) Post 2015: why is the water–energy–land nexus important for the future development agenda? German Development Institute, Briefing Paper 3/2013
Cohen E (2014) The water footprint of urban energy systems: concepts methods and applications for assessing electricity supply risk factors. University of Colorado, Denver
Deleware Valley Regional Planning Commission (DVRPC) (2011) Eating here: Greater Philadelphia's food system plan. Available at www.dvrpc.org/reports/10063.pdf
Denver's Climate Resiliency Committee (2014) City and county of Denver climate adaptation plan. Available at www.denvergovorg/content/dam/denvergov/Portals/771/documents/Climate/Climate_Adaptation_Final%20with%20letter.pdf
Endo A, Tsurita I, Burnett K, Orencio PM (2015) A review of the current state of research on the water, energy, and food nexus. J Hydrol Reg Stud. https://doi.org/10.1016/j.ejrh.2015.11.010
Foran T (2015) Node and regime: interdisciplinary analysis of water–energy–food nexus in the Mekong Region. Water Altern 8(1):655–674
Griffiths-Sattenspiel B, Wilson W (2009) The carbon footprint of water. River Network, Portland. Available at https://www.csu.edu/cerc/researchreports/documents/CarbonFootprintofWater-RiverNetwork-2009.pdf
Hoff H (2011) Understanding the nexus. Background paper for the Bonn 2011 conference: the water, energy and food security nexus. Environment Institute, Stockholm
IEA (2016) World energy outlook 2016. France, Paris
IGES (2013) Water availability for sustainable energy policy: assessment of South and East Asia cases. IGES, Hayama, Japan
Kanagawa Prefecture (2005) Basic policy for Kanagawa water source environment conservation and restoration (in Japanese). Retrieved in Jan 2018 from http://www.pref.kanagawa.jp/
Kanagawa Prefecture (2015) Kanagawa water source environment conservation and restoration—overview of achievements to date (in Japanese). Retrieved in Jan 2018 from http://www.pref.kanagawa.jp/
Kanagawa Prefecture (2016) Third phase of Kanagawa water source environment conservation and restoration 5-year action plan (in Japanese). Retrieved in Jan 2018 from http://www.pref.kanagawa.jp/
Ministry of Power (2015) Fact sheet on agriculture demand side management. Ministry of power, New Delhi, India. Available at http://pib.nic.in/newsite/PrintRelease.aspx?relid=128572
Ramaswami A, Boyer D, Nagpure AS, Fang A, Bogra S, Bakshi B, Cohen E, RaoGhorpade A (2017) An urban systems framework to assess the trans-boundary food-energy-water nexus: implementation in Delhi. India Environ Res, Lett, p 12
The Guardian (2016) Where are the world's most water-stressed cities? Available at https://www.theguardian.com/cities/2016/jul/29/where-world-most-water-stressed-cities-drought
United Nations (2011) World urbanization prospects
Weitz N, Huber-Lee A, Nilsson M, Davis M, Hoff H (2014) Cross-sectoral integration in the sustainable development goals: a nexus approach. Stockholm Environment Institute (SEI), Discussion Paper
World Economic Forum Water Initiative (2011) Water security: the water–food–energy–climate nexus. Island Press, Washington, DC

World Water Assessment Programme (WWAP) (2012) The United Nations world water development report 4: managing water under uncertainty and risk. UNESCO, Paris

WRI (2017) Droughts and blackouts: how water shortages cost India enough energy to power Sri Lanka. Available at http://www.wri.org/blog/2017/07/droughts-and-blackouts-how-water-shortages-cost-india-enough-energy-power-sri-lanka

Zhang F, Polyakov M, Fogarty J, Pannell DJ (2015) The capitalized value of rainwater tanks in the property market of Perth, Australia. J Hydrol 522:317–325

Zhao X, Liu J, Yang H, Duarte R, Tillotson MR, Hubacek K (2016) Burden shifting of water quantity and quality stress from megacity Shanghai. Water Resour Res 52:6916–6927

Chapter 6
Coping with Scarcity and Urban Water Governance: Case of Udon Thani's City Region

Wijitbusaba (Ann) Marome

Abstract This paper is a case study of the limitations to urban climate resilience of the city region of Udon Thani province in Thailand which is rapidly growing and aims to become a regional hub despite being confronted with climate-related challenges such as drought and flood. The research focuses on the formal institutional factors which are obstacles to mainstreaming climate change adaptation into urban management and thus exacerbate Udon Thani's vulnerability to drought and flood. The research examines plans and policies of land use and water supply services at national, regional, and local levels regarding the provincial development strategy and the spatial coordination of public agencies at the same level of governance in order to examine the vertical and horizontal institutional arrangement, legal framework, policy drivers, and other limitations to the mainstreaming of climate resilience into urban development planning. Interviews were also held with relevant public officials and practitioners. The findings reveal that there are limitations caused by formal institutional arrangements and gaps which have resulted in the lack of coordination between agencies at different levels or administration or agencies operating under different ministries. There are also discrepancies of responsibility and information and financial resources which significantly undermine the efficiency of the two urban services, and, consequently, Udon Thani city region's climate resilience and ability to accommodate spatial and socioeconomic growth.

Keywords Urban resilience · Climate change · Water · Thailand · Institution Governance

This paper is part of the research called "Spatial and socioeconomic risk assessment of flooding in Udon Thani Nakorn Municipality and outlying municipalities and Tambon administrative organization" funded by the Thailand Research Fund (TRF).

W. (Ann) Marome (✉)
Faculty of Architecture and Planning, Urban Futures Research Unit of the Thammasat University, Bangkok, Thailand
e-mail: wijitbusaba@ap.tu.ac.th

B. Ray and R. Shaw (eds.), *Urban Drought*, Disaster Risk Reduction,
https://doi.org/10.1007/978-981-10-8947-3_6

6.1 Introduction

Located in the Northeastern Thailand, Udon Thani is province that became a regional transport hub after experiencing rapid economic and infrastructure development in the 1960s because it was the US Air Force's base during the Vietnam War. The Office of the National Economic and Social Development Board's 2007–2011 and 2012–2016 strategic plans have also proposed the development of the logistic systems in Udon Thani's city region, while the border areas of Udon Thani are supported by the Special Economic Zone in Nong Khai. Udon Thani is therefore expected to experience further growth and is addressed in various national and regional strategies. Udon Thani is a dynamic area with high intensity and diverse land use as well as high socioeconomic importance to Thailand. However, like other provinces in the Northeastern region, Udon Thani's development potential is largely threatened by extreme climate-related risks and a freshwater ecosystem such as drought and flood, which undermines not only the rising urban areas but also the province's large agricultural sector.

This case study focuses on the formal institutional factors which are obstacles to mainstreaming climate change adaptation into urban management and thus exacerbate Udon Thani's vulnerability to drought and flood. This examines via analysis of the two urban services of land use and water supply. These components are explored because they are linked to issues of flood, drought, and urban growth. The improvement of land use and water supply is necessary for Udon Thani's development and climate resilience, and the degree to which these components are able to accommodate climate-related risks are indicative of the extent of the mainstreaming of water management into urban development planning. The research area covers the urban areas of the province and is comprised of Udon Thani city municipality, Nong Samrong town municipality, Banjan subdistrict municipality, Nong Bua subdistrict municipality, Na Dee tambon administration organization, and Banjan tambon administration organization. It should be noted that these areas fall under different boundaries of the regional and the local administrative division systems which are inconsistent, thus highlighting one of the various obstacles to the mainstreaming of climate resilience and ensuring policy coherence.

6.2 The Importance of Formal Institutional Factors in Making Urban Climate Resilience in Udon Thani City Region

The Global Climate Risk Index (Kreft et al. 2017) ranks Thailand as the 10th country most affected by long-term impacts of climate change. It is therefore unsurprising that adaptation, with the ultimate goal to enhance climate resilience, is highly prioritized in the national agenda. Thailand's National Adaptation Plan (NAP) has been initiated since 2015 under the framework of Climate Change Master Plan 2015–2050 and is expected to be launched in late 2018. Under the NAP, six priority sectors are identified, namely: water management, agriculture and food security, tourism, human health, natural resources management, and human settlement and security. The formulation of NAP focuses on three main aspects of climate change adaptation: national risk map, database of best practices, and guideline for implementation. Thailand's NAP will establish a framework for mainstreaming adaptation plans into sectoral and local planning. Regarding human settlements, a large number of big cities in Thailand are situated in risk-prone areas. Without proper national and local adaptation plans ad response measures, Thailand and its cities will not be able to achieve their sustainable development goals, and the consequences of climate change will have disastrous impacts upon the well-being and livelihood of the inhabitants.

Following a large flood in 2011 across multiple cities in Thailand, water management became a priority for many agencies such as the city planning department. Nonetheless, the efficiency in the spatial aspect of urban management of cities in Thailand remains limited; this is especially significant due to trans-boundary and multidimensional nature of city's key threat of flooding and drought. Cross-jurisdictional urban management is not supported by a mandated entity, and various basic infrastructures are managed by agencies of different levels, leading to limitations of cross-jurisdictional management and creating gaps in multiple forms, such as basic infrastructure and urban planning. The fragmentation of governance also means that there is limited cooperation between the local governments of cities and the surrounding areas regarding urban planning, drainage system planning, floods and drought prevention. Moreover, the overlaps or discrepancies between climate action plan and policy between local governments and national formal institutions, which undermines the need to link local action to the mainstreaming of climate change adaptations into urban development policy.

Climate change adaptation and resilience must be integrated to development strategies in order to minimize the impact climate related on investments and infrastructure and reduce social vulnerability of the residents. In order to mainstream climate resilience, the following components contributing to city resilience system need to be considered: the physical, socio-cultural and economic, environmental formal institutional setting systems. These four components are closely interrelated and considered as 'system of cities' or 'cities network'. City resilience cannot be considered as a single system of cities but should be viewed in terms of highly interrelated levels of systems of the city, peri-urban areas, cities network and surrounding rural areas. For the institutional setting system, there are four enabling factors to be considered, which are the legislations, decision- and policy-making process, formal and informal institutions, and structure and participation (Ernstson et al. 2010). This paper focuses on institutional setting system of formal institution that is related to the environmental quality system of water resource and physical system of land use. Not only do effective formal institutions enable climate resilience, they also contribute to improving urban governance. Consolidating city resilience will enhance the capability of city systems to function better not only during the time of crisis but also the normal times, and public agencies will have a spatial coordination at the same level of governance. Hence, this research approach will contribute to the

Sustainable Development Goal, especially Goal 11 of making cities inclusive, safe, resilient, and sustainable.

Within formal institutional setting system, another requirement is for mainstreaming climate resilience is to ensure that there is coherence and unity in the approach of local and national authorities (Blanco et al. 2011), especially in the case of Thailand where central government agencies remain in control of many aspects of urban and environmental management at all levels. Climate change adaptation and resilience in Udon Thani can be analyzed using IIED's building blocks for climate mainstreaming (Pervin 2013) which categorizes the components of the process into three 'blocks': the enabling environment building block, the policy and planning building block, and the programmes and project building block. This study focuses on the enabling environment and policy-planning building blocks.

The research includes the examination of primary data and conducted multiple in-depth interviews with officials in Udon Thani, which offered a better understanding to the enabling environment upon which these policies were formulated, and the analysis of plans and policies at national, regional, and local levels regarding the management of flood, drought, land use, and water supply management within the context of larger development aims and projects; specifically, the formal institutional gaps present during the policy design or plan formation, the links between water management and the two urban services, and limitations for operation which includes shortcomings in organizational structure, legal authorization, and other limitations. Extensive literature review was conducted of the policy and plans at multiple levels including the province's strategic development plan, development plans of the Joint Committee of Public and Private panel for Fixing Economic Problems, flood and drought management plans, environmental management plans, comprehensive urban plan, and the strategic plan of the Provincial Waterworks Authority.

Formal institutional analysis was conducted with a focus on policy-driving mechanisms, legal framework, and other components in order to establish an understanding of the nature and capacity of organizations which leads to policy change, as well as the obstacles and limitations to policy design and implementation. The analysis of land use and water supply management in Udon Thani province also focuses on the vertical and horizontal institutional arrangements, which refers to the coordination between public agencies of the national, regional, and local levels, and the spatial coordination of public agencies at the same level of governance, respectively. Moreover, there is a need for improved incorporation of climate change mitigation and water management to various aspects of urban planning and management.

6.3 Formal Institutional Analysis of Land Use and Water Supply Services in Udon Thani's City Region

Udon Thani city region experiences water shortages almost every year. Drought was declared in 2013, and there was flooding in 2001, 2006, and 2011. An analysis of

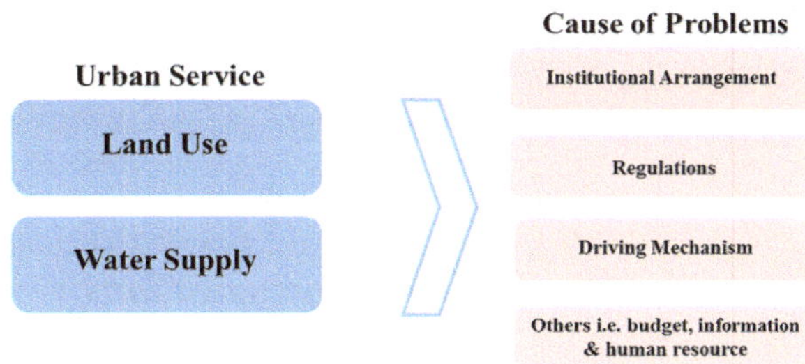

Fig. 6.1 Causes of the problems to urban water governance

plans and policies at national, regional, and local levels regarding the management of flood, drought, land use, and water supply management reveals that when flooding is addressed, the inclusion of drought in local and regional level plans is very limited. Plans addressing flood and drought are usually created by central agencies on an ad hoc basis, or only created during the year when there was severe flooding or drought, thus making it difficult for local authorities to incorporate and implement those plans and climate-related planning in their jurisdictions. This is further exacerbated by how most development or action plans of local authorities cover a period of three years while the central governments' flood and drought plans are created on a one-year basis.

The research team has identified five key limitations to the development of Udon Thani province. Firstly, there is incoherence and contradiction between the provincial authority's strategic plan and the flood and drought management plan. Secondly, there is a lack of continuation of flood and drought management plans which are usually issued by the central government. These plans are created annually and often after an emergency event, so they are more reactive than preventive in nature and lack the essential integration of the water resource management. Thirdly, there is limited information forecasting scenarios of future floods or droughts in the provincial strategic plan, this resulting in a plan that only addresses general environmental issues. Fourthly, there is a lack of holisticness in spatial and infrastructure management, leading to incoherence between urban policies and risk transfers to neighboring areas. Lastly, there is a distinct lack of unity in the planning and development of the two urban service sectors of land use and water supply. This undermines the development capacity and climate resilience of Udon Thani's city region as well as the province's aim to become a regional economic hub (Fig. 6.1).

6.3.1 Incongruence of Plans

There is inconsistency between the strategic plans at the provincial and local levels and the flood and drought plans. The development strategy from both the central and local governments fails to sufficiently take into account Udon Thani's recurring flood (which have occurred in 2001, 2006 and 2011) and its annual water shortage, thus hindering future spatial development plans.

6.3.2 Lack of Continuity and Integration of Flood and Drought Management Plans

Flood and drought management plans are usually issued by the central government and cover a period of only one year. These plans have a tendency to be reactive plans which address only the issues arising from a flood or drought event of the year that it covers. Meanwhile, local plans regarding flooding tend to be plans for drainage system maintenance. This highlights the lack of coordination as well as the lack of climate change forecasting and how it will increase the frequency and severity of flood and drought.

6.3.3 Limitations of Data Forecasting Flood and Drought in the Future

The lack of climate forecasting leads to limitations in guiding the strategic plan of Udon Thani province. Thus, the provincial plan focuses mostly on improving the general environment, i.e., creating a livable city.

6.3.4 Lack of Comprehensiveness in Spatial and Infrastructural Management

In the study area of four municipalities and two Tambons (Udon Thani city municipality, Nong Samrong town municipality, Banjan subdistrict municipality, Nong Bua subdistrict municipality, Na Dee tambon administration organization, and Banjan tambon administration organization), there is a distinct lack of comprehensiveness in spatial management. For example, land use is not consistent with the planned basic infrastructures for the province's future developments and oftentimes utilizes areas that are known to be flood prone. The lack of unity or congruence in flood and drought management means that policies may be transferring risks to other areas or sectors instead of mitigating them.

6.3.5 Lack of Unity in Planning and Developing Important Public Services in the City

The important urban service sectors are land use and water supply, which are drivers for the province's aim in becoming a center of trade and investment. The problems of each sector will be further explored.

Four limitations have been identified in terms of the vertical and horizontal institutional arrangements, legal framework, policy-driving mechanisms, and other components. Regarding land use, there are limitations from the enabling environment that are obstacles for mainstreaming climate change adaptation into the comprehensive plans. The latest provincial comprehensive plan of Udon Thani was based on outdated information, leading to a plan that is inconsistent with existing settlements. The comprehensive plan fails to take into account flood risks, as demonstrated by how the plan-designated settlements in areas with the risk of recurring flood. The shortcomings in land-use planning may be a result of how there was a transfer of responsibilities from the Udon Thani department of Public Works and Town and Country Planning to the local-level organization, the Udon Thani city municipality, to create the Udon Thani comprehensive plan. According to an interview with a city official, this transfer of responsibilities was not supported by the allocation of personnel and budget, leading to inefficiency and delays. This also indicates that there are discrepancies in vertical institutional arrangements. Few local government authorities have the skilled personnel and/or resources, knowledge, and budget to do so, thus demonstrating a gap between assigned responsibilities and available capacity. In the horizontal dimension, the formal institution arrangement of municipalities lacks cross-jurisdiction coordination which is important for mitigating or preventing flood. Civil groups such as the Udon Thani Chamber of Commerce and the Industrial Committee have therefore submitted a request for the comprehensive plan to be reviewed.

In the legal aspect, there are discrepancies regarding the decentralization of authority. Ultimately, it is the central authorities, and not the local government, who make the final decision in the planning process: according to the urban planning act of 1975, the urban planning committee has the power to reject or amend the comprehensive plan proposal. This committee is composed of representatives of central government agencies and does not include representatives of the concerned area, resulting in a lack of integration. Moreover, while land-use planning has been identified as being useful for the province's development strategy, the development plan does not address tools or substantive process for urban planning and offer only vague goals such as creating livable cities.

Similarly, the water supply sector lacks integration with climate change adaptation policies. Currently, the Udon Thani provincial waterworks authority uses only 58% of its total water production capacity, covering all areas in the Amphoe Mueang with the exception of some villages that have their own water supply system. The Udon Thani provincial waterworks authority has the capacity to accommodate the needs of 120,000 people, but the current number of service users is 69,876 persons. The main limitation for water supply services in Udon Thani is the increasing demand for raw water, especially in times of flooding where the raw water sources located in areas of recurrent flooding may become contaminated. In the past, the water supply did not particularly suffer during droughts because the urban area is prioritized over the surrounding agricultural areas when water is allocated, but the conditions will change with the worsening drought and the decreasing volume of water of existing raw water sources. Another issue is the contamination of pesticides and other chemicals as a result of farming in nearby areas which often gets washed into the water sources by the rain.

The Udon Thani provincial waterworks authority recognizes that there is a need to find alternative water sources not only for satisfying rising demands but also to ensure that the water supply service will be less severely affected by flood and drought as the water treatment capacity suffers in the event of flood. The Udon Thani provincial waterworks authority is allocated 33 million m^3 of water from the Huai Luang basin, based on a MoU between the Department of Irrigation, the Provincial Waterworks Authority, and the Udon Thani city municipality. This information is inconsistent with the projection made in the Provincial Waterworks Authority's plan stating that the demand for raw water in 2014 is 47.9 million m^3 (Table 6.1), suggesting that it is possible that there water supply is insufficient in some areas. Similarly, the demand of water predicted by the Department of Irrigation is less than actual use. In 1999, the Department predicted that in 2017, the demand for raw water will be 24.022 million m^3 per year while the amount of water Department allocated to Udon Thani provincial waterworks authority in 2016 was 33 million m^3. The issue of available raw water resources is further complicated by the lack of data on raw water consumption through alternative water supply service providers such as local or village water supply systems. This also includes the areas within Nadee tambon administration organization and Banjan tambon administration organization which are within the research perimeters.

Another reason why the Udon Thani provincial waterworks authority needs to find an alternative, more sustainable source or raw water is its goal in extending its service coverage to more residents in order to offer increased access to more secure and higher-quality water supply.

Water supply in Udon Thani further complicated by how it involves multiple actors with roles in the three stages of the water supply process. The three stages are (1) Finding water sources and allocating and maintaining raw water resources (actors: The Department of Irrigation, Joint Management Committee irrigation and protection of Huai Luang, Provincial Waterworks Authority, and the Department of Water Resources), (2) Water treatment and water supply systems (Provincial Waterworks Authority), and (3) Tap water distribution (Provincial Waterworks Authority, municipalities or other local administration organizations) which are required to provide water supply services in the area. Other actors include the Provincial Electric Authority who provides electricity needed for water supply services, the Udon Thani Office of Natural Resources and the Environment which is the central government entity that oversees the environmental aspects of water supply operations, the Department of Highways who enables the laying of water supply distribution systems in accordance

Table 6.1 Current demand for water supply from the Udon Thani Provincial Waterworks Authority and predictions for the future (from 2015)

Water supply system		Udon Thani's Provincial Waterworks Authority		
		Present (2014)	Stage 1 (2020–2029)	Stage 2 (2030–2039)
Maximum water supply demand	(cubic meter per day)	108,018	186,127	252,083
	(cubic meter per hour)	4,501	7,755	10,503
Productivity	(cubic meter per day)	121,200	190,800	260,400
	(cubic meter per hour)	5,050	7,950	10,850
Future plan of water supply plant	(cubic meter per day)		69,600	69,600
	(cubic meter per hour)		2,900	2,900
The amount of water supply user	(person)		119,406	155,390
Demand of raw water	(million cubic meters per year)	47,995	75,557	103,118

Source Udon Thani Provincial Waterworks Authority master plan (2015)

with new roads, and the local administration organizations who own the land used for pipe laying as well as approving operations and providing policy guidance through their development plans. Moreover, there are additional agencies who are involved in water supply and services during times of emergency or disasters, including the Udon Thani Provincial Agricultural Office, Udon Thani District Office of Groundwater Resources which coordinates with the local administration in times of drought, and the Regional Office of Water resources who are involved with the restoration of water sources in areas afflicted by disaster.

There are multiple issues regarding the formal institutional arrangement of the water supply service. Vertically, the Regional Waterworks Authority is largely bound by central government authorities. However, the Regional Waterworks Authority's desire to find an alternative raw water source in response to anticipated demand of the future is not shared by the central authorities such as Subcommittee on Water Management who are more focused on addressing issues as they arise. Regarding the horizontal institutional arrangement, the Udon Thani Provincial Waterworks Authority lacks the support from and coordination with other entities in finding an alternative raw water source to accommodate rising future needs as well as for planning and operational purposes. There is also a lack of coordination between the Waterworks Authority and other relevant urban services such as power supply and urban planning. Within the financial framework, there is a lack of budgetary support from local authorities despite the fact that they are legally mandated to ensure the provision of water supply in the 1953 Municipality Act because the water supply expenditure has been grouped with other services under the public utilities budget.

6.4 Deliberative Visioning Workshop

A workshop was held after the completion of the research. Participants of the workshop included personnel from Udon Thani Provincial Waterworks Authority, Udon Thani Province Administration Organisation, Udon Thani City Municipality, and the wastewater management agency. Findings from the formal institutional gap analysis conducted were presented to the relevant actors and stakeholders who expressed their agreement. The research team also assessed Udon Thani's capacity to cope with the risk of flood and drought through the creation of three transient scenarios: business as usual, population increase according to the strategic plan of the Provincial Waterworks Authority, and the scenario where all of the province's envisioned development projects of the strategic plan are materialized, leading to significant urban growth. For each scenario, projections are made for the requirements that the land use, and water supply sectors. These scenarios were then used in a visioning workshop with public officials from relevant formal institutions in Udon Thani to stimulate dialog on whether or not their organization has considered the scenario and the possible limitations they will face in the case of flood or drought.

For instance, in Scenario 2, Udon Thani is the center of trade and investment. There was development in the water supply infrastructure, which caused the population density to increase. There are issues that should be considered in the future in the event that Udon Thani is the center of trade and investment and has development in the water supply infrastructure.

During the year of 2020 and 2030—or in 6 and 16 years, respectively (counting from 2014)—of the Provincial Water Authority, Udon Thani was chosen for future analysis due to the fact that those will be the years when the maximum demand for tap water will exceed the Udon Thani's water authority's capacity to supply, if there are no terms of capacity expansion of the operation of a new water station under the master plan.

In 2020, the maximum demand for tap water will be as much as 5,452 m^3 per h, while the maximum capacity of supply is only 5,050 m^3 per h. The new water station will increase tap water capacity by 2,900 m^3 per h, resulting in a total tap water output of 7,950 m^3 per h, which will be able to meet the rising water demands until 2030. That is when the maximum demand for tap water will increase up to 8,021 m^3 per h, thus there will be a need to increase tap water production by another 2,900 m^3 per h.

Through analysis, it is found that the area that Udon Thani waterworks authority covers is the entire district, with exceptions in a few villages where the village water supply is used. As predicted in 2020, there will be 409,362 inhabitants or 149,417 households in said area; and 452,190 inhabitants or 165,045 households in 2030.

In the future, there is a tendency that the demand of the use of water per person will increase, and Udon Thani Provincial Waterworks Authority aims to increase their coverage. Thus, there is a need to increase the capacity of water supply and reduce water loss. It is estimated that the Udon Thani Provincial Waterworks Authority needs to find additional sources of raw water before 2030 when the demand for water of approximately 452,190 will need to be met. Indeed, securing a sustainable raw water source is the most significant challenge of the water supply service. It is predicted that in 2030, the daily demand for water will be about 148,079 m^3 per day, with the maximum demand reaching 260,400 m^3 per day.

In the third scenario, the city develops according to all strategic plans, including that of the Joint Committee of Public and Private panel for Fixing Economic Problems: Udon Thani Ratchapat University fully opens in 2021, the Udon Thani industrial estate opens in 2020, and the urban transport network in 2021. The population growth is expected to experience an increase of 60,000 people, and water supply demand will be 67,008.09 m^3 per day in the next 6 years, 69,878.55 m^3 per day in the next 8 years, and 72,874.29 m^3 per day in the next 10 years. These estimates are inconsistent with the projection of the waterworks authority which include the population growth of 46,237 in 2021, and the water supply of 75,495.12 m^3 per day in 2020 and 111,056.85 m^3 per day in 2030. Water supply will not be at risk in this scenario because its capacity is also expected to increase as the population grows, but it will need to find an additional source of raw water. It should also be noted that the projects included in this scenario have not taken into account the potential impacts of flood and drought.

This end-of-project workshop was also intended to foster a more holistic approach to mainstreaming climate change adaptation and to encourage participants to consider the temporal aspect of operational planning. However, these scenarios are not predictions for the future—they are simply a tool for understanding the climate resilience of existing policies; these scenarios do not accurately reflect the reality on the ground due to limitations of the availability of data such as the number of informal residents. Two scenarios are based on plans and strategies that are built upon potentially inaccurate data themselves, as suggested by discrepancies in estimates on the same subject such as population size by different plans.

The output of this deliberative workshop demonstrates that the relevant actors have some awareness of climate-related issues but find it difficult to translate these ideas into practice due to limitations in political will, authority, formal institutional arrangement, and budgets. Participants also agree that measures such as reducing water loss through leakage and excessive usage is no longer sufficient and must be complemented by more significant measures such as to increase water procurement and production capacity. Likewise, for land-use planning, it is no longer sufficient to simply avoid flood-prone areas but also to implement adaptive measures to ensure that built structures can withstand flood and that they will not exacerbate climate risks. Incidentally, these are also the measures recommended by UNDP-UNEP Poverty-Environment Initiative (2011) for mainstreaming climate change adaptation into development. However, the workshop also highlighted limited risk assessment as well as the lack of adequate coordination between organizations which needs to be addressed immediately as the impacts of climate change will exacerbate the challenges of the ongoing urban development.

The output of the deliberative workshop stresses the important to mainstream climate adaptation into long-term planning such as urban development strategies, drawing on the findings of the research team, highlights some of the main policy directions the city's future strategic framework should follow. A long-term adaptation strategy that is collective and participatory, guided by adequate data and a deeper recognition of the complex social dimensions of flooding, is crucial. Together with substantial national and municipal governance reform, this approach will help to address these challenges and preserve city's immense social and economic vitality.

6.5 Conclusion

This study examined and demonstrated limitations to the mainstreaming of climate resilience in two urban services (land use and water supply) and urban development planning by focusing on the policy and planning and the enabling environment 'building block' of the building blocks of for climate mainstreaming. Existing policies and plans in the national, provincial, and local levels lack coherence and there is very limited integration of climate change adaptation and development. Consequently, the urban services of water supply and land use are not adapted to immediate climate risks. There are limitations in the formal institutional arrangement aspect in the vertical and horizontal dimensions: for example, the local authorities are legally mandated to create a comprehensive plan but lack the ability to do so, thus the responsibility falls upon the central authority that does not have a thorough understanding of the area. There are also insufficient channels for coordination with the local jurisdictions and utilities service providers. Similarly, the water supply service lacks vertical coordination between entities of different levels of government as well as horizontal coordination between relevant agencies operating under different ministries. This resulted in incoherence in planning, which is further complicated by the process of raw water allocation.

There are also significant discrepancies between development plans and the allocation of power and resources (i.e., the legal and financial frameworks) needed to execute certain policies which are further confounded by ineffective formal institutional arrangements. These policy limitations are a result of how the government is only partially decentralized and how decision-makers did not perceive climate resilience as being inherently linked to development policies and the insufficiency of information services. The case of Udon Thani is not unique: the province's situation is similar to that of many other emerging provinces in Thailand. Thus, there is a need for further research regarding the mainstreaming of climate resilience in urban development policies in Thailand in order to ensure a more sustainable future.

References

Blanco H, McCarney P, Parnell S, Schmidt M, Seto KC (2011) The role of urban land in climate change. In: Rosenzweig C, Solecki WD, Hammer SA, Mehrotra S (eds) Climate change and cities: first assessment report of the urban climate change research network. Cambridge University Press, Cambridge, UK, pp 217–248. Retrieved 23 July 2015. From: http://uccrn.org/files/2014/02/ARC3-Chapter-8.pdf

Cynthia Rosenzweig and CiudadesSostenibles y CambioClimatico UCCRN (2014) Building the knowledge base for city action on climate. Retrieved 23 July 2015. From: http://seminario.pucp.edu.pe/ciudades-sostenibles/wp-content/uploads/sites/6/2013/07/5-UCCRN-Rosenzweig-Lima-Sep-17-2014.pdf

Department of Land Development, Ministry of Agriculture and Cooperatives (2008) Data on land use in Udon Thani province 2008. Accessed 19 Sept 2015 via http://traffregion.otp.go.th/mis/LandUse/landuse_info.aspx?rid=10&pid=41&zid=0#

Department of Water Resources, Ministry of Natural Resources and Environment (2015) Flood rehabilitation project (1st). Accessed 19 Sept 2015 via http://pm.water.go.th/floodhab551/finish.aspx?org=3

Department of Public Works and Town & Country Planning, Ministry of Interior (2015) Comprehensive plan, maps and plans annexing ministerial law download services. Accessed 19 Sept 2015 via http://www.dpt.go.th/urbanplanning/page5.php

Department of Public Works and Town & Country Planning, Ministry of Interior (2015) Northeastern regional plan, part 3: visions and regional development strategy. Accessed 19 Sept 2015 via http://www.dpt.go.th/nrp/images/stories/pdf/ReportOf6Region/Northeast_report/Northeast03.pdf

District 7 regional waterworks authority (2015) Organization missions. Accessed 19 Sept 2015 via http://reg7.pwa.co.th/view_content.php?id=28

Ernstson H, van der Leeuw SE, Redman CL et al (2010) Urban transitions: on urban resilience and human-dominated ecosystems. AMBIO 39:531. https://doi.org/10.1007/s13280-010-0081-9

Kreft S, Eckstein D, Melchior I (2017) Global climate risk index 2017: who suffers most from extreme weather events? Weather-related loss events in 2015 and 1996 to 2015, Germanwatch

Pervin M (2013) A framework for mainstreaming climate resilience into development planning. Working paper, International Institute for Environment and Development

Phuttharak T (2013) The regional rapid growth city and urbanization in Thailand. Retrieved 19 Sept 2015. From: http://ibac-conference.org/ISS%20&%20MLB%202013/Papers/ISS%202013/B6261.docx.pdf

Udon Thani department of Public Works and Town & Country Planning (2012) Vision and mission. Accessed 19 Sept 2015. http://www.dpt.go.th/udonthani/main/index.php?option=com_content&view=article&id=2&Itemid=3

Udon Thani Regional Waterworks Authority (2015) Final report on the strategic plan or the Udon Thani regional waterworks authority. Udon Thani

UNDP-UNEP Poverty-Environment Initiative (2011) Mainstreaming climate change adaptation into development planning: a guide for practitioners

World Bank (2016) Institutional analysis from the PSIA workshop. Retrieved 19 Feb 2016. From: http://siteresources.worldbank.org/INTPSIA/Resources/490023l121114603600/12996_workshop_instanalysis.pdf

Chapter 7
Bridging the Distances Between Far Water and Near Water: Case of Yanagawa City from Japan

Shimpei Iwasaki

Abstract Urbanization requires expanding the necessary infrastructure for water and sanitation to make city water safer and keep the environment clean. Instead of stable water supply and drainage, the modernized water infrastructure development weakened the relationship between water and human, which has brought a shift from near water to far water-based lifestyle. In response to this, traditional wisdom which takes into account social-ecological system has been under-evaluated or ignored so that urban water management may become vulnerable due to their lower awareness to the water. With this recognition, this chapter presents a case of successful urban water management in Japan by describing the historical process of canal restoration in Yanagawa City. The case study highlights traditional water management which was linked to 'near water' while exploring the underlying causes of 'far water'. The research provides insights into bridging the distances between near water and far water by creating 'new near water' in the urbanized world.

Keywords Near water · Far water · New near water · Horiwari · Yanagawa City

7.1 Introduction

Urbanization which is the progressive increase of the number of people living in urban areas is common all over the world. In 2014, there are around 54% of the world's population residing in urban areas, in comparison with 30% of it in 1950 (United Nations 2014). In response to the rapid increase of global population, it is expected to bring further changes in the size and growth of urbanization in the future. Cities where a large number of people live together require meeting water demands of the people into a small area (McDonald et al. 2014). At the same time, water quality problems such as human sanitation, and destruction and degradation of

S. Iwasaki (✉)
Department of Environmental Science, International College of Arts and Sciences, Fukuoka Women's University, 1-1-1, Kasumigaoka, Higashi-Ku, Fukuoka 813-8529, Japan
e-mail: iwasaki@fwu.ac.jp

B. Ray and R. Shaw (eds.), *Urban Drought*, Disaster Risk Reduction,
https://doi.org/10.1007/978-981-10-8947-3_7

water ecosystem also emerge as a result of urbanization that increases the variety and amount of pollutants and nutrients in receiving water bodies (Zhou 2014). To make city water safer and keep the environment clean, sustainable urbanization requires expanding the necessary infrastructure for water and sanitation.

In Japan, the infrastructure development for water and sanitation such as dams (for water storage), weirs (for withdrawal of river water), water channels (for water transportation), water treatment plants (for purification of the drawn water), and sewage treatment plants (for purification of water used at homes, plants and other uses) has been accelerated with a sense of urgency from the mid-1950s to cope with the surge in strong water demand during the high economic growth period (MLITT 2014). It needs to be mentioned that before the introduction of modernized water infrastructure, the distance of water–human relationship was physically, socially, and psychologically quite close even in urban areas. Such a close relationship between water and human was named as 'near water' by Kada (2000). The lifestyle was closely tied to near water where the people used to have easy access to water from sources such as wells, reservoirs, rivers, and natural springs, especially for drinking, daily household work such as cooking and washing, transportation, and livelihood works (agriculture and other industries). The local water was used for everyday life, but the water infrastructure development has brought a shift from near water to far water-based lifestyle. For instance, many places replaced their canals and rivers with underground concrete sewers in part due to heavy accumulation of smelly nutrients and wastewater. Furthermore, taking into account hydrological hazards, the government also applied a precautionary approach to safety by building tall dikes along the water's edge, thereby cutting off the access to the local water. As a result of this, the close relationship between water and human has been lost in many parts of Japan, which ignored traditional wisdom related to water management. Improper water management will become more vulnerable for their lifestyle and livelihoods.

With this recognition, this chapter presents a case of Yanagawa City where innovative efforts between city officials and local people on water management (canal restoration) are showcased. In the following section, the chapter firstly profiles the case study site. Second, the research highlights 'near water' based lifestyle and its water management. Third, it explores the underlying causes of 'far water'. Fourth, the chapter describes the process of the canal restoration to revitalize the water environment in urban areas by bridging the distances between far water and near water. Finally, the research provides insights into building the relationship between water and human in the urbanized world.

7.2 Profile of Case Study Site

Yanagawa City, which is the southwest part of Fukuoka Prefecture in Kyushu island, Japan, is situated between 33°05′ and 33°11′ North latitude and 130°28′ and 130°21′ East longitudes, according to Google Earth (Fig. 7.1). It has a population of 67,490 in March 2017 (Yanagawa City 2017). The population has been decreasing after peaking

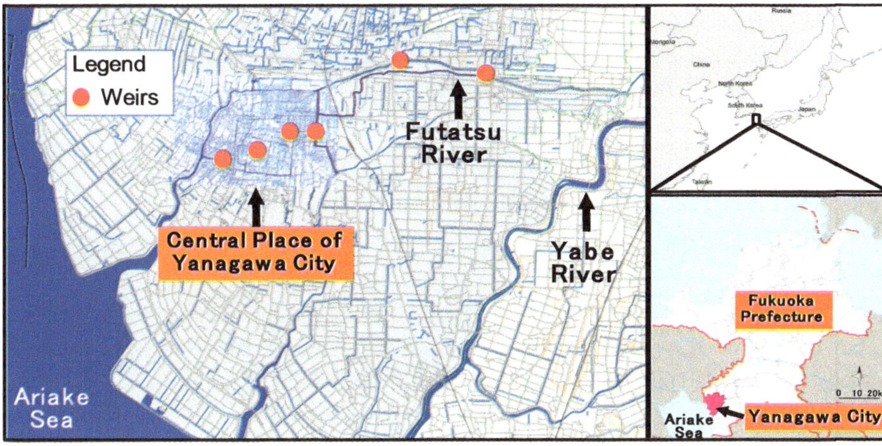

Fig. 7.1 Map of Yanagawa City. *Source* Modified from Mizkan Water Cultural Center (2017)

in 1960 (86, 888), but the number of households has been constantly increasing due to trends toward the nuclear family and aging. The area lying at a height of 0–6 m above sea level is on the mud flat land near the mouths of Chikugo River and Yabe River flowing into the Ariake Sea.

Long times ago (Yayoi period dated 300 BCE to 300 CE), the ancestors had made great efforts to replace the mud land with human settlement and agricultural fields. In the land reclamation process, they built houses on slightly higher grounds which were circled with ditches, in order to keep out flooding and high tides. When it rains, stored water in the ditches was used for irrigation, enabling them to make a living in the area. The land use was further affected by the national laws during seventh and eighth centuries, which included the Jorisei System of land division and the Handenshuju law of land distribution, leading to the foundation of the lay of the land in the region (Tanabe 1991). Then, the city has been built and developed as a castle town since the late sixteenth century.

Yanagawa City, situated at the fringes of Ariake Sea, receives a sole supply of freshwater from Futatsu River. Futatsu River is manmade one which was constructed under the supervision of the lords. Massive people were involved in the construction work and struggled to dig to divert water from the mainstream of Yabe River, in order to cope with irrigation as well as protection of human from flooding. The water from Futatsu River is the lifeline of the city, especially for consumption, agriculture, and other uses of freshwater.

Owing to the traditional water infrastructure development which takes into account social-ecological system, there are a large number of narrow canals even in the central place of Yanagawa City. The network of waterways flowing through Yanagawa City has been called as ‘Horiwari’. A famous film director, Isao Takahata, coined the term in his film entitled 'YanagawaHoriwariMonogatari' which was made in 1987 (Mizunokai 2014). The people are able to enjoy various benefits of living close to Horiwari. Hence, Yanagawa is often called as 'the city of water' in Japan.

7.3 Near Water: Close Relationship Between Water and Human

The Horiwari provides major six important environmental functions to make a living in Yanagawa City; (i) water supply, (ii) water purification, (iii) food security, (iv) transportation, (v) flood regulation, and (vi) subsidence regulation.

Due to saline groundwater, freshwater flowing through the canals used to be a fundamental necessity of life for drinking and other uses of daily work. The people residing in the city required cooperation for sharing the freshwater. If they bring pollution to the canal water in an indiscriminate manner, an undrinkable tragedy may occur in the absence of management as Hardin (1968) espouses. On this account, a given form of social control among users was developed. In 1896, a prefectural ordinance entitled 'River Drinking Water Regulations' was formally established to govern the behavior of users. To make the water safer, for instance, no washing activity was allowed during the period from 1 am to 8 am according to the Article 14. The wastewater was treated by pouring it into a hole dug in the backyard or vegetable fields to ensure clean water. Contaminated water and garbage to the canals shall not be allowed at all times. Based on the strict rules, the people used to draw the water from the canals at the cleanest time of day in the early morning.

Although human activities bring pollution to the canals to some extent, the nature purifies the quality of water (water purification function). Contaminants are filtered through plants. Bacteria also can break down contaminants and then the water environment promotes algae growth. Related to this, the people are able to catch fishery resources which largely depend on richness of plankton as well as quality of water in the canals. When it came to autumn season,[1] the people tried to clean the water environment by drying up the water flow of canals. Then, they cut freshwater weeds and dredge the sludge. These were used as fertilizer in the agricultural fields. Besides, a plenty of fishery resources were caught in the process of water removal so that everyone including small children was actively involved in the event with joy. The elaborations played important roles in purifying the quality of water, providing foods, and recycling wastes naturally. The function of food security includes not only fishery sector but also agricultural work by carrying the water from the city to the fields for irrigation purpose.

Living at the water's edge was useful to carry loads to upstream areas. Before land transport development, most shipping and transportation activities took place as observed in many cities of Japan. Taking advantage of the water transport system,

[1]Taking into account heavy flow of nutrients to the aquaculture grounds for seaweeds in the Ariake Sea, the date of clear-out in the canals has been recently changed from autumn to winter season.

most of the canals were run behind the houses. The boat traffic was common for the Yanagawa people.

As for flood regulation, the ancestors developed various types of traditional unique water infrastructure. The network of waterways in the canals was not enough to stop flooding so that V-shaped dams called as ‘motase’ was built at various road overpasses. The shape under the bridges is narrowest at the bottom. The V-shaped dams serve as a basis for making the water flow through the network, keeping the canals full and maintaining the water level. When the water level is low, the V-shaped dams adjust to make the water flow quickly. The increased water flow can generate oxygen in a whirl, contributing to water quality improvement as well. On the other hands, when it rains heavily, the canals become wide enough to store the water until the tide recedes for its drainage. Apart from the motase, there are many types of sluices and dams to control the water level from both sides of upriver water flow and tide backflows.

It is important to be noted that the fine sediment making up the Ariake clay formation has been deposited in the land of Yanagawa City. Hiromatsu (1977) insisted that land subsidence may occur if the moisture ratio reduces up from 70 to 50% with an assumption that the layer thickness is 10 m, which will cause four-meter depth of land sinking. The constant stored water in the network is expected to minimize subsidence damage. In this sense, Horiwari is fundamental to sustain their life and lifestyle in a sustainable manner.

Putting them all together, the near water structure of water–human relationship can be summarized as shown in Fig. 7.2. Under the traditional water management of Horiwari, the people directly depended on the canals for drinking, foods, and transportation. Their lifestyle was closely tied to the water for daily use. To ensure these functions, they promoted a collective action to improve the water quality. What is more, the water infrastructure development took into account vulnerability of natural hazards from the viewpoint of flood and subsistence regulations. These functions served as a basis for life security to make a living, though the water management through Horiwari required strong commitment to maintenance of water infrastructure among the local people.

7.4 Far Water: Loss of Close Relationship Between Water and Human

The close water–human relationship in Yanagawa City has begun to lose after changes in their lifestyle. In the process of urbanization, mobility and accessibility have increased tremendously. Instead of boats, road traffics and railways spread and became the dominant means of transportation. On the other hands, the canals around Yanagawa castle have been used for tourism by launching a new business of river cruising in 1955 (Sakae 2005). Since then, the occupants of waterways have been shifted from residents to tourists who came from outside place. It is important

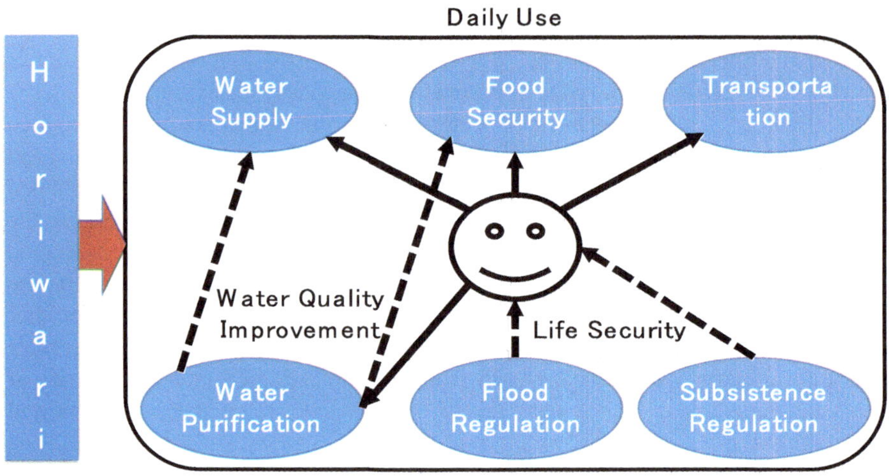

Fig. 7.2 Near water structure of traditional water–human relationship

to be noted that the installation of waterworks enabled the people living in the canals to use tap water. Water has been sealed up in pipes and concrete from a distant dam which was built upstream on the Yabe River. Accordingly, the wastewater from such rooms as kitchens and bathroom was drained into the canals. When the residents engaged in water drawing or treatment in the backyard along the canals, they might have become increasingly nervous tourists' eyes from the boats. The entry of outsiders in the network of waterways might keep the local people away from the water and then make them discharge the wastewater through domestic drainpipes without any treatment. Furthermore, their lifestyle pattern of daily consumption has been also changed quantitatively and qualitatively, in response to rapid economic growth since 1960s. The modern mindset of mass consumption and throwaway prevails, thereby resulting in increased amounts of garbage. There has been an increasing trend in the proportion of papers and plastics-related waste. Now that they depend on tap water from the distant dam through pipelines, their incentives to make the canal water safety for consumption have been lost. The indiscriminate consumption and discharge of the canal water caused worsened pollution. Furthermore, the construction of the dam reduced water volume of Yabe River flowing through Yanagawa City via Futatsu River (Sakae 2005). The lower volume of the water flow is expected to weaken the ability of water purification function which the nature performs. It is also important to be noted that use of chemical fertilizers became popular for agriculture so that there has been seldom any need for sludge or weeds. The conversion reduced the collective action to dry up Horiwari to remove and collect sludge and weeds, resulting in accumulation of pollutants and nutrients in the canals. Hence, the water canals had begun to be filled with sludge and garbage. The people were severely plagued by unpleasant odors, increased insect pests such as mosquitos, and dirty landscape. The canal water was stagnated and some of them no longer flowed.

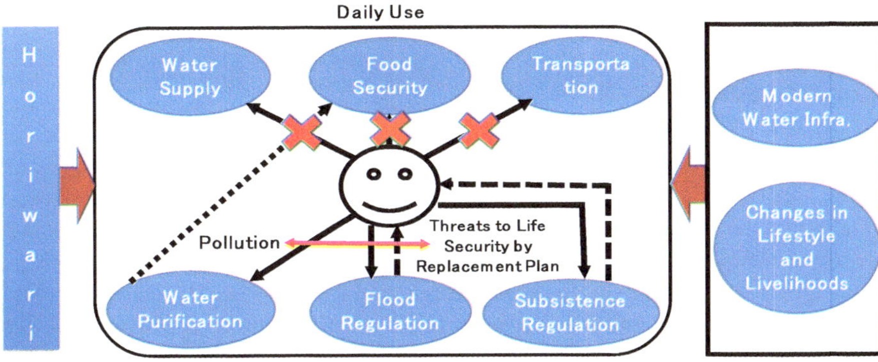

Fig. 7.3 Far water structure of water–human relationship

Under the circumstances, there have been increasing calls to fill up some of the canal waters and turn them into roads, etc. (Tanabe 1991), as many cities in Japan had already replaced canals and rivers with underground concrete sewers. In response to the urgent challenges, the city government implemented environmental actions such as promotion of raw sewage treatment tank, dredging operation, and establishment of environmental department under the authority. The dredging operation from 1968 to 1971 was implemented in the city centering around the covered course of Yanagawa boat cruising and to some extent improved the condition temporally. As years passed, however, the water canals were stagnated and filled with sludge and garbage. In those days, many residents showed an attitude of indifference to cope with the pollution, though some of them working in certain shopping street struggled to clean, dredge, and sanitize the surrounding water environment. Even when the government dredged in the canals, many residents thought that the project was dedicated to clean the environment for tourists, rather for them.

It can be mentioned that the people living in Yanagawa City had easy access to the water canals physically, but the distance of water–human relationship was socially and psychologically quite kept away. Figure 7.3 summarized the far water structure of its relationship. Although the farmers still depended on the water from the canals for irrigation, installment in tap-water supply system and changes in their lifestyle and livelihoods have resulted in disappearance of water areas for daily use such as drinking, less catch of fishery resources, and few opportunities for boat navigation except tourism. The nature of water purification had not been functioned, worsening water pollution. To shut to the stench, dominant strategy was land reclamation of canals with underground concrete sewers. In 1977, the Yanagawa City government finally proposed a project entitled 'urban sewer system plan', which would be scheduled to reclaim most of the canals (5,500 m) except major waterways centering around the castle from April 1, 1977. Horiwari having ecological functions of flood and subsistence regulations were on the brink at that time.

7.5 Canal Restoration in Yanagawa City

The urban sewer system plan was about to be implemented, but there was a person who strongly objected to the plan with the idea of significant ecological functions of Horiwari as mentioned in Sect. 7.3. His name is Tsutae Hiromatsu (1938–2002), an engineer officer of Yanagawa City government. When the plan was formally proposed to the public, he just took over as an urban sewer section chief under environmental department of Yanagawa City. Hiromatsu having expertize of water supply system as an engineer and local experience of lifestyle and livelihoods with the surrounding nature from childhood gained a sense of crisis in relation to loss of Horiwari. On this account, he made a direct plea to the city mayor to stop the plan. Although the plan had been already sanctioned among the stakeholders, it is worth noting that the mayor decided to give him time for six months to propose an alternative plan. The decision was a turning point of canal restoration which was headed by Hiromatsu.

Based on field observation and collection of data related to Horiwari, Hiromatsu made one booklet (Hiromatsu 1977) under the slogan of water restoration what it used to be in the city. Based on his booklet, he submitted a new plan entitled ‘River Clean Plan’ to the stakeholders including city council members. The plan consisted of three pillars; (i) canal restoration by dredging, (ii) wastewater treatment by promoting introduction of a series of septic tanks, (iii) participatory water management. Based on the historical and scientific characteristics of Horiwari, the alternative plan was plausible and feasible so that his plan was successfully accepted by the mayor.

The new plan has been implemented during the period from 1978 to 1983. In the beginning stage, there were opponents, especially from those who were strongly involved in the initial plan. The key to succeed water restoration project was active participation of local people, but most of the people had less interest in revitalizing the water canals. As written in Sect. 7.4, the close relationship between water and human was cut, resulting from modernized water infrastructure development and changes in their lifestyle and livelihoods. Hiromatsu faced with difficulties in finding the way how to promote involvement of canal restoration in a collective manner. To tackle with the challenge, two types of meetings were held. First, Hiromatsu discussed about the project proposal with representatives from each school area and other relevant stakeholders. Before getting onto the point for discussion to understand the essence of the project, his elaboration was made on addressing their precious memories linked to Horiwari. Participants were asked to recollect their good memories, which inspired a feeling of enthusiasm to revive invaluable water environment in Yanagawa City. Their memories which were largely tied to near water-based lifestyle led to agreement of voluntary cooperation from the leaders. Then, more than 100 local meetings with the cooperation of the leaders were held to address the residents for involvement of water management in the same manner. Throughout the discussions, the people changed their mind to a great extent. For instance, those who illegally occupied places in the canals voluntarily engaged in removal of their belongings. In the process of dredging operation, Hiromatsu and other officers were concerned about ensuring sites for dump sludge, but some volunteers offered them for free.

The planned dredging operation covering 26 km was supposed to be completed for 4 years. In this regard, however, their strong commitment to canal restoration in combination with introduction of powerful jet hose for sludge removal succeed to complete it only for 1 and a half years on a budget of around 50% of the planned total cost. Therefore, the project further implemented dredging operation to cover additional 10 km. Throughout these efforts, there has been drastic improvement to the canals.

Apart from the large-scale dredging operation, nested structure of canal management institutions was established in 1980. In 2017, Yanagawa City is divided into 19 area-wise canal committees. Then, each area-wise canal committee was further divided into canal division committees, amounting to 127 canal division committees in total. Several members of each canal division committee are elected on the basis of administrative boundary. Then, some of them are further elected as members of area-wise canal committees. These committees are in charge of maintenance of the canals by themselves. The city government support them by providing instruments for the cleanup, collection and disposal of sludge and waste materials, and information dissemination to the stakeholders. With the supports from the government, the committee-centered water management is encouraged to dredge in the canals, monitor illegal activities, adjust to water for use, and disaster prevention.

In addition to canal management institutions, Yanagawa City government also made efforts on shortening the distance between water and human. Promenades and water docks along the canals were built. This interactive waterscape attracts a lot of people even living near the canals to refresh their mind with joy. On the other hands, learning center and a series of environmental education program under the theme of Horiwari and the surrounding environment in Yanagawa City have been promoted. Understanding the ecosystem of the canals and lessons learned from the past experience is expected to be fundamental to strengthen the attitude of water management, especially for young children. It will be difficult for young generation to develop close relationship between water and human in their daily life as the old experienced. Instead of life experience, environmental education is expected to be essential as a tool to promote a feeling of enthusiasm to sustain the canal water in an appropriate manner.

7.6 Bridging the Distance Between Far Water and Near Water Toward Resilient Water Management

Putting them all together, this chapter highlighted the historical process of canal restoration in Yanagawa City by using near and far water concept. From the national context, water resource system including tap-water supply and drainage has been mainstreamed, especially urban areas including Yanagawa City. The tap-water supply and drainage is readily available in one's household so that the water use has been shifted from common-pool to personalized resources. In other words, modernized

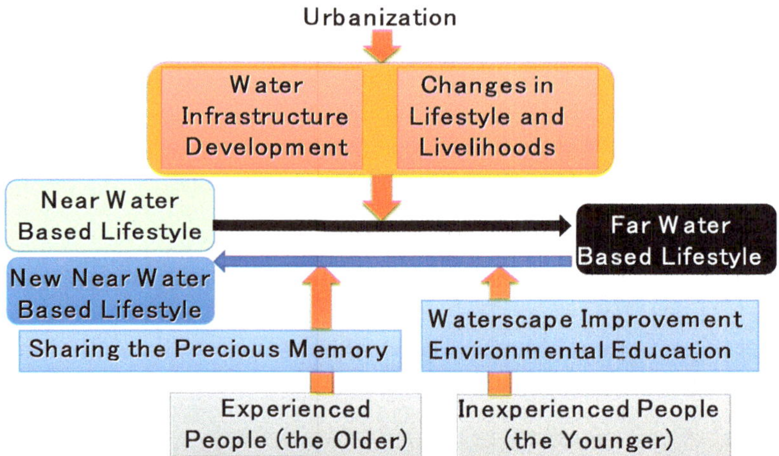

Fig. 7.4 Approaches from far water to new near water-based lifestyle

water development resulted in the social separation of water users and managers (Moritaki 2003). As the case of Yanagawa City experienced water has been sealed up in pipes and concrete from distant dams so that there is no need to ensure clean water for drinking purpose in the canal area. The waterworks were commensurate with the convenience and comfort of water consumption for them. However, it is worth noting that conventional engineering did not take into account traditional water management system where the local people used to be actively engaged in water maintenance while using water for everyday life. With the firm commitment to the tap-water supply and drainage, the social separation of water user and manager has been accelerated, causing various water-related issues as shown in the case of Yanagawa City. To cope with water pollution, many cities in Japan had already replaced canals and rivers with underground concrete sewers. The choice of water-based infrastructure further weakened the relationship between water and human at the society level. As a result of this, it can be said that most of the cities in Japan have been shifted from near water to far water urban areas. The far water-based lifestyle might not only stimulate indiscriminate use of water quantitatively and qualitatively, but also undermine traditional water management institutions taking into account socio-ecological system.

In the case of Yanagawa City, it is worth noting that tremendous efforts headed by Hiromatsu succeeded to restore Horiwari having significant ecological functions as shown in Fig. 7.4. The city chose to strengthen the water–human relationship by initiating the 'river clean plan' with active involvement of local people. It seems that the key to mobilize their strong commitment to clean the canals was sharing the precious memory among those who had experienced the near water-based lifestyle in the old days. The old experience might inspire a feeling of enthusiasm to rebuilt the relationship.

For those who have been no such experience of the relationship, it is very difficult to maintain Horiwari system such as regular cleanups. Their natural and common practice of far water-based lifestyle diminished incentives to make the canal water clean. Instead of the experience-based approach, the case study revealed that waterscape design and environmental education has been encouraged as a means to create a new near water-based lifestyle, especially for younger generation. It is difficult for the people residing in the city to return to their old life which was linked to near water-based lifestyle, but the water–human relationship can be rebuilt by adjusting to changing socio-ecological systems in the local context. Nowadays, waterfront urban renewal projects became common-place in the process of urban regeneration in many cities of the world, which is expected to shorten the physical distance to some extent. In addition to it, elaborations were further required to promote the sense of unity among the stakeholders, in order to revitalize resilient water management. Given the population decline with the aging in Yanagawa City, active involvement of younger generation for water management institutions will be one of the major challenges to maintain the canal water in a collective manner. This chapter just focuses on the case of Yanagawa City, but creating the new near water-based lifestyle varies from place to place. Further research is called for addressing the new relationship between water and human in the urbanized world.

Acknowledgements The author is extremely grateful to Yanagawa City government officers for taking the time for interviews.

References

Hardin G (1968) The tragedy of the commons. Science 162:1243–1248

Hiromatsu T (1977) Kyodo no kawaniseiryu wo torimodosou: Yanagawasiminhenoyobikake (in Japanese)

Kada Y (2000) Tooimizu, chikaimizu: gendaishakainiokerukankyo no zibunka. In: Kada Y, Kunihiro Y, Tsutida A (eds) Kyokansurukankyogaku: tiiki no hitobitonimanabu. MinervaShobou, pp 24–37 (in Japanese)

McDonald RI, Weber K, Padowski J, Flörke M, Schneider C, Green PA, Gleeson T, Eckman S, Lehner B, Boucher T (2014) Water on an urban planet: urbanization and the reach of urban water infrastructure. Glob Environ Change 27:96–105

Mizkan Water Cultural Center (2017) Osusumeshiryo-shu. http://www.mizu.gr.jp/archives/recommended.html. Accessed on 13 April 2017

Mizunokai (2014) Yanagawamachizukurioendankai. Mizunokai, Yanagawa (in Japanese)

Moritaki K (2003) Magari kado nikita mizushigenkaihatsu: "tikai mizu" tai "tooi mizu". J Water Env Issues 16:1–8

Sakae S (2005) Yanagawahoriwarisaiseikaramirurekishitekikousatsu. Dissertation, Prefectural University of Kumamoto (in Japanese)

Tanabe K (1991) Yanagawa: man, land and water. Fukuoka Style 1:52–55

The Ministry of Land, Infrastructure, Transport and Tourism (MLITT) (2014) Water in Japan. https://www.mlit.go.jp/common/001044443.pdf.Accessed. Accessed on 13 April 2017

United Nations (United Nations Department of Economic and Social Affairs, Population Division) (2014) World urbanization prospects: the 2014 revision. ST/ESA/SER.A/352. https://esa.un.org/unpd/wup/Publications/Files/WUP2014-Report.pdf. Accessed on 13 April 2017

Yanagawa City (2017) Juuminkihondaicho zinkou setaisuu. http://www.city.yanagawa.fukuoka.jp/shisei/tokei/tokeijoho01.html (in Japanese). Accessed on 13 April 2017

Zhou Q (2014) A review of sustainable urban drainage systems considering the climate change and urbanization impacts. Water 6(4):976–992

Chapter 8
Resilience to Water-Related Disasters: Risks, Vulnerabilities and Coping Strategies in Dhaka

Md. Anwarul Abedin and Mohammad Golam Kibria

Abstract Climate change has become one of the major developmental concerns of countries in the Bay of Bengal region, which is already causing setbacks such as flooding, waterlogging, and drinking water problem. Around 3.49 billion people or more than half the world's population lives in urban area, of which 827.6 million are slum dwellers. Dhaka the mega city of Bangladesh covering an area of 360 km^2 contains more than 20.0 million people that noticeably creates huge burden on water-related disaster, especially flooding. Dhaka receives about 2,000 mm of rainfall annually, of which almost 80% falls during the monsoon. Water crisis is a major problem in Dhaka city due to overexploitation of groundwater and limitations in water supply throughout the city. On the other hand, floods are also one of the main natural hazards affecting the city and are associated with river water overflow, groundwater depletion due to overexploitation of water during dry period, and rainwater stagnation. The drainage capacity of the city has also decreased alarmingly due to the development of unauthorized settlements. Illegal occupation of drainage canals and wetlands by land grabbers has further contributed to the problem. Furthermore, the risk of drought and subsequently flooding is aggravated through rapid urbanization and concurrent encroachment on retention areas, as well as increasing problems with both the natural and man-made drainage system. Therefore, this paper explores the risk and vulnerabilities of water-related disasters and their management framework, which has now been accepted by the government as well as causes of water crisis in Dhaka city with their possible remediation strategies. It is envisaged that the concepts put forward in this framework will be of immense value in planning future strategies to cope with the problem of flooding and water crisis in Dhaka city.

Md. Anwarul Abedin (✉) · M. G. Kibria
Department of Soil Science, Bangladesh Agricultural University, Mymensingh 2202, Bangladesh
e-mail: m.a.abedin@bau.edu.bd

M. G. Kibria
e-mail: kibria.ss@bau.edu.bd

B. Ray and R. Shaw (eds.), *Urban Drought*, Disaster Risk Reduction,
https://doi.org/10.1007/978-981-10-8947-3_8

8.1 Introduction

The urban areas are becoming more vulnerable to natural disasters with almost half of the world's population and thus turning the big cities into the main point of destruction as well as loss of assets. The cities are the most potential disaster hot spot areas due to rapid urbanization and global climate change with a number of human settlements those make cities more vulnerable to disasters. There are several water-related disasters which occur in Dhaka city like flooding, scarcity of fresh drinking water during dry period, poor drainage system, intensive use of groundwater.

The scarcity of fresh drinking water is an emerging problem in Dhaka city. The population of Dhaka city is increasing day by day due to rapid urbanization. Groundwater level in Dhaka city is also going down due to withdrawn of water frequently, and availability of fresh drinking water becomes a major problem now. Surface water like ponds and rivers is already polluted by wastewater from different industries. That is why surface water cannot be used now without any treatment. Besides, poor drainage creates another problem in Dhaka city, and due to this, waterlogging condition is very common in Dhaka city. And in rainy season urban flooding is very common in Dhaka city due to this poor drainage system.

Groundwater is the main water supply source of Dhaka city, and around 87% of the total supplied water is gained through the process of groundwater extraction. This indicates that Dhaka city water supply is solely dependent on this source. The groundwater table in Dhaka city decreases at a very high rate due to continuous extraction of groundwater. During the last seven years, a sharp decline in groundwater table in Dhaka city was observed at the rate of 2.81 m per year, which caused the down of total around 20 m during the stated periods. From this data, the study predicts that if the process of groundwater extraction continues at the present rate, the groundwater table will go down to 120 m by 2050. Despite sufficient rainfall in Dhaka city, the groundwater recharge is not sufficient to meet the present demand. It was found that the potential groundwater recharge of Dhaka city is only 1.33 m per year (1.48 m per year groundwater recharge deficit), but as the extraction rate is higher than the recharge rate, the decline in groundwater table is extensive. Other than this, surface water is a very good source of water, but the water reservoirs around the Dhaka city become polluted, and their amount and volume also becoming limited due to rapid urbanization, industrial wastes, illegal construction, and encroachment (Azim Uddin and Baten 2011).

Dhaka is the ninth largest city in the world with a rapid urbanization rate, an urban hotspot for climate risks, and a city which faces the recurring phenomena of urban flooding and waterlogging following intense rainfall nearly every year. Dhaka is situated beside Buriganga River and on the lower reaches of the Ganges–Brahmaputra Delta. Tropical monsoon climate is predominant in Dhaka city and experiencing different natural hazards due to river flooding as well as water crisis. Recent major floods have been worse in depth and extent of inundation and duration, especially in fringe areas, where many of the city's poor reside. Rapid, unplanned urbanization and the gradual filling up of low-lying floodplains, rivers, canals, and other water bodies traditionally used to drain or retain water during rainfall have exacerbated the problem; Dhaka's average annual rainfall is 2,000 mm, 70% of it occurring during the summer monsoon. This chapter overall aims to provide the history of major water-related disasters in Dhaka city and their impacts, water management strategies, implementation of framework to reduce disaster risks, role of local decisionmakers in effective planning approach for minimizing the risk of urban drought, waterlogging and poor drainage system in Dhaka.

8.2 History of Major Water-Related Disasters in Dhaka

Bangladesh is situated under a normal tropical monsoon climate with high temperature, relative humidity, and heavy rainfall (1200–6000 mm), and depending on the location, the dry season covers from June to September and rainy season covers from December to March. Dhaka city is also experiencing this monsoon climate and facing some water-related disasters in both of these seasons. Floods occur in rainy season, and water crisis is common in dry season in Dhaka city (Fig. 8.1).

Water crisis is a very common problem in Dhaka city because people in this city are totally dependent on groundwater, and in dry season, water becomes unavailable due to depletion of groundwater level. Many researchers working on water management system in Dhaka city stated that over-exploitation of groundwater could be threatening for the future water supply system. Dhaka is considered as mega city with

Fig. 8.1 Peoples are holding empty pitchers upside down, indicating the scarcity of water in Dhaka city. *Source* Enamul Haq (2017)

a population of over 15 million people and experiencing different water management problems common to other major cities.

A regularly disregarded outcome of these activities in the urban communities of Dhaka city is the related to the impact on the water quality and supply to surrounding urban or provincial groups, where water assets are lacking and surface contamination from lethal metals, natural materials, and different toxic elements is boundless. Toxic element contamination in the groundwater is a problem in Dhaka city because of deposition of untreated industrial wastes directly into the water bodies, irrational use of fertilizers for agricultural purposes, etc. Arsenic contamination in the groundwater of Dhaka city becomes prominent when water is extracted from a shallow depth (less than 200 feet), which threatens human health of millions of people living in Dhaka city. To avoid the groundwater contamination of arsenic in Dhaka city, water is extracted from deep groundwater source (greater than 500 feet). Due to extensive dependency on the groundwater, Dhaka city is experiencing water crisis because of groundwater depletion and malfunction of deep tube wells. Besides, due to the problem of load shedding, the process extraction of groundwater and purification also hampers and results in the limited supply of water in Dhaka city.

Floods are very common in Dhaka city that mainly occur due to river water overflow and rainwater stagnation. With the increase of urbanization, floods are occurring more frequently than before in Dhaka city. There were several floods occurred in Dhaka city from 1954 to 2007, and the city experienced heavy flooding in 1978, 1988, 1998, 2004, and 2007, where the effects of flooding were catastrophic (Rabbani 2009). Figure 8.2 illustrates the flooding area in Dhaka city during 1978 and 1988. Almost all of the population in the entire Dhaka city was affected by those heavy floods. The consequence was devastating in the slum areas of Dhaka city due to drinking water crisis, diseases outbreak, and poor management strategies. Around 30% of the total population in Dhaka city live in slum areas, those are normally situated along the water's edge, thus become exposed to floods particularly (Shaw 2013).

The consequences are more adverse after a flood occurs. Different parts of Dhaka city remain waterlogged for a couple of days after occurring flood. Many roads in Dhaka city remain waterlogged and become inaccessible for couple of hours (8 h) after heavy rainfall, and depending on the situation, water becomes stagnant in the road for the period of 12 h or even more (Shaw 2013). The problem of waterlogging becomes more prominent in Dhaka city from May to October, which means during the monsoon season. According to the findings of different researches, the increase of river water level surrounding the city is the main reason of urban flooding, which occurs due to poor water management system (Faisal et al. 2003; Shaw 2013). Fast and spontaneous urban development with an unplanned way and an uncontrolled land boom in Dhaka city causes severe encroachment of natural drainage system and retention areas, ruining the natural flow or stream of water in river and causing considerable waterlogging and flooding in relatively each year of the previous decade (Parvin et al. 2013).

Dhaka, the capital of Bangladesh, and other major cities are vulnerable to both geological and climatic disasters due to its geographical location. These cities are

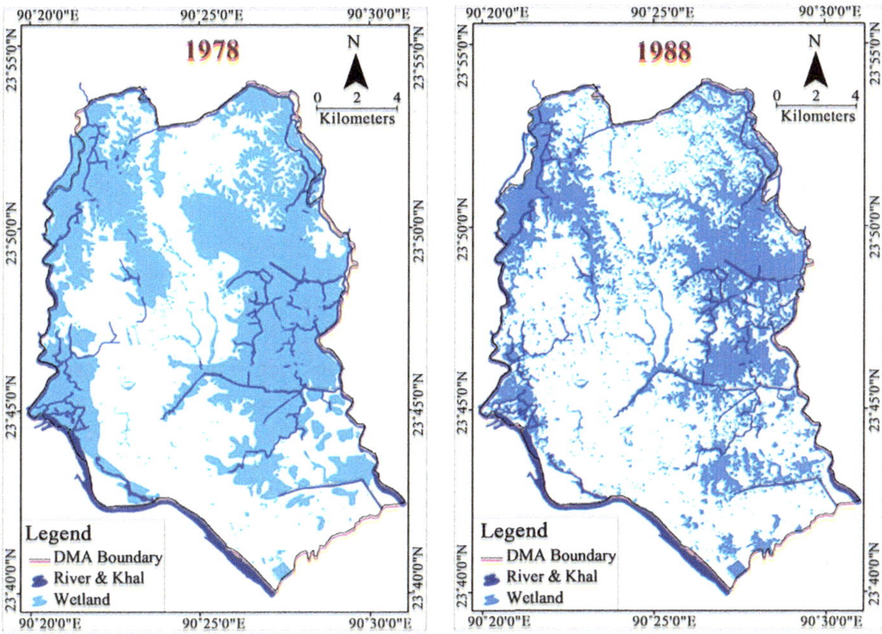

Fig. 8.2 Flooding area in Dhaka city during 1978 and 1988. *Source* Mahmud et al. (2011)

Table 8.1 Major floods in Dhaka city with their records

No.	Year	Flood area (%)	Sum of damage (million Taka)	Death	Injured people	Homeless	Sufferers
1	1974	36.6	28,490	28,700	0	2,000,000	38,000,000
2	1987	39.9	35,000	2,680	0	0	29,700,000
3	1988	84	1,00,000	2,440	0	28,000,000	73,000,000
4	1998	68	1,00,000	1,000	50	0	15,000,050

Source Bangladesh compendium of environmental statistics (BBS 1999)

the most climate hot spot areas of Bangladesh and remain vulnerable to different natural disasters due to rapid and unplanned urbanization. The urbanization rate in Bangladesh is one of the most noteworthy on the planet, and this quick pace is relied upon to proceed for a long time to come (Bashar and Rashid 2012). Table 8.1 figures out the impact of major floods in Dhaka city during 1974–1998.

Floods and prolonged waterlogging are exceptionally normal in Dhaka city. Recently, this city experienced four noteworthy floods in the previous 20 years occurred in 1988, 1998, 2004, and 2007. In terms of inundation and duration of flood water in the city periphery regions, the floods of 1998 and 2004 were the worst floods occurred in the previous time. As many people in Dhaka city live in slum areas, around 30% of the total people in Dhaka city experienced a severe catastrophic effects of floods. After flooding in Dhaka city, poor people in slum areas experienced different health problems due to the lack of fresh drinking water and suffered from lack of food. People face severe problem to arrange their foods, because all the roads and fields remain under water and people need to swim or travel by boat to move anywhere. Women, children, and elder people living in slum areas are extremely vulnerable to flood risks.

Dhaka City Corporation (DCC) worked with different national and private organizations and operated relief and rescue operations as well as rehabilitation works through established flood control room and shaped administration groups. Government and NGOs completed alleviation circulation and restorations exercises, and DCC found a way to restore the damaged road, embankments, and other facilities.

8.3 Sudden Water Crisis Affects Parts of Dhaka

Sudden water crisis is also observed in Dhaka city, which is another most important water-related disaster in urban areas. In Dhaka city, the acute water crisis occurs due to continuous heat wave during summer for more than 2 weeks with almost no rain. The water crisis takes a substantial toll on the general population battered by the sweltering warmth with around 37 °C temperature or more in Dhaka city. There are also some problems with the authorities of water management of Dhaka city. They claim that due to the declining groundwater table and technical problems at some water extraction pumps in the city, water supply becomes limited during dry season. The situation becomes more worsen when load shedding or power cuts occur. In spite of sufficient power generating capacity and supply limit, the people in Dhaka city experience severe load shedding for the two or three days continuously, as a large portion of the fuel oil-based power plants stays close because of a disturbance in fuel supply. Considering this situation, the Power Development Board stated that the load shedding has been occurred due to the ongoing waterways transport workers' strike. Other than these reasons, the water crisis in Dhaka city mainly occurs due to groundwater depletion during dry period with almost no rainfall.

Among the different reasons for gradual groundwater level depletion in Dhaka city, excessive groundwater extraction with an increasing rate from the aquifer is the most significant one. Rapid and unplanned urbanization and construction of structures, flood protection dams, and embankments aggravate the acute problem of water crisis in Dhaka city. These human interferences are restricting natural water flow and making the existing surface water reservoir unfit for use (Rahman and Alam 2005). The natural drainage system of floodwater as well as groundwater recharge is managed by different water reservoir sources such as cannels, lakes. Due to rapid urbanization, the number of these water sources is decreasing day by day, which creates the problem of flooding as well as the crisis for safe drinking water. Other than these, unauthorized groundwater extraction system in Dhaka city also aggravates the problem of declining groundwater level, thereby causing water crisis during summer. Therefore, people in Dhaka city experience the problem of illegal encroachment over the last four decades.

Consequences of Groundwater Depletion

(a) The most serious outcome of continuous groundwater extraction is letting the groundwater table down. On the off chance that groundwater levels decline too far, at that point, the well may need to extend the well depth to pump water from the aquifer. Finally, it may cause severe water crisis due to inadequate supply of water.
(b) Water extraction cost from the groundwater increases with the diminishing groundwater level. If the groundwater table goes down, the pump needs to lift water from deeper source than before, which increases the total cost to get it ready for pipeline supply. Therefore, the pump will consume more energy to lift water from deeper source and ultimately increases the cost involved in water supply system.
(c) Water flow in rivers, streams, and lakes from groundwater to streambed through seepage can be reduced greatly when the groundwater table decreases. A lot of the water streaming in waterways originates from leakage of groundwater into the stream bed. Groundwater pumping can modify water movement between an aquifer and a water body by either capturing groundwater stream that is released into the surface water body under regular conditions or expanding the rate of water movement from the surface water body into an aquifer. Groundwater pumping is bringing down the groundwater levels underneath the profundity which is required by the stream side or wetland vegetation to survive, which ultimately causes the loss of vegetation and wildlife habitat.
(d) Land subsidence is another long-term consequence of groundwater extraction. Due to illegal and indiscriminate extraction of ground water from the main aquifer, the respective land subsequently loses the support from below and makes the land to become unfit to use and may contribute to land subsidence.
(e) Water quality deteriorates because of indiscriminate groundwater extraction from the main aquifer. Naturally, the boundary of fresh water and salt water is relatively stable, but due to overpumping of freshwater, saltwater may migrate inland and go upward. This may result in the contamination of freshwater with saltwater, and then, the water becomes incompatible to drink and use.

8.4 Water Management System in Dhaka City

Being a developing country with lots of population, water management system of Bangladesh especially in Dhaka is not well advanced and undated. From the history, it was found that piped drinking water system in Dhaka city was installed in 1874. This framework was served by a water treatment plant in Chadnighat close to the bank of Buriganga. Department of Public Health Engineering of the Pakistani government mainly managed the drinking water supply system during Pakistan period, and later, Dhaka Water Supply and Sewerage Authority (DWASA) was established in 1963. From then, this authority is mainly managing the drinking water supply system in Dhaka city. Besides, the management of drinking water supply system, DWASA started to handle storm water drainage system of Dhaka city in 1989, and the next year, DWASA extended their service coverage to include Narayanganj city. In the early 1990s, World Bank agreed to provide loan for improving water supply management system of Dhaka city if DWASA works in collaboration with public–private partnership with an international organization. But when the proposal was rejected by the Bangladesh government, it asked that income charging and gathering ought to be outsourced to a privately owned business for no less than one administration region on a pilot basis and that DWASA ought to be changed into a financially arranged utility. In 1997, outsourcing in one service area was done, but the performance of pilot project was not satisfactory and thus it was stopped. Finally, DWASA was transformed into a service-oriented commercial organization that was reorganized by the Dhaka WASA Act, 1996. From then, DWASA is working to manage water supply system in Dhaka city as a government organization.

8.5 Factors Contributing to Water-Related Disasters

The occurrence of floods in an area depends on the number of factors, and the chance of flooding can vary with the change of those factors. The chance of flooding in an area can be predicted from the change in water accumulation in watershed from precipitation, total water-bearing capacity of drainage channels, the elevation of an area from sea level and land subsidence. Other than flooding, water crisis is another most important water-related disaster predominant in Dhaka city. These water-related disasters subsequently occur in Dhaka city due to a combination of number of factors, which are discussed below:

8.5.1 Overexploitation of Ground Water

Around 82% of the total water supply system in Dhaka city is solely dependent on the groundwater, and total 577 deep tube wells are used to pump water from this source. On the other hand, only 18% are provided through four surface water treatment plants. Due to this sole dependency on groundwater, the level of groundwater is dropping at the rate of 2–3 m each year (Azharul Haq 2006). The water table in Dhaka city dropped by 50 m within last four decades, and now, the closest underground water table is around 60 m below from the surface. Due to this problem, recently, people in Dhaka city experience the severe problem of water crisis during dry season.

8.5.2 Unplanned Urbanization

Being the overpopulated city, unplanned urbanization is very common in Dhaka city. To accommodate the growing population in Dhaka city, there is an increasing pressure on land, which destroys the agricultural land as well as using the watershed to develop infrastructure on them. This rapid and unplanned urbanization in Dhaka city increases the flooding intensity during rainy season by restricting natural water flow and poor drainage system also aggravates the problem of flooding. For instance, Dhaka city that is completely served by storm channels and where 60% of the land surface is secured by streets and structures, surges are very nearly six times more various than before urbanization (Pipkin and Cummings 1983).

8.5.3 Riverbed Aggradation

Floods become more prominent when the amount of water exceeds the level of water-bearing capacity of water reservoirs. This mainly occurs due to riverbed aggradation, which is more pronounced in all over Bangladesh. Due to riverbed aggradation, water carrying capacity of rivers decreases and overflow of river bank occurs. This current increment in riverbed aggradation levels more likely increased the flooding rate in Dhaka city.

8.5.4 Soil Erosion

The top soils are removed from the land through the process of soil erosion, especially surface runoff. This erosion is more prominent in agricultural field because the topsoil is loose due to plowing which makes it vulnerable to be eroded. It can create flooding in two different ways. Firstly, through the soil erosion process, sediments are deposited on the riverbeds, thereby reducing the depth of river and water-bearing capacity and increasing flooding propensity. Secondly, when the topsoil becomes eroded, the land elevation from riverbeds decreases and in turn increases the chance of flooding.

8.5.5 Deforestation in the Upstream Region

A rapid increase in population in Dhaka city has resulted in an acceleration of deforestation in Dhaka city. Increasing deforestation rate to meet the demand of growing population accelerates the soil erosion process in urban areas. Landslides and severe soil erosion occur during rainy season, which become deposited in the riverbeds and then contribute to severe flood propensity in Dhaka city.

8.6 Impact of Water-Related Disasters in Dhaka City

Safe drinking water crisis has become one of the major and common problems in Bangladesh during dry season, especially in Dhaka city. According to the report of DWASA, the Dhaka city requires around 2.2 billion L of water in a day, but only 1.9–2 billion L is available. This report clearly indicated the severe water crisis problem in Dhaka city. Scarcity of fresh drinking water is very common in Dhaka city, especially in dry season (Fig. 8.3). Besides in rainy season, surface water becomes more polluted due to continuous waterlogging condition becomes totally unfit for drinking purpose.

During dry season, load shedding and a drop in water table from March to May in every year create the severe problem of water crisis. In that time, DWASA becomes unable to supply water due to inadequate water extraction from groundwater source. In some parts of Dhaka city, water crisis becomes very severe during summer, especially in April, when people experience the problem of water crisis for several days continuously. Furthermore, people of Dhaka city reported that the water supplied was not suitable to drink and the situation is still continuing. All these taken together resulted in protests from the people of Dhaka city. People continuously stands in a line with containers to collect water from a very few sources (Fig. 8.3). Many people become victims of various waterborne diseases due to safe drinking water crisis throughout the year and especially from April to August every year. Scientists are suggesting to use surface water to meet the increasing water demand of Dhaka city using different water treatment plants.

On the other hand, Dhaka city is experiencing the problem of waterlogging and flooding during monsoon period (June–September), when around 70% of the total annual rainfall occurs. Due to unplanned urbanization and poor drainage system, the people of Dhaka city are suffering from the extreme problem of waterlogging during rainy season, even after a week from raining. The rivers and canals remain interconnected naturally, but the network gets disturbed due to unauthorized development. Rivers surrounding the Dhaka city include the Tongi Khal to the north, the Balu and Lakshya to the east, the Dhaleswari to the south, and the Buriganga, Bangshai, and

Fig. 8.3 Few glimpses of focusing fresh drinking water scarcity in Dhaka city. *Source* Authors

Fig. 8.4 Waterlogging condition in Dhaka city due to poor drainage. *Source* Authors

Fig. 8.5 Poor drainage system in Dhaka city. *Source* Authors

Turag to the west. But due to unplanned urbanization, waterlogging becomes very common in Dhaka city even after little rain and remains stagnant for a couple of days (Fig. 8.4).

The problem of waterlogging and flooding is very prominent in Dhaka city due to poor drainage system. Unplanned urbanization, illegal encroachment, dumping of solid waste, and lack of coordination between local community people with government and non-government organizations create the problem of waterlogging and flooding. This problem is very severe in slum areas where the management system is very poor and unplanned. Almost all the surface water reservoirs are got polluted, and only some surviving khals and canals are acting as water flow pathways, but those are really in poor conditions. Thus, the drainage system of Dhaka city remains in poor condition, which in turn causes waterlogging and floods (Fig. 8.5).

In many areas of Dhaka city, the drainage system is not good. Almost well channels are open and become blocked due to accumulation of garbage figure. As a result, water movement becomes impossible through those channels and ultimately causes waterlogging condition in Dhaka city.

Fig. 8.6 Disruption of traffic movement and normal life. *Source* Authors

8.6.1 Traffic Movement and Normal Life Disruption

Waterlogging disrupts the normal life, and poor peoples mostly suffer due to this problem. When the roads remain under water, normal traffic movement gets hampered, which in turn results in traffic jam in city area. Besides, the movement becomes so risky, and many accidents occurred due to the waterlogging problem in roads, and the risk is highest for the pedestrians (Fig. 8.6). Puddles are formed in roads, where water cannot drain out easily and remains stagnant for a couple of days. Poor people are suffering more due to waterlogging problem because they normally live in low-lying areas with poor management system.

8.6.2 Damage of Infrastructures

Waterlogging damages the roads, substructures of buildings in low-lying areas, construction site during the rainy season. The roads are mostly damaged due to waterlogging problem, and when the water drains out from roads after several days, the roads become unfit to travel, getting more risky for the people (Fig. 8.7).

The brick foundation also gets damaged because their longevity reduces due to dampness of the building. Due to heavy flooding, some buildings in low-lying areas go under water and the resident needs to move to other places creating severe problem for poor people. Besides, those buildings cannot be used further even when the water drained out.

Fig. 8.7 Damage of roads and structures due to waterlogging in Dhaka city. *Source* Authors

8.6.3 Environmental Impact

Waterlogging problem in Dhaka city is very prominent due to improper water management plan and system with inadequate management. Due to this problem of improper water management, the trees and vegetation in Dhaka city become damaged by stagnant water for longer time. Disease outbreak is one of the most serious impact of waterlogging in Dhaka city. Urban runoff water gets mixed with sewage from overflowing latrines and sewers in poorly drained low-lying areas, causing water pollution associated with waterborne diseases. The poor people of Dhaka city need to rely on polluted surface or shallow groundwater sources, as they do not have access to portable water during the period of monsoon.

8.6.4 Economic Problem

The rapid and unplanned urbanization in Dhaka city is not only affecting the physical, social and environmental systems but also creating significant economic problem. Due to damage of natural drainage system by uncontrolled urbanization, the economic loss of individual people and community people becomes more prominent in Dhaka city. The cost of construction and maintenance of infrastructures increases, and life span of structures goes down rapidly, and roads as well as underground pipes and connections become damaged due to waterlogging problem in Dhaka city. This problem in turn increases the cost involved to replace and repair those structures with a huge economic loss.

8.7 Role of City Government in Enhancing Dhaka's Resilience

The ability to withstand hard times or disasters is termed as resilience. If a city or town gets destroyed by any disaster and people in that city are able to restore the previous condition as the things were before quickly, then it is called resilient to disasters. Due to rapid urbanization in Dhaka, the city is becoming more vulnerable to different water-related disasters which are irreversible to restore. The disaster risk in Dhaka city is increasing not only for the rapid urbanization, but also for poor governance and inappropriate planning. In this situation, city government can play an important role in disaster risk reduction (DRR) and enhancing disaster resilience by implementing proper management plan with the involvement of community people. It is not possible to maintain a sustainable development process in Dhaka city without enhancing the city's resilience capacity to disasters by prioritizing disaster risk reduction (DRR).

Disasters risk reduction has attained more concern of development initiators, designers, and scholars due to paradigm shift, compared to disaster response and recovery under the system of disaster management (DM). In this approach of DRR and resilience enhancement, city governments have the primary responsibility for implementing all advanced management action plans. DRR and enhancing disaster resilience must be an integral part of planning and investment, as the city government has the main authorities.

City corporation can install water pumping station and surface water treatment plants to ensure proper water supply throughout the city to mitigate the problem of water crisis. Besides, Dhaka city corporation (DCC) should establish and implement strict rules to protect the natural drainage system of Dhaka city by restricting unauthorized constructions to avoid waterlogging problem. They can establish a good communication network between control room and the City Disaster Management Committee headed by Mayor to inspect the disaster management with their responses The city government should work with DESA, WASA, Fire service, Red Crescent, and other sectors during disasters to increase the resilience capacity of Dhaka city.

8.8 Building Community Resiliency: Linkages Between Individual, Community, and Local Government in the Urban Context

The increase in community resilience to different water-related disasters occurring in Dhaka city can mitigate the economic and environmental losses. Resilience can be established when individual people, community, and local government will work together and act like a social network. This community resilience is essential to reduce the disaster risk which is also known as next generation of sustainability. Community resilience capacity includes sustainability's nested triad of environment, society, and economy. This resilience to risk becomes more effective when stakeholders are engaged in the implementation plan and the primary stakeholders are the people

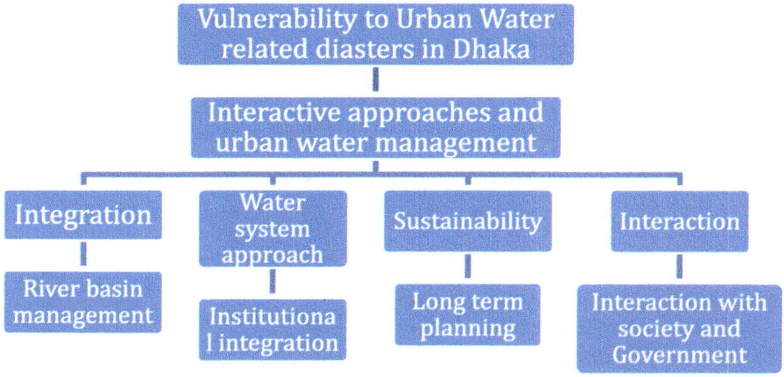

Fig. 8.8 Interactive urban water-related disaster management. *Source* Authors

who live there. Water-related disasters, most commonly water crisis (urban drought) and flood, in Dhaka city can be managed by interactive urban water-related disaster management plans (Fig. 8.8). Interaction between local community and government is essential to reduce risk in Dhaka city from flooding.

Resilient communities are capable to restore the damages from adverse situations by active participation of community people for bringing changes in economic and social level. Resilience in a community can be characterized by having an unrestricted access to information with a good communication networks during disaster and can call upon a wide range of resources. For mitigating the risk from different water-related disasters, advanced disaster risk management (DRM) strategies both at community and local levels are essential. Community disaster risk management (C-DRM) and local disaster risk management (L-DRM) refer to processes and procedures promoted at different, but complimentary sub-national/sub-regional levels with an aim to reduce and prevent disaster risks at community level.

8.9 Implementation of Plans and Policies Toward Resilience

In order to mitigate the risk of different water-related disasters occurring in Dhaka city, the water management system requires substantial development with advance technologies and management plans. Being the capital of Bangladesh, Dhaka is the main hub for growing economy. People are migrating to Dhaka city gradually to earn their livelihood, which in turn makes the city more crowded and leads to unplanned development with high rate of urbanization. Due to rapid and unplanned urbanization in Dhaka, this city becomes more vulnerable to disaster risk with different water-related disasters like waterlogging and water crisis with a highest frequency. Therefore, it is essential to implement an advanced plan with the involvement of community people to reduce the water crisis and waterlogging problem in Dhaka city.

Implementation plan for urban water-related disasters management includes:

- Integration of spatial dimension plan is essential with sectorial planning and investment both in government and community level.
- Authority must be accountable to improve the planning and management practices with adequate supervision.
- Low-lying areas and areas with less rainfall are more sensitive to waterlogging and water crisis, and these areas require special management plans.
- Laws and codes should be built immediately to restrict any illegal construction, landfill with strict implementation.
- Decentralization of Dhaka city is essential to restrict people migration to Dhaka city from rural areas so that rural people can get access to each and every facility from remote areas of Bangladesh. This action in turn can reduce the rate of unplanned urbanization.
- Sharing database within and among government and non-government organizations with an integrated approach.
- Risk sharing and risk transformation through the implementation of different financial tools and instruments.
- Implementation of efficient and judicial groundwater abstraction strategies should be followed during dry period to minimize the risk of urban drought.
- Change of lifestyle concerning water use during dry season.

8.10 Conclusions and Recommendations

This study synthesized the relevant information from different available sources (plans, programs, reports, and scientific literature), and these could be helpful for local and government policy-makers for mitigating water-related disasters in Dhaka city. In this study, an interactive water-related disaster management framework has been developed which includes risk mitigation, developing resilience and implementation. Safe water supply is very important for sustainable development, and water crisis can directly affect the development process as well as the ecosystem. The present water supply system in Dhaka city solely dependent on groundwater sources and we couldn't find other suitable options of supply water system yet. Through DWASA has started to use the surface water through few water treatment plants, these are not sufficient to meet the demand. So, more construction of water treatment plants is essential as well as use of rainwater needs to be introduced. Illegal encroachment and unplanned urbanization should be stopped to avoid waterlogging and flooding problem in Dhaka city. Local community people can play an important role in mitigating water-related disasters in Dhaka city. A city government or local government can act to implement the decentralization process and stop unplanned development in an area.

Possible management practices to mitigate water crisis in Dhaka city

Some management strategies can be implemented to solve water crisis problem in Dhaka city. They are:

1. Raising awareness among the community people for efficient use of water for reducing the water misuse
2. Seeking alternative sources of water other than groundwater source
3. Establishment of water treatment plant to use surface water to eliminate dirt and ill-smelling
4. Efficient human resources to manage water supply system throughout the Dhaka city
5. Increase power generation for continuous supply of water.

References

Azharul Haq K (2006) Water management in Dhaka. Water Res Dev 22(2):291–311

Azim Uddin AFM, Baten MA (2011) Water supply of Dhaka city: murky future-the issue of access and inequality. Unnayan Onneshan-the innovators

Bashar T, Rashid S (2012) Urban microfinance and urban poverty in Bangladesh. J Asia Pac Econ 17(1):151–170

BBS (1999) Bangladesh Bureau of Statistics, Bangladesh Compendium of Environmental Statistics

Enamul Haq SK (1 June 2017) Water crisis all around Dhaka. The Daily Star. Retrieved from https://www.thedailystar.net/city/water-crisis-all-around-1413664

Faisal IM, Kabir MR, Nishat A (2003) The disastrous flood of 1998 and long-term mitigation strategies for Dhaka City. Nat Hazards 28:85–99

Mahmud MS, Masrur A, Ishtiaque A, Haider F, Habiba U (2011) Remote sensing and GIS based spatio-temporal change analysis of wetland in Dhaka city. Bangladesh. J Water Res Prot 11:781–787

Parvin GA, Ahsan SR, Shaw R (2013) Urban risk reduction approaches in Bangladesh

Pipkin BW, Cummings D (1983) Environmental geology—practical exercises. Belmont Star Publishing Company, California, p 215

Rabbani MDG (2009) Environmental risks in Dhaka: present initiatives and the future improvements. In: Shaw R, Srinivas H, Sharma A (eds) Urban risk reduction: an Asian perspective. Emerald, United Kingdom, pp 319–338

Rahman AA, Alam M (2005) Dhaka city—state of environment (SOE) UNEP, Dhaka. Online available at http://www.rrcap.unep.org/pub/soe/dhaka-soe-05/3-2Dhaka-water.pdf

Report of the Task Forces (RTF) on Bangladesh Development Strategies for the 1990s—environment policy (vol 4), 1991. University Press Limited, Dhaka, 265 pp

Shaw R (2013) Urban disaster risk reduction framework: assessing urban resilience in world vision project sites in Bangladesh, China, and Indonesia—final report, 1 April 2013. World Vision International. Singapore

Chapter 9
Resilience Perspective for Planning Urban Water Infrastructures: A Case of Nagpur City

Sameer Deshkar

Abstract The role of urban infrastructures in building resilience is acknowledged in most of the literature pertaining to climate change and disasters. Basic infrastructures and services, like water supply and sanitation, not only play a crucial role in delivering services to communities and supporting economic growth, but also are vital for determining community's resilience against adverse or stressed scenarios. In cities from developing countries like India, such conditions might emanate either due to climate variability or even due to lack of effective management of services itself. While much emphasis over last two decades has been on the engineering and management aspects for infrastructure development in the million-plus cities in India, resilience perspective calls for an integrated approach involving local communities and the natural resources. This chapter discusses some of the challenges for planning urban water infrastructures particularly from resilience perspective and presents a case study of Nagpur city in India.

Keywords Urban resilience · Water infrastructure · Community resilience Water-sensitive planning

9.1 Urbanization, Water Infrastructure, and Disaster Risk Nexus

Water is the very basic necessity for progression of humankind. The food, health, and livelihood security that are fundamental for survival and growth of human population primarily depend on water security. Rise and fall of some of the historic cities, like Fatehpur Sikri in India, too were in tandem with the developments in regional and local hydrological regimes. In the recent times too, the way water resources are managed is considered to be a guiding factor for the sustainability of human settlements

S. Deshkar (✉)
Department of Architecture and Planning, Visvesvaraya National Institute of Technology, Nagpur, India
e-mail: sdeshkar@gmail.com

B. Ray and R. Shaw (eds.), *Urban Drought*, Disaster Risk Reduction,
https://doi.org/10.1007/978-981-10-8947-3_9

under changing and complex environmental scenarios. The natural systems at large and water resources in specific across the world are put to immense risks owing to rapid urbanization and erratic climate fluctuations. It is predicted that by 2025, half of the world's population will be living in water-stressed areas (WHO 2016) with the majority of them inhabiting developing countries in tropical climate regions that witness unpredictable rainfall and frequent meteorological drafts.

With nearly 90% urban population growth occurring in developing countries (UN DESA 2015), the development challenge heavily puts pressure on water infrastructures in these countries. Most of the mega-urban centers in developing countries like India are characterized by a large-scale inward migration and haphazard outward sprawl that makes provisioning of adequate and safe water to its inhabitants a daunting challenge for the city administrations. The water woes are already becoming common features affecting not just cities and their hinterlands but larger regions supporting medium and several small urban centers pulling populations from all directions. The rising intermediate and small towns, in numbers, population size as well as their spatial expanse, are closing the gaps between cities and large metros. Misra (1998) calls these urban centers as micropolises, which he rightly presumed to be the backbone of urban India in the coming years. But at the same time, the pressures for economic development and industrialization of these regions are exacerbating tremendous demand for the vital natural resources, water in particular. The trends of migration and development patterns within urban regions could shift significantly owing to rising occurrences of severe stressed conditions of environmental services, water scarcity, flooding, and subsequent rise in health expenditure from water-borne diseases and epidemics (Revi 2008).

From intense energy production for industrial processes, building constructions, agriculture sector, and food processing to water-intensive land uses and consumptive lifestyles in cities, the demand for water resources is growing exponentially. These precious resources are threatened under changing development priorities marked by inadequate environmental policies and their ineffective governance in urban centers. The increased pressures on water resources can easily create or modify existing hazards or can generate completely new hazards (UNDP 2004; Coppola 2011). For example, reclamation of wetlands for supporting growing housing demands in cities can reduce the hydrological holding capacities of the land and may result in flooding in areas, which were previously safer. Poorly managed, the diminishing water resources in urban centers as well can lead to conflicts over their sharing between societies and regions. For instance, the conflicts turned violent over the sharing of water resources between interbasin states of Andhra Pradesh and Karnataka in India in the recent past put cities in region at further risks and were resolved only through geopolitical intermediation.

The existing water infrastructure networks continue to get pressurized not only due to rising development demands but the loss of precious urban land from being misused or consumed on account of unregulated, haphazard urban development constrains the space for developing new infrastructures or implementing non-conventional water conservation methods like water percolation ponds. The deficient urban basic infrastructure situation not only adversely affects the economic productivity but also often leads to unhygienic living conditions, specifically in slum-like developments in cities. In 2008, Prüss-Üstün et al. reported the unsafe drinking water and poor sanitation practices to be among the major environmental factors that cause more than 10% of worldwide deaths. Poor quality water and sanitation in developing countries has been a major concern for the public health with people living significantly reduced years of healthy life in contrast to those in high-income countries. The looming dangers from climate change to the global and local water regimes are equally alarming with several regions across India already facing acute water shortage in summers and severe floods in rainy seasons.

Developing adequate infrastructures and effective management mechanisms for water resources becomes the key for addressing several overwhelming issues with agriculture, energy, livelihood, and poverty reduction along with adaptation and mitigation of climate change. However, with the existing trends, water challenges in future can be predicted to become more complex owing to low competence and capacities of water management processes and governing institutions, ineffective implementation of environmental regulations, volatile social and economic conditions, and emergence of technologies, etc. The conventional approaches for provisioning and regulating water infrastructures in urban regions, therefore, have to look beyond just the augmentation of demand–supply gaps and with a resilience perspective where the impending vulnerabilities and various stressors, natural as well as man made, are addressed holistically.

9.2 Urban Resilience and Infrastructures

The 2011 Tohoku earthquake in Japan and the tsunamis that followed made the role of critical infrastructure (CI) very evident in sustaining the societies in adverse scenarios (Bach et al. 2013). The susceptibility of urban infrastructures such as water, electricity, and telecommunications, to various natural as well as man-made hazards, makes the urban communities, which are heavily dependent on these systems for their daily chores, increasingly vulnerable. A need for achieving resilience-integrated urban development has been emphasized through various public policies and studies. From a concept, resiliency is being developed into a tool to evaluate the capacities of urban areas to adapt and evolve against the increasing adversities arising from natural as well as man-made calamities. For sustaining the urban growth under rising conditions of uncertainties due to increasing complexities of urban functions, overlapping infrastructures, and rising concentration of population as well as economic assets, the idea of risk resiliency is of high significance.

Among several parameters influencing the disaster risk resiliency of urban areas, basic infrastructures and critical services are most significant as they not only play role in reducing vulnerabilities but also generate higher levels of community's quality of life. The status of basic infrastructures in terms of their availability and robustness is also considered crucial in determining the resilience of the urban systems (Mayunga 2007; Chang 2009) and communities against disasters. Revi (2008) also highlights various vulnerabilities arising from inequitable distribution of infrastructures across different city regions. However, these studies emphasize either on supply augmentation or look at only infrastructure mitigation as a loss reduction strategy. Secondly, there are conditions such as network complexities and environmental disruptions to which when exposed, the infrastructure systems also are susceptible to failures. The failure of infrastructure systems immediately affects the closely dependent human systems in urban areas through reduced accessibility and availability of life support services. Therefore, analysis of the status of accessibility and availability of infrastructures may hint at the potentials and challenges for achieving community resiliency at a particular urban area (Deshkar et al. 2011). Bruneau et al. 2003 considered the reduced failure probabilities and consequences of failures of built structures as key elements in determining the seismic resilience of communities and proposed technical, organisational, social, and economic (TOSE) dimensions to understand the resilience of critical infrastructures like electricity, water supply, health care, and emergency services.

However, Elmqvist (2014) argues that resilience is not a locational but a systemic condition and that efforts to establish a desired state of resilience at a place may catalyze to reduce resilience elsewhere or create "undesired resilience" (emphasis added). From this perspective, putting more stress on infrastructure building in urban centers by extracting natural resources from rural hinterlands could create robust systems in urban areas but at the same time result in vulnerable communities in the rural. The common perspective that emphasizes on the efficiency of urban systems like transport, energy, and governance for building sustainability is questionable as over-emphasis on efficiency may potentially undermine resilience through reduction in redundancy. An increased dependence on limited resources and very less available alternatives would generate more vulnerability in the urban systems, and the insufficient overlap in functions might lead the entire system to unstable conditions (Elmqvist 2014). There is also a debate as to increasing efficiency of existing services, and creating newer infrastructures may lead to adding more GHGs and thereby contribute toward the climate change. How then the non-normative concept of resilience is applied in determining the normative objectives of urban development through building infrastructures is challenging.

9.2.1 Measuring Infrastructure-Based Community Resilience

Different evaluation frameworks and measures to achieve resiliency of urban systems have been proposed (Prasad et al. 2009; Bam et al. 2009) which are so far limited only to megacities. The resiliency evaluation frameworks suggested in these studies consider knowledge and capacities of local governing authorities for dealing with the adverse situations resulting from climate change. However, the understanding of risk resilience also depends on the risk perception, which is governed by the characteristics of the local community. The sociocultural and socioeconomic attributes of the community together with the characteristics of physical developments and access to life support services are vital components that influence the risk perception. The idea of resilience is also nested within the quality of life of local communities (Deshkar et al. 2011) and that for gaining community resilience integrated with physical developments, the QoL perspective can be valuable. The experiences by local communities of adversities from failing infrastructures and their responses to such situations may vary in urban areas, and therefore, for a comprehensive evaluation of risk resiliency, the community's evaluation of components of resiliency is crucial. The importance of capacities of local community in resilience has been highlighted through other studies (Adger 2003; Manyena 2006; Mayunga 2007; Cutter and Finch 2008), which focus more on socioeconomic capacities. However, the community's capacities that may be induced or influenced by other factors such as the characteristics of existing infrastructures remain less addressed. It is opined that the capacities at local levels play a significant role in evaluating the fiscal demands for resilience and the bottom-up process is crucial for effective mobilization of resources for infrastructure building for resilience (ICLEI 2011).

The resiliency evaluation based on robustness of infrastructures needs to be holistic. Since the local population may have varying access to the complex, networked infrastructure systems on which they rely, the robustness evaluation could vary dramatically within the same community (Longstaff et al. 2010). As communities are the end beneficiaries of infrastructure provisions, their evaluation of infrastructure services and perceptions for resiliency can be vital for urban local bodies to decide which parameters of infrastructure provisioning are significant in specific localities. The existing resiliency evaluations provide citywide blanket policies for infrastructures but often are standalone in nature and are not integrated with existing infrastructure development policies. Deshkar and Adane (2016) proposed a framework for prioritization of infrastructure provisions for generating community resilience and the areas for physical intervention based on the local community's evaluation for services. The measures taken up by communities to adapt to the situations of non-availability of infrastructure services can significantly influence the way services are provided in urban areas and also determine necessary infrastructure improvements, in which community engagements could prove to be vital.

9.2.2 Challenges for Infrastructure Development for Gaining Resilience

9.2.2.1 Interdependency of Infrastructures

Infrastructure is a broad term encompassing various sectors such as energy, transportation, telecommunications, water and sanitation, financial systems, agriculture, and public health systems. All these infrastructures are complex systems in themselves and are made functional through highly overlapping conditions of economic market, techno-legal regimes, and social conditions. At the same time, these systems are intricately interdependent on each other; e.g., electricity that runs telecommunication networks depends on water systems and fuels for its generation, while the modern transportation systems require robust information systems for efficient and safe movements, etc. Interconnected infrastructure systems, therefore, are beyond just a theoretical construe and have been guiding numerous policy documents.

Interdependencies in critical infrastructures pose a serious challenge for managing risks to these systems since cities rely on interdependent and interconnected infrastructure systems for the well-being of their habitants and for their economic performances. Any disturbance to the conditions in which infrastructure systems remain interconnected may lead to what is known as "cascading events" or a "domino's effect." Urban systems and processes that are connected to the infrastructures can severely get affected due to a single disruption that is potential enough to trigger a series of disturbances in all interconnected systems (Benoit et al. 2003). In the much same way, even recovery or restoration of an infrastructure affected by any disruption within or from outside of the system depends on the availability of another supporting infrastructure system. This non-availability may occur either because of non-provisioning of infrastructure itself or due to a common cause failure where both the interdependent infrastructures get disrupted due to a common factor such as a natural disaster. Thus, all the developmental processes involving economies and societies, which rely on interdependent infrastructure systems, can seriously get affected due to a short disruption in one system followed by the cascading events of disruptions in other networks.

9.2.2.2 Climate Change Adaptation and Water Infrastructure

Approaches for developing cities and their basic infrastructures are already witnessing paradigmatic shifts due to projected climate change impacts on urban areas. The climatic and environmental compositions are predicted to change seriously and rapidly. Realization of any of the projected conditions might as well lead to the collapse of the witnessing urban areas, which would primarily be due to the failure of their infrastructure systems. Glimpses of such potential devastations were seen during worst-ever floods in Mumbai and Chennai in India during the years 2005 and 2015, respectively. The infrastructure provisioning in the Indian cities and rural regions is poised to witness an immense pressure under the projected implications of climate change such as drought, river and inland flooding and extreme rainfall events, cyclonic storms, storm surge and coastal flooding, mean and extreme sea-level rise, and environmental health risks (Deshkar and Adane 2016). Studies focusing on the implications of climate change also highlight the economic implications due to disruptions in infrastructures those could be similar or even greater in magnitude. Such implications could be more concentrated in urban regions since much of the infrastructures are typically located in urban areas. Water supply systems would be affected by climate change-induced excessive demands for drinking and cooling systems, failure of local water sources, and increased demand for regional water supplies. More frequent floods could also present a significant threat if these lead to the contamination of floodwaters with fecal material.

Efforts to integrate adaptation and mitigation measures for climate variability and extreme events are being taken while developing water, sanitation, transport, and energy infrastructures and their management. Climate change adaptation in infrastructure systems typically involves increase in reserve margins and back-up capacity, and systemic modifications that can accommodate more extreme conditions for operations. The concept of green infrastructure (GI) is also been explored for which it offers non-structural benefits to the society such as noise reduction, air filtering, microclimate regulation recreational and cultural values along with the mitigation of climate-induced extremes. Green infrastructures can be beneficial for various hydrological functions that include water carriage and permeation, reduction in soil and water pollution, storm protection in coastal regions, surface flow reduction and erosion control through vegetative barriers, water capture and storage with potential for reuse (Bartens et al. 2009). However, formulating any adaptation and mitigation measure with infrastructure systems for any city is challenging in the absence of any clarity in specific trends and likely future impacts due to climate variability. Secondly, even if predicted, existing large infrastructure deficits and crunch of physical space for the implementation of various measures make planning for such adaptation a daunting task. The poor quality of infrastructure and the lack of maintenance characteristic of typical developing cities are already seen causing catastrophic failures and devastations to public life during or after extreme weather events. In such circumstances, retrofitting existing urban infrastructures to cater to rising pressures of demands as well as designing futuristic urban services that are climate resilient can potentially enhance the prosperities from infrastructure development. This necessitates including adaptation and preparedness in sustainable infrastructure planning and development.

9.2.3 Water-Sensitive Urban Planning

Major cities in the world are facing greater challenges for managing water not only in terms of protecting freshwater resources for drinking purposes but also managing huge quantum of wastewater generated as part of urban processes. Growing water shortages makes it imperative to look for alternate ways we consume these vital resources and the ways we develop our cities. The option for city planners of becoming smart water resource managers is apt in the face of changing environmental conditions than remaining as resource users and making cities increasingly dependent on rural areas for water sources and simultaneously growing their urban footprint and potentially exploiting all available water resources.

Water-sensitive urban planning looks for an integrated and comprehensive approach to urban water services, taking water supply, wastewater, and storm water as components of an integrated physical system. The approach advocates utilizing all possible available water resources in a way in which water supply, storm water, and wastewater are managed independently and are separated from land-use planning and economic development (Bahri 2012; Biswas 2008). The idea of water-sensitive urban planning is being propagated in various countries and regions through different concepts such as water-sensitive urban design (WSUD) in Australia, low-impact development (LID) in the USA, and sustainable urban drainage systems (SUDS) in the UK. Water-sensitive urban planning is an integrated approach that connects the urban water cycle with urban planning and design so as to achieve together the goals of minimizing environmental degradation and improving aesthetic appeal of urban spaces (Asit 2008).

Similarly, the concept of integrated urban water management (IUWM) redesigns and restructures a city's relationships to water resources and creates new opportunities in which it can be best utilized. The IUWM is envisaged "as a process that promotes the coordinated development and management of water, land, and related resources, with an objective to maximize the resultant economic and social welfare in an equitable manner without compromising the sustainability of vital ecosystems" (Bahri 2012). The concept of IUWM seems more promising from resiliency perspective also as it looks for developing redundancies in water sources through conservation of water and identifying alternative sources, addresses robustness by integrating water storage, distribution, treatment, recycling, and disposal as part of single resource management cycle, and encourages resourcefulness by recognizing interrelationship between water resources, land use and energy and participation by all stakeholders. All these approaches call for an integration of urban development and basin management to achieve the objectives of economic, social, and environmental sustainability.

9.3 Managing Urban Water Infrastructure in India: Some Issues for Building Resilience

9.3.1 Status of Service Provisioning and Financing

Good infrastructures and basic services play a vital role in improving quality of life at a place as well as at larger scales; it stimulates the country's economic growth. However, with exploding population and already crunched economy, infrastructure development still remains a major challenge in developing countries like India. Adding more physical infrastructures assumes a priority for city governments than developing economically and environmentally sustainable services. With ever-increasing urban population and sprawling city limits, inadequate coverage, intermittent supplies, and poor quality have become inherent features of infrastructures in many cities of India.

In India, although 97% of the urban population is reported to have access to improved water sources (WHO-UNICEF 2015), there are huge gaps with regard to the quantum of water available to urban residents. There are remarkable disproportions in access, desired quality and quantity of water supply availability not just within adjoining urban areas but also within different parts of the same city. In Delhi, nearly 30% of the population receives only 25 L or less per capita water supply compared to as good as 200 lpcd that is supplied to other parts of the city (GoI/MUD 2011). India's large cities are worst in terms of hours of availability of water per day varying between 4 and 5 h, while for other smaller cities it ranges from 3 to 8 h per day. Though drinking water forms a smaller proportion of total water resources, issues like leakages in distribution systems, which range between 25 and 50% of the total water supply, exacerbate pressures that may mount up to a water-scarce scenario. The average share of non-revenue water (NRW) constitutes as high as 50% of the levels of system losses in Indian cities, which not only accounts for huge financial but also severe environmental losses in urban areas (UNICEF 2013). While some improvements and progress have been made in terms of coverage of network, the service is generally of poor quality and is not sustainable.

By and large the storm water drainage service for many cities in India including big metros is grossly underestimated and far from satisfaction level. It is a common experience that the major circulation within the city often comes to a standstill due to flooding of roads due to even occasional heavy rain showers. The existing plight of drainage system can be attributed to the lack of coordination among various development agencies within the city. Around 46% of urban households have water-sealed toilets, but the share of urban households having connections with public sewer system is only 28%. In India, nearly 300 urban centers have a sewerage system, out of which only 70 have sewage treatment facilities (GoI/MUD 2011).

In the recent fiscal year, setting the priority on health and hygiene, Indian government reported an investment of nearly 3.6 million US$ toward achieving objectives of Water, Sanitation and Hygiene (WASH) under SDG through development assistance (WHO 2017). But owing to the lack of financial sustainability for reaching the unserved and maintaining the services, implementation of WASH has been inconsistent in nature. Secondly, in spite of the well-laid guidelines for developing water infrastructures, participation by service receivers and local communities has been very moderate in India, which affects the effective implementation of the initiatives. With an average 40% population in Indian cities living in slums, balancing between the needs of infrastructures for city development and those for enhancing the quality of life of population residing in urban areas becomes crucial. Decision making with regard to the required levels of infrastructure services and their locations for provisions still remains a critical gap for city administration.

On the other hand, considering the socioeconomic dynamics, pricing of urban infrastructure services excludes O&M costs, which when coupled with poor tax recovery makes developing water infrastructures in cities heavily reliant on government subsidies and thereby an unsustainable practice. The existing pricing mechanism for the basic urban services works through the levy of various tax and non-tax components. The recovery made through these components is normally expected to meet the cost of infrastructure provision. However, in many instances, the cost recovery falls way short of even meeting the operation and maintenance cost and all other costs like capital cost, operational or working cost, cost of replacement are still left uncovered. The gap between income and expenditure widens and results in deficit burden and financial liabilities for the municipal bodies.

The irony is that where local development authorities are facing difficulties in sustainable provisioning of basic amenities to citizens, no effective financial or policy mechanisms are adopted to address environmental pollution or hazard management till it assumes a crisis level. The devastations from Mumbai floods of July 2005 or Chennai floods of 2015 made the gaps in addressing these issues very conspicuous and brought the discussions on India's crippled development path to the forefront on global forums. In Maharashtra State alone, a single drought (2003) and flood (2005) captivated nearly 13% higher budget than the entire planned expenditure on irrigation, agriculture, and rural development for the duration of 2002–2007 (World Bank 2008).

The scarcity of basic municipal services forces urban communities to adopt unsustainable and illegal practices of resource consumption. Illegal and unmetered water connections, over-extraction of municipal water by connecting motor pumps, etc., are observed commonly in several cities across the country. This not only creates unrealistic service gaps and inequitable distribution of services, but also could have serious implications for resilience building. At the same time, the populations living in outskirt areas unserved by basic amenities are exposed to various vulnerabilities. However, the coping mechanisms against extreme conditions of non-availability of services adopted by these local communities go unaddressed while developing infrastructure services, which rely on standards and service-level benchmarks set at the national level.

9.3.2 *Standardization and Service-Level Benchmarking*

Standardization is in principle a way of providing solution for repetitive applications to problems essentially in the domains of science, technology, and economics aimed at achieving optimum degree of order in the concerned aspects. It also acts as a base for newer technology and management innovations targeted at achieving efficiency in the processes and products. Technical standards and specifications propagated through various codes, regulations, and guidelines have an effect on almost every part of modern economy. In case of water supply management, as seen in the previous sections, the legitimacy and effectiveness of existing mechanisms, which mostly follow top-down approach, are being increasingly questioned. Secondly, under the fast urbanizing world together with changing global environmental conditions, it is nearly certain that competition for water will intensify in decades to come. The supply-side management, which primarily emphasizes on techniques of water resource generation such as construction of dams, groundwater management, and rainwater harvesting, and demand-driven approaches that focus on managing demands from various sectors such as agriculture, irrigation, and industry, are crucial for balancing the water budget and the development priorities. The pace of urban population increase and emerging gaps in service provisioning make the need of standards more valid in case of water supply and sewerage infrastructures.

However, effective standardization requires the active involvement of business, industry associations, standards bodies, and government agencies, agencies involved with consumer safety and environmental protection (Kanika 2009). The process fails to acknowledge the socioeconomic dynamics and locational parameters including weather events that could alter the way water infrastructure services are provided. The challenge faced by the current standardization regime is to be able to address ground realities and take cognizance of local sociocultural aspects particularly those linked with natural resource conservation.

On the other hand, understanding the growing concern for improving efficiency in the delivery of basic services in urban areas, the Government of India launched a series of initiatives, policies, and programs which aimed at enabling the urban local bodies to meet the unprecedented challenges that are currently being faced. To facilitate the critical reforms in the urban sector, the Ministry of Urban Development, Government of India, in the year 2009 implemented National Service-Level Benchmarks (SLBs) in four key sectors, viz. water supply, sewerage, solid waste management, and storm water drainage (GoI/MUD 2009). Acknowledging that the investments in constructing new city infrastructure has not always resulted in corresponding and desired improvements in levels of service delivery, a need for a shift in focus in the direction of service delivery was realized, especially the case in water supply and sanitation. Measuring service levels of civic agencies infers measuring results and reveals on the institutional capacity and financial performances.

The objectives of SLBs, however, were to facilitate comparison between different cities/sectors and to monitor the changes in the performance of the urban local bodies with regard to infrastructure development. The indicators and their benchmarks for the evaluation of each sector were done through a consultative process involving various experts and based on the experiences of the government agencies with past initiatives on infrastructure improvement, which mostly focused only on augmenting demand–supply gaps or increasing the coverage areas. The SLBs were also developed with a baseline data collected from a sample of 28 cities, spread across 14 states and different city sizes. It is evident that neither the additional requirements for adaptation against extreme climate events nor the financial requirements for creating supporting infrastructures for strengthening resilience of local communities were taken into account while formulating these SLBs as the majority of the previous government schemes pertaining to infrastructure development emphasized on the issues of coverage and supply augmentations. Secondly, with existing mechanisms of service deliveries and infrastructure development in urban India and the rapidly changing resource consumption patterns, the SLBs so defined seem only ideal and far from being realistic (Table 9.1).

Table 9.1 Service-level benchmarks for water supply

Sr No.	Indicator	MoUD benchmark
1	Coverage of water supply connections	100%
2	Per capita supply of water	135 lpcd
3	Extent of metering of water connections	100%
4	Extent of metering of water connections	20%
5	Continuity of water supply	24 h
6	Efficiency in redressal of customer complaints	100%
7	Quality of water supplied	80%
8	Cost recovery in water supply services	100%
9	Efficiency in the collection of water supply-related charges	90%

Source Handbook on service-level benchmarking, MoUD, GoI

9.4 Case of Nagpur City in Central India

Nagpur is the third largest city in the state of Maharashtra. Located between latitude 21° 5′N and longitude 79° 5′E (Fig. 9.1), the city is spread over an area of 217.56 km^2. With a population of nearly 2.5 million, Nagpur is the 13th largest urban agglomeration in India and one of the largest cities in Vidarbha region. It serves as the seat of the winter session of the Maharashtra State government. In addition to its importance for political and geographical reasons, it is also a major education center. It is known in India as the "City of Oranges" because of the large production of oranges in the region. Located in the center of India, the city is well connected to other major cities through an extensive network of highways, railways, and air traffic.

The Nagpur Metropolitan Region is also one of the fastest growing industrial and economic hubs in central India with the majority of the industrial hubs such as Butibori, Kamptee, Hingna, Wadi, Khapri, and Kalmeshwar located within a range of 25–40 kms from the city center. Development of the present-day airport at Nagpur as a Multimodal International Hub (MIHAN), planned as a multidimensional, multidisciplinary project of international standards, along with formation of a special economic zone (SEZ) has catalyzed a growth in real estate sector in recent past through increased housing demands. However, through increased migration is also burdening the water infrastructures which are already limping to meet existing requirements.

Fig. 9.1 Location map of Nagpur. *Source* https://d-maps.com/carte.php?num_car=24865&lang=en

9.4.1 Hydrometeorology

Nagpur comes under a subtropical hot and dry climate zone with dry conditions prevailing for most of the year. Summer season ranges from March to June and is extremely hot with average temperatures up to 43 °C, which often reaches to peaks of 48 °C, during the month of May. Monsoon season in the city prevails from June to September. The winter season is observed from end of November to end of February, during which the temperatures may drop to 10 °C and below. The average annual rainfall witnessed in this city is 1112 mm. This includes the monsoon rainfall of 962.5 mm. The mean total of rainy days during the year is observed to be 58, of which 48 are witnessed during monsoon season and 10 during non-monsoon season (NMC 2014).

However, in the recent times the rainfall has been erratic in nature with more than 100 mm being received in shorter spans of few hours. On May 22, 2013, the city recorded its hottest ever temperature at 47.9 °C, and just 2 months later, it received a total of 652.7 mm rainfall over the month of July, making it Nagpur's second wettest July ever. Local floods during heavy downpour, mainly in the areas adjacent to river channels and low lying areas, have become regular feature in the city.

9.4.2 *Physiography and Drainage*

Nagpur is one of the prominent cities situated on Deccan Plateau of the Indian Peninsula and is covered by Deccan trap formation in western part, while metamorphic and the crystalline series occupy eastern part of the city. The city that is located at a mean altitude of 310 m above sea level is a landmark on the eastern edge of the rolling Mahadagarh Hills, extensions of the more prominent Satpura Ranges spanned almost over Central India. Nagpur city is dotted with many remnants of these hills, viz. Seminary Hill, Starkey Point Hill, and Ramnagar Hill on western edge while Sitabuldi Hill located at the center of the city. On a rim of the high elevation of these hills on western edge are located the major tanks from which two rivers originate and flow eastward. The tiny streams, viz. the Pili River and Nag River, and their tributaries originate on the western plateau and flow toward southeast (Fig. 9.2). Nagpur city falls into three distinctive drainage basins spread in west to east direction. The central zone comprises of one-fourth of the city municipal area being drained by the Nag River. The northern zone of the city, which is a less than half of the city municipal area, drains into the Pili River while the southern part of the city drains into the Pohra River.

9.4.2.1 Surface Water

Nag and Pili streams are the two prominent streams passing within the boundary of the Nagpur Municipal area. The other rivers flowing within the Nagpur Municipal Corporation (NMC) are the Pohra and Pili rivers. The Nag River flows through major part of the city, and it carries mostly sewage and solid waste generated from adjoining areas. The Pilli River passes mostly through outskirt areas of city, and it also carries domestic waste generated from the localities besides it. The two rivers confluence at Kanhan River and eventually join the Gosikhurd reservoir. Over the years, these rivers have been carrying heavy sewage load from the adjoining localities and converted into wastewater channel or nallahs. The water quality in the rivers is contaminated due to untreated sewerage and wastewater being directly let into the river and its

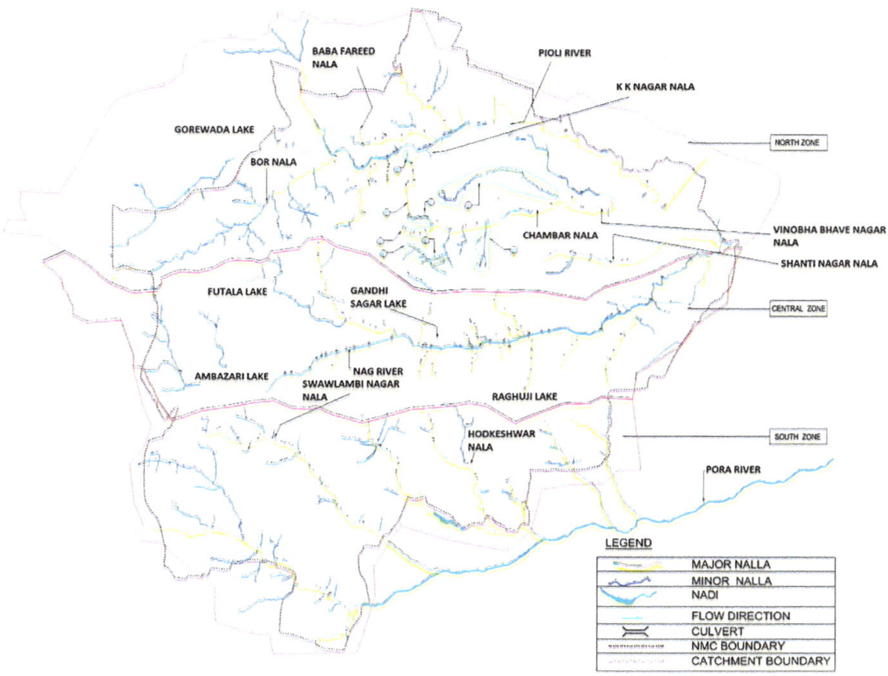

Fig. 9.2 Drainage pattern and basin map of Nagpur. *Source* NMC (2011)

tributaries. The Nag River traverses across the city from west to east. Situation becomes most serious in the monsoon season, when these drains are flooded and water/wastewater enters into the nearby locality due to unembanked edges of these rivers. Astonishingly, some of the major residential developments and channelization have been proposed along the drains (NMC 2012), which would not only affect the quality and quantity of the water in the future but would also pose threat to the nearby developments due to flooding.

Natural water bodies, which are mostly located on the elevated western rim of the city, assume prime importance for the sustenance of the city. There are around 11 number of lakes situated in and around Nagpur city, the significant being Futala, Ambazari, Sonegaon, Shukrawari, Sakkardara, Naik, and Lendi Talav. The surrounding agricultural and forest areas and also the faunal diversity depend on these water bodies. Nagpur Municipal Corporation (NMC) supplies around 96% of water mainly from some of these surface reservoirs. However, currently, the city meets demand just sufficiently and the low competent supply system creates a deficit situation during extreme summers.

The lakes in the city are mostly being used for domestic and recreational purposes. During the festival seasons, a few of the lakes are used for the immersion of idols and the local authorities make provisions for the same. On the bank of these lakes, some of the slums and informal settlements are located which are using lake water for washing, bathing, and domestic activities. The NMC in recent past has taken actions to protect and conserve these lakes from getting polluted due to local activities. All the lakes situated in the city area are found to be bacteriologically unsafe for drinking point of view (Puri et al. 2011).

9.4.2.2 Groundwater

The depth of groundwater table in Nagpur city ranges between 1.65 and 1.95 m in central part of the city and extends up to 16 m in the peripheral areas. Crystalline (Granite Gneiss) and Deccan Trap Basalt, which do not possess any primary porosity due to its massive and compact nature, are the main consolidated rock formation occurring within the city limits (NMC 2014). Owing to this and a lack of effective water conservation measures, there is a probability of drought in every 5.5 years in the city and its surrounding region. Similarly, water scarcity is also a regular feature in the city both during monsoon and summers, which mainly is characterized by inadequate availability and poor quality of water. In the recent times, with sudden seasonal variations, the occurrence and impacts of these calamities have been intensive in nature.

It is observed that in the older parts of city almost every residential plot has a provision of dug well. However, these dug wells are seldom used. In the newer parts of the city, provision of bore wells has been made in almost every residential plot. These are often used to supplement the water demand. It is generally observed that the majority of people living in the fringes area of Nagpur have no or poor municipal water supply connectivity and are dependent on community wells, which during summer period go dried up due to low water level. The water demand is also projected to increase exponentially with vast developmental activities and population growth. This necessitates prioritizing the use of groundwater source for potable or non-potable usage. Contemplating the scarcity of surface as well as groundwater during summers and deficient water supply in the outskirt areas of city, managing the groundwater in these areas becomes very crucial.

9.4.3 Urban Water Management

The urban water management of Nagpur city includes supply of water from different sources, collection of wastewater through sewer lines, and collection of storm water. Little amount of the collected sewage is sent to the single available treatment plant, and the rest is directed to the water bodies. As highlighted in the previous section, Nagpur has been more vulnerable to water-related adversities in the form of flash floods and severe water scarcity.

9.4.3.1 Water Supply

The provision of water supply to the city residents is managed by Water Works Department of NMC and is being operated via private operator, M/s Orange City Water Services Ltd., under the public–private partnership (PPP) mode. Gorewada Tank, Kanhan River, and Pench Canal are the three major surface water sources from where water is drawn to supply to the city.

Ambazari and Shukrawari lakes were some of the oldest water resources developed by the rulers of the city in the nineteenth century. Ambazari continued to the city for nearly a century, but after the establishment of industrial units in its catchment that polluted the lake, the extraction of water for drinking purpose was stopped in the year 1980. The Gorewada Tank, which was developed in the year 1911, still remains a significant source of water to the city. However, with the sprawl of city limits and rise in population leading to subsequent rise in water demand, River Kanhan located 15 km away from city was considered for augmenting the water supply.

In the year 1961, the city lost its status as a capital city of Madhya Pradesh and became an integral part of Maharashtra State as its second capital. It saw an upsurge in its population through natural growth as well as due to merging of surrounding villages into city limits. However, correspondingly no new water sources were developed; rather the existing source from Ambazari Lake was discontinued. As a consequence, the per capita water supply dropped considerably from 124 to 103 lpcd in span of two decades (Table 9.2).

Meanwhile in 1976, a storage dam across River Pench for hydroelectric project at Totaladoh and pickup dam at Navegaon Khairi were executed by the Irrigation Department, Government of Maharashtra. The water resources from Pench River

Table 9.2 Development of drinking water sources

Year	City Population	Water supply in mld	lpcd rate	Sources
1921	1,45,000	16.50	114	Ambazari + Gorewada
1941	3,02,000	45.00	149	Ambazari + Gorewada + Kanhan
1961	6,44,000	80.00	124	Ambazari + Gorewada + Aug. 1, Kanhan
1981	12,17,000	125.00	103	Ambazari (discarded), Gorewada + 3 Times Aug. from Kanhan
2001	21,50,000	370.00	172	Gorewada + Kanhan + Pench-I + Pench-II
2004	23,50,000	470.00	200	Gorewada + Kanhan + Pench-I + Pench-II + Pench-III
2011	24,47,000	651.00	266[a]	Gorewada + Kanhan + Pench-I + Pench-II + Pench-III + Pench-IV

[a]Includes water losses
Source NMC (2011)

Table 9.3 Estimation of water demand till 2041

Indication/year	2013	2021	2031	2041
Per capita water supply demand	135	135	135	135
Water losses (%)	53%	45%	30%	20%
Per capita water demand—(LPCD + losses)	207	196	176	162

Source MoUD (2015)

are commissioned in a phase-wise manner, and currently, Pench and Kanhan rivers form the major sources of water for the city.

However, the actual supply of raw water, which is quite high from the sources, becomes less than 50% on reaching the consumer's end. So, with the same pattern of losses, the water requirement of city's future will need new resources. But it is reported that the raw water sources, primarily meant for irrigation purposes reservoirs, are increasingly being allocated for drinking and industrial purposes (NMC 2011). The estimated water demand based on per capita supply demand (135 lpcd) and the average losses indicate a reduction in per capita water supply by the year 2041 (Table 9.3) assuming that the water supply losses would decrease from 53% in 2013 to 20% in 2041. However, what mechanisms would be adopted to reduce these losses and how the success of those mechanisms would be ensured are yet uncertain.

Network coverage is another issue with sustainable water supply in the city. About 85% of the households in Nagpur city receive water supply through a distribution network of 2,050 km. The remaining households that are not connected to the distribution network receive water supply through water tankers. While 63% of water connections in the city are metered, it was observed that 35% of metered connections are non-functional. Against this background, the target of 100% metering of water supply connections by NMC as part of JNNURM reform implementation is an ongoing herculean effort.

9.4.3.2 Storm Water Management

The city has about 917 kms of drains which are divided into three zones; i.e., only 35% of city roads are covered with storm water drains. The storm water is carried by three rivers in the city, viz. the Pilli River, Nag River, and Pohra River. The storm water from some of the areas in west Nagpur, south Nagpur, central and east Nagpur is finally drained into Nag River through major and minor drains (nallahs), and Pohra River collects the storm water from the southern part of the city.

The average annual rainfall observed in the city is 1112 mm. As there are only 35% roads with storm water network, most of the rainwater is not reused or recycled. In areas not covered by underground sewerage, the rainwater enters into the open sewer lines that lead to water bodies and river channels. In the absence of inadequate storm water drains, the excess rainwater during heavy rainfall causes water logging in areas those are low lying and near to water bodies. The practice of rooftop rainwater harvesting is not followed in most of the areas, and thus, the rainwater flows through the roads and goes to waste. With increasing imperviousness in the city causes the rainwater run through it and very less infiltration into the ground.

9.4.4 The Abortive Paradigm in Urban Water Management: 24 x 7 Supply Project

Continued growth in urban population, depleting water bodies, and inadequate capacities of existing water infrastructures to cater to newer developments forced the Nagpur city administration to tap far located Pench water resources for augmenting the rising water demands in the city. However, the Pench project, which was implemented in phased manner and involved huge investment through soft loans, witnessed major roadblock during the third phase of the augmentation. Pench-III project, which was commissioned in the year 2003, increased the input volume by 120 mld, but the actual water sales by local authorities dropped by 20 million rupees in a single year. This was discovered to be on the account of major water losses through leakages, which lead not only to economic crisis for the local municipal body but also an unrest among the citizens who underwent inconvenience for years during construction of infrastructures but ended having no improvement in water supply. The authorities had difficult task in explaining to public and media for reasons behind persisting water problems despite average water supply in city being 200 lpcd.

Against this background, the Nagpur Municipal Corporation undertook Water Loss Control Program (WLCP), which was to be implemented in four-phase manner with an ambitious 24 × 7 water supply as its end target. Being first of its kind, the 24 × 7 water supply by NMC became a much-hyped project at national level under the ambitious Jawaharlal Nehru National Urban Renewal Mission (JNNURM) targeted at improving urban infrastructures. The project was initially looked upon as a pathbreaking initiative that adopted public–private partnership (PPP) for providing continuous water supply instead of the age-old urban service provisioning and management mechanisms that had burdened local municipal authorities, financially as well as technically.

9.4.4.1 Public–Private Partnership in 24 x 7 Water Supply

Public–private partnership (PPP) in urban water infrastructures was explored as a viable financial option in Indian cities. However, due to the problems of tariff setting and lack of political will, many of these projects, which mainly aimed at augmenting bulk water supply systems in cities (WSP 2011), were abandoned (ADB 2011). In spite of risks of capital, revenue as well as political in nature, the private players are expected to raise huge capital and also establish efficiency in operations and delivery to enhance the financial potential in service delivery sector. From the earlier failures of PPP in large-scale project, states and cities initiated projects at local levels with the involvement of the private sector for improving the operation and maintenance (O&M) efficiency of urban services. In case of 24 × 7 water supply experiment in Nagpur as well, the role of the private entities was limited to O&M and partially in planning and construction of infrastructures.

In 2008, NMC formed a special-purpose vehicle (SPV) by the name of Nagpur Environmental Services Private Limited (NESPL) as a 100% subsidiary owned by NMC that provided autonomy to the water supply department for managing water infrastructures and supply within the city. The NMC implemented a 24 × 7 water supply system in a few selected wards in the city as a pilot project that provided clean drinking water with adequate pressure and an uninterrupted continuous water supply to more than 16,000 consumers, including 5,000 slum households (NIUA 2015). Improved water pressures and coverage, increased billed water volume, and reduction in non-revenue water were projected as some of the prominent achievements of the pilot scheme. Based on these experiences at pilot level, NMC upscaled the initiative at citywide level by entrusting Orange City Water Private Limited (OCW), a private entity formed in 2011 under PPP, to take up 24 × 7 water supply project. Under the provisions of contract, the private operator was expected to enhance the coverage of water connections through compulsory provision of metered connections and replacement of old and non-workable meters. Also, the private operator was made responsible to regularize the illegal connections through obligatory provision of metered connections and billing of the supply of water, which will help to reduce the commercial losses and improve tax collection efficiency.

However, the project, which was expected to start showing results in the year 2012, is still in its infant phase of implementation. The cost of the ambitious 24 × 7 water supply project has reportedly escalated by Rs. 38 crore (5.8 million USD) due to inordinate delay in completion (Anjaya 2017), which further may burden water consumers. The reasons extended by the concerned authorities for delay range from discontinuation of JNNURM scheme, change of government, lack of funds, and non-completion of Pench-IV water supply project on time. With no clear argument on any of these, it only puts a question as to whether the visualized paradigm was ever possible and inclusive or was beyond the capacity of the authorities.

9.4.4.2 Lessons Learnt

A critique into the status of the project indicates that the agencies failed to appreciate the involvement of stakeholders including citizens and local population. Even at the pilot stage, both NMC and OCW had reported their difficulties in convincing the citizens to implement 24 × 7 in the pilot zone (NIUA 2017). NMC shared that a long process of stakeholder consultations, door-to-door campaign, pamphlets, and other means of communication were used to build consensus among citizens of the city. But it failed to gain the trust and credibility due to discontinued stakeholder engagement and communication during later phases of the implementation. Inadequate customer awareness and spike in billings following implementation of 24×7 supply during the pilot project did contribute to negative vibes for the project.

Nagpur Municipal Corporation took almost 2.5 years to design a detailed contract covering all aspects at various levels of service delivery. Delays on administrative fronts, inability to implement revised tariff rates, increased burdens on NMC's coffers toward operator's payment, electricity and raw water costs, and poor recovery marred the project even before it could prove its merits. The whole paradigm was also driven by impending pressures for increasing infrastructure coverage and water demand augmentation through heavy investments in newer construction technique and materials. The opportunity to redefine its innovative urban agenda was missed by the local authorities by not assessing the ground realities holistically and including community's perceptions, existing adaptation practices and environmental challenges. Also, the project emphasized only on drinking water and ignored the problems of drainage systems. Water stagnation and flooding during monsoon rains continue even in the pilot area, and soon there could be excavation works from making drainage that could conflict with 24×7 water supply lines.

Continuing with the decades-old priorities for urban infrastructure improvements set by the government, another mission named Atal Mission for Rejuvenation and Urban Transformation (AMRUT) was launched in June 2015 with the objective of achieving the SLBs for basic urban infrastructures set by MoUD. The State Government of Maharashtra prioritized water supply followed by sewerage under the mission, and the City of Nagpur is yet again attempting to achieve universal coverage of household connections for the city over a period of three years.

9.5 Conclusion

The National Water Policy (GoI Ministry of Water Resource 2012) of Government of India highlighted the concerns for deepening water crisis owing to the anticipated implications of climate change as well as the continued failure of city governments to deliver safe drinking water access to its residents. In 2010, to address the problems and challenges associated with climate change, Government of India launched the National Mission on Sustainable Habitat. Recognizing the future challenges due to changing climatic conditions may cause issues at regional level such as droughts or floods; the focus was set on urban areas for mitigating environmental risks emanating from climate change and developing the cities to be more resilient and environmentally sustainable for the future. Opportunities certainly existed for integrating these concerns into ongoing urban reforms, particularly through infrastructure regimes, but seem to have been missed by various cities.

Secondly, planning for infrastructure development in India often takes a piecemeal approach and often land use and local community requirements and characters are ignored while planning the city infrastructures. Instead, it emphasizes on a planning process suiting to private market situations, limited public resources and capitals, forcing cities to scuffle with prioritizing interventions such as water, or education, or housing and phasing out areas of the city for their implementation built on various city-specific conditions, such as rapid and haphazard growth and deadly environmental issues. The financial requirement of the infrastructure building is huge compared to the low availability of resources with local bodies that creates a wide gap between emerging needs and restricted provision. At the same time, the needs for climate finance is going to be another challenge in coming decades and innovative funding mechanisms by leveraging private finances along with government sources will have to be worked by the local authorities. Various approaches, including technological deployments, toward disaster risk mitigation and adaptation in developing countries need to be taken up in a way that do not deter potentials for realizing growth and reducing poverty (World Bank 2010).

Many cities are burdened with additional responsibilities of providing services for slums and squatters that comprise around 30–40% of the city population and, also, rising informal sector population needs. Under such conditions, any effort on improving infrastructures and urban services only looks as a piecemeal project. As a silver line, however, the situation is gradually changing and many urban local bodies are looking for necessary corrective measures. There is a major shift in the approach about providing basic urban services like water, sewerage, solid waste disposal, road, street lighting as these services are no longer perceived as free public goods. Planners and local governments in most of the developing countries are progressively exploring the concept of privatization of basic urban services for all its merits.

Water shortage is a reality in many cities across India. In case of the Nagpur city, it has been witnessing water crisis on regular basis on account of harsh climatic conditions, which are turning erratic in the recent times as well as due to poor management of its water infrastructures and resources. In spite of a sophisticated water distribution system, frequent repairs to its infrastructures expose local communities to vulnerabilities due to non-availability of water. Protection and conservation of water bodies and other sources of water should be the priority for city government. Nothing seemed wrong logically with the citywide PPP and a well-articulated post-implementation water audit. But the greatest lesson for the urban local bodies would be not to rely on public–private partnerships as cure-all solutions to their current problems and challenges, but instead consider it as a strategic opportunity to advance on performance while remaining dedicated on achieving a resilient future development.

References

Adger WN (2003) Building resilience to promote sustainability. IHDP Update 2:1–3

Anjaya A (April 25th 2017) 24 × 7 water project cost escalates by 38 crore. Times of India, Nagpur edition. Available at: http://timesofindia.indiatimes.com/city/nagpur/24x7-water-project-cost-escalates-by-38cr/articleshow/58350900.cms. Accessed on 25 June 2017

Asian Development Bank (ADB) (2011) Tool Kit for public—private partnerships in urban water supply for the state of Maharashtra. Asian Development Bank, Manila, India

Biswas AK (2008) Integrated water resources management: is it working? Int J Water Resour Dev 24(1):5–22

Bach C, Gupta AK, Nair SS, Birkmann J (2013) Critical infrastructures and disaster risk reduction. National Institute of Disaster Management and Deutsche Gesellschaft für internationale Zusammenarbeit GmbH (GIZ), New Delhi, 72 p

Bahri A (2012) Integrated urban water management. Technical Background Paper no. 16 Global Water Partnership, Stockholm

Bam H N Razafindrabe, Parvin GA, Surjan A, Takeuchi Y, Shaw R (2009) Climate disaster resilience: focus on coastal urban cities in Asia. Asian J Environ Disaster Manag 1(1):101–116

Bartens J, The Mersey Forest Team (2009) Green infrastructure and Hydrology. Available at http://www.greeninfrastructurenw.co.uk/resources/GI_&_Hydrology_Report_May_2009.pdf. Accessed on 15 June 2017

Benoit R, Jean PS, Mathias G, Frederic P, Marie S (2003) A new structural approach for the study of domino effects between life support networks. In Alcira K, Margaret A, Anne C (eds) Building safer cities: the future of disaster risk, disaster management series no. 3. The World Bank, pp 245–272

Bruneau M, Chang SE, Eguchi RT, Lee GC, O'Rourke TD, Reinhorn AM, Shinozuka M, Tierney K, Wallace WA, von Winterfeldt D (2003) A framework to quantitatively assess and enhance the seismic resilience of communities. Earthq Spectra 19(4):733752

Chang SE (2009) Infrastructure resilience to disasters. The Bridge 39(4) (Journal of the National Academy of Engineering, Washington, D.C.)

Coppola DP (2011) Introduction to international disaster management. Elsevier, Butterworth-Heinemann

Cutter SL, Christina F (2008) Temporal and spatial changes in social vulnerability to natural hazards. Proc US Nat Acad Sci 105(7):2301–2306

Deshkar S, Adane V (2016) Community resilience approach for prioritizing infrastructure development in urban areas. In: Shaw R, Rahman A, Surjan A, Parvin GA (eds) Urban disasters and resilience in Asia. Elsevier, New York

Deshkar S, Hayashi Y, Mori Y (2011) An alternative approach for planning the resilient cities in developing countries. Int J Urban Sci 15(1):1–14

Elmqvist T (2014). Urban resilience thinking. Solutions J 5(5):26–30. Available at https://www.thesolutionsjournal.com/article/urban-resilience-thinking/. Accessed on 10 June 2017

Government of India, Ministry of Urban Development (GoI/MUD) (2009) Handbook on service level benchmarking. New Delhi

Government of India, Ministry of Urban Development (GoI/MUD) (2011) Report on Indian urban infrastructure and services for estimating the investment requirements for urban infrastructure services. High Power Committee, New Delhi

Government of India, Minstry of Water Resources (2012) National water policy. New Delhi

ICLEI (2011) Financing the resilient city: a demand driven approach to development, disaster risk reduction and climate adaptation. An ICLEI White Paper, ICLEI Global Report

Kanika K (2009) Standards: a necessary adjunct to sustainable water management. In: Gupta R, Ralegaonkar RV, Ghare AD, Katpatal YB (eds) Sustainable development of urban infrastructure, Dattsons

Longstaff PH, Armstrong NJ, Perrin KA, Parker WM, Hidek M (2010) Community resilience: a function of resources and adaptability, White Paper, Institute for National Security and Counter Terrorism, Syracus University

Manyena SB (2006) The concept of resilience revisited. Disasters 30(4):433–450

Mayunga JS (2007) Understanding and applying the concept of community disaster resilience: a capital-based approach. Draft working paper prepared for the summer academy, Megacities as hotspots of risk: social vulnerability and resilience building, Munich, Germany, 22–28 July

Misra RP (1998) Urban India: historical roots and contemporary scenario. In Million Cities of India, Misra RP, Misra K (eds) p 52. Sustainable Development Foundation, Delhi

Ministry of Urban Development (MoUD) (2015) City Development Plan for Nagpur City-2041

Nagpur Municipal Corporation (NMC) (2014) Environmental Status Report, Nagpur
Nagpur Municipal Corporation (NMC) (2012) Nag River Rejuvanation Plan, Nagpur
Nagpur Municipal Corporation (NMC) (2011) Draft City Sanitation Plan, Nagpur
National Institute of Urban Affairs (NIUA) (2015) Urban water supply and sanitation in Indian cities, Compendium of good practices. Available at https://pearl.niua.org/sites/default/files/books/GP-IN2_WATSAN.pdf. Accessed on 25 June 2017
Prasad N, Ranghieri F, Shah F, Trohanis Z, Kessler E, Sinha R (2009) Climate resilient cities: a primer on reducing vulnerabilities to disasters. World Bank
Prüss-Üstün A, Bos R, Gore F, Bartram J (2008) Safer water, better health: costs, benefits and sustainability of interventions to protect and promote health. World Health Organization, Geneva
Puri PJ, Yenkie MK, Sangal SP, Gandhare NV, Sarote GB (2011) Study regarding lake water pollution with heavy metals in Nagpur city. Int J Chem Environ Pharm Res 2(1):34–39
Revi A (2008) Climate change risk: an adaptation and mitigation agenda for Indian cities environment and urbanization. Int Inst Environ Dev (IIED) 20(1):207–229
United Nations Development Programme (2004) Reducing disaster risk: a challenge for development. Bureau for Crisis Prevention and Recovery, New York
UNICEF, FAO and SaciWATERs (2013) Water in India: situation and prospects
United Nations Department of Economic and Social Affairs, Population Division (2015) World urbanization prospects: the 2014 revision. (ST/ESA/SER.A/366). Available at https://esa.un.org/unpd/wup/Publications/Files/WUP2014-Report.pdf
Water and Sanitation Program (WSP) (2011) Trends in private sector participation in the Indian water sector: a critical review. Water and Sanitation Program (WSP), The World Bank, Washington D.C.
WHO-UNICEF (2015) Estimates on the use of improved drinking water sources. Joint Monitoring Programme for Water Supply and Sanitation. Available at https://www.wssinfo.org/documents/?tx_displaycontroller[type]=country_files/. Accessed on 06 July 2017
World Health Organization (2017) Financing universal water, sanitation and hygiene under the sustainable development goals. UN-Water Global Analysis and Assessment of Sanitation and Drinking Water (GLAAS) Report. Geneva
World Health Organisation (2016) Factsheet on drinking water. Available at http://www.who.int/mediacentre/factsheets/fs391/en/. Accessed on 10 June 2017
World Bank (2008) Climate change impacts in drought and flood affected areas: case studies in India. Available at http://documents.worldbank.org/curated/en/34787069/pdf/439460ESW0P0841sclosed0July03002008.pdf. Accessed on 06 July 2017
World Bank (2010) Development and climate change. World Development Report

Chapter 10
Urban Droughts in India: Case Study of Delhi

Shyamli Singh and Vinod K. Sharma

Abstract Delhi pioneers in tabling its climate change adaptation action plan in year 2009–2012 with a **Water Mission** jacketing water conservation, recycling, and distribution of water. The action plan deals with river water sharing pact with other neighboring states, decentralization of wastewater treatment system, connections, and treatment of drain water through interception sewer project. Revival of water bodies and recharging of rainwater, Delhi is having acute water shortage every year and needs long-term strategy for mitigating urban drought. There is a need to include urban drought as one of natural disaster and have its separate guidelines and preparedness plan as in California (USA) and other developed countries. The chapter deals with existing situation of water availability, well known facts establishing climate change such as rise in temperature, intensity of rainfall, storms and cloud bursts. The climate-related extreme events, viz floods and droughts show an increased occurrence and magnitude too. As Indian economy is agrarian, more emphasis is laid on floods and droughts. Ministry of Agriculture, the nodal ministry for droughts lays emphasis on rural area and agriculture. The urban water scarcity leading to "urban drought" is the responsibility of urban planners and urban municipal authorities. To add to the woes of the already scarce water and sanitation supply, the climate change is slowly but certainly skulking into effect, by the means of altering pattern of rainfall and diminishing groundwater resources. 16.78 million and a population density of 11,320/km^2 is housed in Delhi, Capital of India. The water demand is further increasing because of floating population and tourists. To add to the despairs, the population is gradually increasing every year with continuous migration from rural areas of neighboring states. Delhi is situated on the bank on river Yamuna, flanked by Indo-Gangetic alluvial plains in the North and East, the Thar Desert positions in the west, and the south is laced by the Aravalli. Out of the annual rainfall of 65–72 cm, 75% precipitation is experienced within a span of three months. Summer months from April to June show a maximum temperature of 40–45 °C. Water in summer season particularly in the urban slums need urban drought regulation with legal support, guidelines and regulation on water usage. There is need of sensitization

S. Singh (✉) · V. K. Sharma
Indian Institute of Public Administration, I.P. Estate, New Delhi, India
e-mail: Shyamli.env@gmail.com

B. Ray and R. Shaw (eds.), *Urban Drought*, Disaster Risk Reduction,
https://doi.org/10.1007/978-981-10-8947-3_10

of urban authorities, involving communities and academic institutions, and resident welfare organizations for public awareness. Similar plans can be shared with other metropolitan cities of India to provide water security to the urban population. Mainstreaming disaster risk reduction (DRR) is an imperative pledge of India thus various ministries, viz Ministry of Urban Development, Ministry of Water Resources, and Ministry of Agriculture the nodal Ministry can conjointly develop an approach for urban drought mitigation. India is committed to mainstream disaster risk reduction (DRR) and hence, Ministry of Urban Development, Ministry of Water Resources, and Ministry of Agriculture the nodal Ministry can conjointly develop a strategy for urban drought mitigation.

Keywords Climate change · Urban drought mitigation · Urban planning
Disaster · Water and sanitation · Capacity building

10.1 Introduction

Urban drought and scarce water supply have a larger impact on socioeconomic development. It has been realized beyond doubt that for the economy to achieve new laurels, a resilient foundation of water security is a must. The continuous urban sprawl and boost of economic activities in the urban locales are on upsurge making water a more priced commodity. The development of sustainable cities marks the necessity of robust water supply.

Indian developmental scenario witnesses the ever-increasing water demand in both urban and rural areas. To add to the woes further, Climate Change too plea for the apprehensions all over the globe and demands instant responsiveness. Conferring to the Intergovernmental Panel on Climate Change (IPCC) as the temperatures continue to increase globally there will be a rise in sea levels while the snow cover shall shrink extensively. As a result increase in precipitation will be witnessed in certain area while some areas might experience increase in the frequency of droughts. Global warming directly influences the precipitation patterns as enhanced temperatures lead to higher evaporation rates and thus lead to drying of surface which may trigger the occurrence of droughts and may influence the intensity and duration of drought. Evidences from various studies link droughts to Climate change. Climate change may be triggered due to indiscriminate deforestation which leads to less vegetated surfaces across the city and pollution of all kinds. As urbanization takes the city by storm the conversion of forest/agrarian land into hard surfaces by construction is a taking a toll. This has led to increase in the amount of surface runoff and decrease in the amount of water infiltration which largely affects groundwater table recharge. Developing countries experience pronounced effects of climate change primarily because they fall short of resources and human/infrastructural capacity to covenant with its impacts. Climate change, although gradually will have marked impacts on Indian economy with indicators as reducing agricultural yield potential, escalating severe weather episodes and wrecking low-lying coastal areas. It could significantly alter both the spatial and temporal distribution and quality of natural resources, thereby adversely disturbing livelihood security of masses (State of Indian Agriculture 2016).

10.2 Drought and Urban Drought

The phenomenon of drought has intensity, duration and spatial expanse as its signature. The climatic intermittence is an added dimension in the present scenario. Declaration of drought condition id confirmed when the rainfall received is deficient with respect to the statistical multiyear average for a spatial region, which might extend for a season, a year or even more (National Disaster Management Plan 2016). This brings about far-reaching implications traversing across many sectors of the economy which is felt far and beyond the actual area of physical drought.

According to the statistics provided meteorologically, ±19% deviation of rainfall from the long-term mean is considered "normal" in India. Dearth in the range 20–59% represents "moderate" drought, and more than 60% is "severe" drought (Samra 2004).

Six types of drought have been identified, viz meteorological, climatological, atmospheric, agricultural, hydrologic, and water management (Subrahmanyam 1967).

10.2.1 Drought: Causes, Effects and Impacts

Urban droughts may become a bigger peril in coming years as water both quality and quantity will increase as way of life in urban area is altering very fast. With the impact of Climate Change water scarcity in urban areas is already seen. In Delhi, water was one of the main agenda in state elections and still important aspect of governance.

All major consequences comprising both direct and indirect comprising of economic, social and environmental with major underpinning as drought (Gupta et al. 2011). As in most disasters, some of the influences are inevitable but timely and planned mediations can scale it down. Drought as a disaster threat is a perennial feature in the Indian context. Droughts in India have a signature, featured by incongruities in the arrangement and strength of rainfall from the Southwest monsoon. El Niño-associated droughts have been known to implicate episodic declines in the Indian agrarian output (Caviedes 2001). India had experienced 22 large-scale droughts until 2003; five of which were severe (Agrawal 2003). Peninsular and western parts of the country are typically drought prone added to it are a few pockets in central, eastern, northern and southern parts. The key reasons for the restricted drought in these expanses are snowball effects of altering precipitation pattern, excessive water utilization and ecologically incongruous agriculture practices. It is researched that the usual high-pressure air mass over the southern Indian Ocean, an ENSO-related oceanic low-pressure convergence center forms, continually pulling dry air from Central Asia, desiccating India during what should have been tropical summer monsoon season. The air flow reversal primes for India's droughts. The magnitude of severity of drought is contingent upon the extent of ENSO sea surface temperature rise in the central Pacific (United Nations 2016). Studies show that El Nino events have a precedence over drought in over 43% cases. The urban drought and the response strategy is lacking, the prominence of the response strategies are evolved upon net sown area or the crop yield (Gupta et al. 2011). Droughts have a primary impact which is laced with reduced agricultural produce, depleted levels of water, loss of biological habitats due to scarce water and high mortality. The direct impacts of drought are further connoted with multiplier effect through economy and society by posing threat to biodiversity, besides economic loss from Agriculture and being a threat to Food security. The indirect repercussions of drought are increase in agricultural loans, increase in agrarian unemployment, reduced purchasing power and the like (National Disaster Management Plan 2016). They are a risk to water and sanitation of the affected area. The cause and effect of drought is depicted in Fig. 10.1.

10.3 Case Study: Delhi

Delhi, the capital of India with a population of 16.78 million and population density of 11,320/km^2 situated on banks of river Yamuna, bound by Indo-Gangetic alluvial plains in the North and East, by Thar Desert in the west and Aravalli hills range in the south. The annual rainfall received by the NCT is 65–72 cm, where 75% precipitation is within three months. The Maximum temperature (40–45 °C) is observed in summer months from April May and June. Delhi has a total area of 1483 km^2; 783 km^2 is labeled as rural, and the remaining 700 km^2 as urban. This distribution makes it the largest city in terms of area in the country. Delhi incorporates three governing bodies: the Municipal Corporation of Delhi (MCD), the New Delhi Municipal Council (NDMC) and Delhi Cantonment Board (DCB). Of the total NCT area, MCD occupies 1,397.3 km^2, whereas NDMC and DCB occupy 42.7 and 43 km^{2,} respectively (Economic Survey of India 2007).

After attaining independence, the rapid incursion of migrants led to rise in population from 920,000 in 1941 to 1.74 million in 1951, recording a decadal growth of approximately 90%. In the subsequent forty years, the decadal growth has been above 50%. Population of the Metro city is 26,454,000 (United Nations 2016), whereas population of the urban agglomerate is 16,349,831 (City Population 2017).

Delhi boundaries the states of Haryana in the north, west and south and Uttar Pradesh in the east. Situated on the banks of River Yamuna, Delhi ridge is a paramount geographical feature. Running from southern border of NCT in southwest to northeast direction, ridge progresses toward the western banks of Yamuna River with Wazirabad in its vicinity (Delhi Jal Board 2007). The main watershed in NCT of Delhi is constituted of ridge. The city's topography generates a natural drainage system carrying rain and stormwater from higher elevation of Yamuna River in the west.

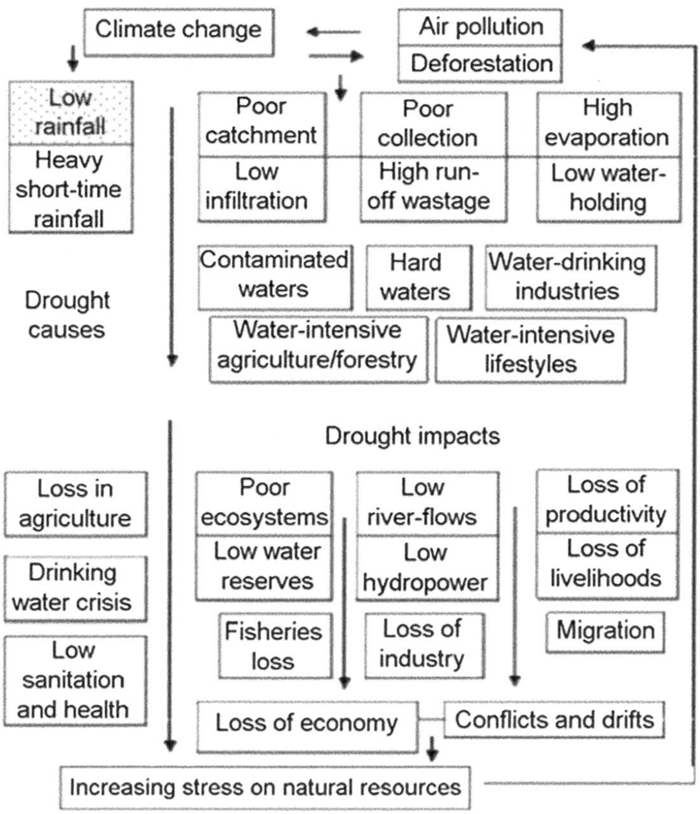

Fig. 10.1 Drought: causes and effects

Whereas the low-lying side of Delhi in the east, originally a part of the floodplain which is considered to be derelict yet it supports 20% of the total population in the trans - Yamuna region. The sprawl of population has led to the consumption of water for different usages and has led to decline of the water table. The depletion of water resource is one of the major causes of onsets of droughts. With consistent warming trends, it is anticipated that there would be severe implication on crop yields like that of wheat due to increase in temperatures, water stress and less rainfall. Urban planning, development of preparedness (contingency/Management) plans and public sensitization would have to be the driving force to mitigate the issue at hand. Table 10.1 shows some of the droughts that occurred in Delhi.

Table 10.1 History of droughts in Delhi

Date	Location (state, regions or districts)	Numbers
July/2002	Haryana, Delhi, Karnataka, Uttar Pradesh, Madhya Pradesh, Rajasthan, Punjab, Kerala, Nagaland, Orissa, Chhattisgarh, Himachal Pradesh, Gujarat, Maharashtra, Andhra Pradesh, Tamil Nadu states	300,000,000 affected; damage: US$910,721,000
May/2001	New Delhi, Rajasthan, Gujarat, Orissa	20 deaths
1875–2004	Haryana, Delhi and Chandigarh	22 droughts

10.3.1 Existing Situation of Water Availability

Urban India today faces numerous challenges, one of the most pronounced challenges is the acute shortage of water, especially during summer when the water bodies due to lack of groundwater recharge and increase in surface evaporation dry up. New Delhi, formally known as the National Capital Territory (NCT) of Delhi, is densely populated housing about 6343 slums with approximately 1 million households as in 2012 (National Sample Survey Office 2014). The Delhi Jal Board (DJB), a public water utility, shoulders the responsibility for production and distribution of drinking water in the NCT of Delhi. Surface water and groundwater caters to the groundwater. Yamuna, Bhakra and Ganga rivers supply as the source of surface water. The extraction of water is both direct from the river at Delhi and also indirect via the western Yamuna canal thus delivering water to Delhi. The upper Ganga and the Bhakra (Narwana branch) canal, respectively act as a conduit for raw water from the other two rivers. The mounting demand of water supply is met by further augmenting supplies from groundwater. Sixteen wells taking water from the Yamuna river channel bed the so-called Ranney wells are instrumental in facilitating groundwater. While the census indicates that 83% used treated tap water as a primary potable water source, only half of the slum households have any water source within their house premises, which reflects the inadequate availability and overdependence on unreliable shared sources. As per DJB reports about 82% of households in Delhi are supplied through piped water supply and warranted average availability of 50 gallons per capita per day of filtered water. Delhi uses an average of 835 MGD raw water daily from a supply of about 906 MGD (as of 2014) (Water policy for Delhi 2015). It is projected that in 2021, Delhi will have a minimum water demand of at least1, 174 MGD (Master Plan for Delhi- 2021).

On an average day, in Delhi, the residents rely on a combination of informal and mostly proscribed techniques and practices to meet their water demand. Albeit Delhi boasts of a very high level of water through the piped infrastructure Delhi's more elite neighborhoods average only 0–2 h of running water per day (Zerah 2000; Sagane 2000). The data available officially has estimated that 200 lpcd is provided but that is mixed with a combination of wasted or missed water, unequal distribution, unkempt water and sewage requirement (Delhi Jal Board 2007; Zerah 2000; Delhi Human Development Report (HDR) 2006; Kandra et al. 2004). Since Independence, Delhi's urbanization has factored on consistent marginalization and economically threatened residents. According to the survey conducted in 1995, it was estimated that each household spends around Rs. 2000 per annum to survive through the unpredictable water supply which is close to about five times more this additional burden furthers them to even more stress (Zerah 2000).

The water demand of Delhi is rapidly increasing because of burgeoning population and tourist inflow. As Delhi is a metropolitan city it attracts migrants in search of employment which increases the water supply burden on the city. Although Delhi pioneers in tabling its climate change adaptation action plan in year 2009–2012 with a Water Mission jacketing water conservation, recycling and distribution of water yet it needs a specific mitigation plan for urban droughts. The climate change adaptation action plan deals with river water sharing pact with other neighboring states, decentralization of wastewater treatment system, connections and treatment of drain water through interception sewer project. Revival of urban water bodies and recharging of rainwater by increasing rates of infiltration is the need of the hour as Delhi faces acute water shortage every year and needs a long-term strategy for mitigating urban drought. There is a need to include urban drought as one of natural disaster and have its separate guidelines and preparedness plan as in California (USA) and other developed countries.

10.3.2 Strategizing Drought Mitigation

Drought is a climatic event that cannot be prohibited, but intrusions and readiness to drought can help to be well organized to manage drought; develop more irrepressible ecosystems; advance resilience to recuperate from drought; and alleviate the influences of droughts. The inclusive approach to improvise the water supply in Delhi focusses upon the identified inadequacies. The strategies for alleviation of the impact approach entails shifting the thinking away from the traditional focus on a small number of research components to take an integrated approach aiming to address challenges. Such an approach should involve a counteractive strategy. DJB Operational Zones is a gray blotch that requires improvement to ensure smooth distribution. A detailed study on the distribution system should be crafted including reorganization into distinct, District Metered Areas (DMAs) so that 24/7 supplies can be familiarized on an incremental basis. Gradual drop of Non-Revenue Water along with DMAs as primer, thus facilitating enhanced financial efficiency and making more water available for consumption thus mitigating the scarcity of water supplies. New management concept with engrained accountability fixation on zone managers should be introduced for supply, distribution, metering, billing, leak detection, and repair. Provisioning of rationalized and equitable distribution of water supplies must be ensured through revamping the entire transmission and distribution system.

10.3.2.1 Economic Instruments and Drought Management

The urban water supply system during drought period relies purely on provisional conservation practices which are by and large non-market programs. These measures hardly change the fringe prices thus such measures are not instrumental in bringing out any significant change. Honolulu Board of Water Supply (1982) studies show that raising the tariff is an effective alternative. A single residential family water demand is a function of tariff, income, precipitation, number of people per family and a dummy variable connoting water constraints program. Short-run resistances recommend that a price hike of less than 40% would be instrumental in achieving 10% reductions in water use during the period of drought occurrence. Thus it comes out very clearly that it the economic measure rather than non-economic measures which bears productive results in reduction of water even during drought situations (Moncure James 1987).

10.3.2.2 Research Instruments and Drought Management

There is an urgent need to revamp and chisel the research instruments to facilitate effective drought management. Strengthened scientific research would ensure further reinforcement of the observational network for drought monitoring to bridge the gap between the existing and desired meteorological and hydrological monitoring network. The improvised research methodologies call for improvement in information and communication technologies in a cohesive manner to embark upon the multifaceted challenge of drought at various spatial and temporal scales. Capacity enhancement for medium and long-range drought forecasting by necessitating better synchronization among ministries and departments. A mechanism should be developed for context-specific and need-based forecasting including local language for better comprehension.

10.3.2.3 Paradigm Shift in Urban Water Management: Toward Demand Management

It has been recognized that there has been continuous increase in the incidences of water shortage and issues of water quality are more rampant. There is a need to focus on the transition from supply-side management (SSM) to Demand Side Management (DSM). The continuous expansion of infrastructure and developing new water resources is expensive and unsustainable. Thus this supply-side management should give way to the more sustainable DSM. It targets the soft path laying on enhancing the water productivity rather than looking for additional supplies (Gleick 2003). The soft path looks at water as the conduit rather than the end product thus ensuring the increase in water productivity rather than additional supplies (Wolff and Gleick 2002; Brooks 2003). The soft path relies on commissioning efficient technologies, education, regulation and the rational use of economic instruments to be applied to expand the efficiency of existing extractions in an impartial manner.

10.3.2.4 Reuse of Water

In the current context of severe urban drought, the recycled water with a substantial degree of treatment should be employed for various purposes as industrial cooling water, landscaping, groundwater recharge. This direct consumptive practice of recycled water and gray water can be put to offset certain futuristic water demands (Lund 1995). This use of water for other purposes would help in easing off the stress from the existing water resources.

10.3.2.5 Rainwater Harvesting

Rainwater harvesting is a hydrological cycle of water comprising of human-related water usage. The rainwater roof catchment system (RRCS) which is widely used throughout the world have several advantages attached to its end such as the collected water is relatively pure. The entire system of rainwater harvesting is stand-alone system thus making it suitable for scattered residential area, this system can be developed by using indigenous material. The system is not energy intensive is easy to maintain thus it is very sustainable (Schiller and Latham 1987). Since June, 2001 The Ministry of Urban Development, Government of India has made rainwater harvesting mandatory in all new buildings with a roof area of more than 100 sq. and in all plots with an area of more than 1000 m^2 that are being developed.

The national capital territory, (NCT), of Delhi receives 611 mm of rainfall on an average annually, and the number of rainy days is as low as 20–30. (A rainy day is specified as a day with more than or equal to 2.5 mm of rainfall) (Centre for Science and Environment 2017). The geology of Delhi comprises Alwar quartzite and alluvium whose vertical hydraulic conductivity, (permeability), is high compared to the horizontal permeability. This makes the conditions favorable for artificial recharge. Thus most of the urban rainwater harvesting efforts revolve around recharge of aquifers which is the best option available taking into consideration the rainfall pattern and availability. Since March 31, 2002 The Central Ground Water Authority (CGWA) has made rainwater harvesting mandatory in all institutions and residential colonies in notified areas (South and Southwest Delhi and adjoining areas like Faridabad, Gurgaon and Ghaziabad).

10.4 Initiatives

There is a need for separate urban drought regulation because the problem is as real as it gets and without appropriate data, it is difficult to plan mitigation strategies. The existing norms are outdated and are in urgent need of being revised. An inventory of urban water bodies needs to be maintained along with the details of quantum of water in their capacity. Deepening existing bore wells or merely providing tankers will not suffice the cause. Soil and water asset management, through enhanced progress and management of fragile catchment areas and river basins. Redirecting research toward more appropriate water conservation methods, rangeland and livestock management and emphasis on drought policy formulation. Reviewing institutional arrangements and physical infrastructure under the lens that drought impact risk reduction can be managed within the scope of long-term development planning including active early-warning systems and highly placed functional implementation and coordination structures.

The stress of drought strategies has always been on short-term mitigation measures rather than on long-term prevention programs. Recent years showcase the trends that new policies have been emerging in which preparedness, rehabilitation, prevention and planning are the strategic elements. Attempts in the current drought management strategies focus on treating drought as a potentially serious disaster and to incorporate it into program management cycles aiming at mitigation and prevention. There has been an acceptance that Drought is a normal phenomenon that would have several recurrences. Thus new policies resort to shift the responsibility of the impacts generated during a drought over the user. This transient shift of burden calls for redefining drought relief programs, for example, long-term projects such as building of dams, roads, buildings may be catalyzed during drought in the form of cash to be replaced by food to integrate relief measures. Diagnosing the gaps in water provision, particularly for the urban poor, the government has some initiatives to improve Delhi's water supply. The key proposed initiatives include free water for all metered connections which consume up to 20 kiloliters of water per month; an extension of piped water supply to unauthorized colonies over the next three years; reducing development charges for water connections to enable residents of unauthorized colonies to afford connections; using information technology for effective, continuous, and strict monitoring of water tankers and to enable people to track the water tankers, find out estimated time of arrival in their colony, and quantity of water. Pilot water ATMs are being set up for serving the urban poor which has received a mixed response. The resettlement colonies are selected for the treatment kiosks and ATMs under the design finance-build-operate-transfer (DFBOT) model of 2012. Designed to be complementary and affordable potable water sources to municipal piped supply, USWEs (Urban Small Water Enterprises) could be compromised by ambitious piped water plans. USWEs can be an important part of the solution to address insufficient potable water supply in Delhi slums. In slums where there is still a gap in water provision, USWEs can play a role, particularly for potable water. To be successful and reach their potential, playing a complementary role to piped water and providing treated, reliable, and affordable water to underserved slum populations.

10.5 Shortcomings in Existing Mitigation Plans

National Disaster Management Plan closes a critical gap in our disaster management system. Albeit most states and districts have prepared their plans, in absence of the national plan which is supposedly the guiding force the process at the sub-national level was absent leading to the exclusion of a good and robust action plan. The National Disaster Management Plan is bereft of the robustness. It may fulfill the ceremonial prerequisite of law of having a plan but it may not be effective in achieving its grandiose vision of building resilience. The drought resilient roadmap with special emphasis to urban sector fails to lay down a clear and practical roadmap. The plan is understated in its approach as it is too generic and lacks the provisioning of a time frame for undertaking activities. The time frame vaguely charters the short, medium, mid- and long-term basis. Another key limitation is the absence of clue on fund mobilization for the various freebies promised. Therefore, there is a dire need to develop a national plan appended by national roadmaps for disaster resilience with clear goals, targets, timeframe, and ideas about how resources shall be mobilized for its effective implementation.

Instead of focusing attention to retain water in drought-proof areas the efforts of the previous mitigation plans have been directed toward development of large-scale irrigation systems and drinking water programs. The importance of the above-mentioned programs is undeniable but to sustain the demands of the population today and tomorrow, strategic allocation and management of resources will have to be an important tool.

10.6 Good Practices

To ensure proper management of droughts some critical parameters have to be kept in mind such as availability of drinking water post the calamity, sustenance of economic activity and availability of food at reasonable price in the affected region.

There should be a predetermined and emergency action plan in place to assist the government agency which is the first to reach out for help to the affected areas. Nodal agencies should make use of climate modeling in a more effective manner in order to provide a warning before the disaster strikes. This would significantly reduce the impact of droughts and can speed up the initiation of drought relief measures. As State response teams are closer to the place of casualty they should have transparent and efficient coordination the central government.

Social security schemes or programs which ensure sustaining the wholesomeness of the hydrological system should be introduced. It is fairly common to see that funds apportioned for drought relief activities are slashed and brought down to paltry when under scrutiny for days. There needs to be a mechanism in place which guarantees advance release of funds under drought mitigation related schemes to the victims. In association with NGOs in each area, the idea of water management by the community shall be encouraged. This can be done be educating the public on methods like rainwater harvesting. In urban areas including Delhi, education has played a very positive role. By educating young children and nurturing them as green ambassadors of efficient water management and conservation in day to day life the message reached among parents and common masses. This has further helped in achieving the goal of water sustainability.

10.7 Linking Sustainable Development Goal and Droughts

The Sustainable development goals pose 169 targets as proposed in 17 objectives, closer observation and analysis reflects that there exists an alignment between drought and health in around 41 targets. An explicit strategy needs to be in place to implement the SDGs, this alone can ensure the adaptation and resilience building in the current face of climate change. Goal 13 of SDG which focusses entirely on climate change demonstrates the reinforcement. Countering to water scarcity and drought depends upon integrated policy and planning and across multiple sectors, especially water, agriculture and energy the same is reflected in Goal 6 which emphasizes on ensuring access to water and sanitation for all. The urban plight especially with respect to the efficient use of water management are highlighted in Goal 2, stating end of hunger, achieve food security and improved nutrition and promote sustainable agriculture. Goal 11 which emphasizes upon making cities inclusive, safe, resilient and sustainable. Goal 12 which states to ensure sustainable consumption and production patterns. Goal 15 to sustainably manage forests, combat desertification, halt and reverse land degradation, halt biodiversity loss. It is imperious to note that without coordinated and collaborative approaches at linking Climate Action and SDGs programs in terms of national implementation strategies, institutional coordination, financing, monitoring, evaluation and reporting, any little gain made in one could easily be eroded by the inaction in the other.

References

Agrawal PK (2003) Impact of climate change on Indian agriculture. J Plant Biol 30:189–198
Brooks David B (2003) Another path not taken: a methodological exploration of water soft paths for Canada and elsewhere. Report to Environment Canada, Friends of the Earth Canada, Ottawa, ON
Caviedes C (2001) El Niño in history: storming through the ages. University Press of Florida. ISBN 0-8130-2099-9
Centre for Science and Environment (2017) http://www.cseindia.org. Accessed 7 Sept 2017
City Population (2017) Delhi (India): Union Territory, major agglomerations & towns—population statistics in maps and charts. Retrieved 28 Feb 2017 from https://www.citypopulation.de/India-Delhi.html
Delhi Human Development Report (HDR) (2006) Oxford University Press, New Delhi
Delhi Jal Board (2007) http://www.delhijalboard.nic.in/. Accessed 12 Oct 2007

Economic Survey of India (2007) The New Delhi municipal council act, 1994. New Delhi Municipal Council. Retrieved 3 July 2007
Gleick PH (2003) Global freshwater resources: soft path solutions for the twenty-first century. Science 302:1524–1528
Gupta AK, Tyagi P, Sehgal VK (2011) Drought disaster challenges and mitigation in India: strategic appraisal. J Curr Sci 100(12):1795–1806
Kandra H, Johri R, Kapur A (2004) Water auditing—tracking unaccounted-for water in Delhi. Waterlines 22(3):19–21
Lund I (1995) Optimization of transfers in urban water supply planning. J Water Resour Plan Manag 121(1). http://ascelibrary.org/doi/abs/10.1061/(ASCE)07339496(1995)121:1(41)#sthash.WOM6q8re.dpuf, https://doi.org/10.1061/(ASCE)0733-9496(1995)121:1(41)
Master Plan for Delhi-2021 (2010) Delhi Development Authority, New Delhi
Moncure James ET (1987) Urban water pricing and drought management. J Water Resour Res 23(3):393–398. Department of Economics and Water Research Centre, University of Hawaii, Manoa, Honolulu
National Disaster Management Plan (2016) National Disaster Management Authority, Government of India
National Sample Survey Office (2014) Ministry of Statistics and Programme Implementation, Government of India
Sagane R (2000) Water management in mega-cities in India: Mumbai, Delhi, Calcutta, and Chennai. In: Uitoo J, Biswas A (eds) Water for urban areas: challenges and perspectives. United Nations press, New York, pp 84–111
Samra JS (2004) Review and analysis of drought monitoring, declaration and management in India. Working Paper 84. International Water Management Institute, Colombo, Sri Lanka
Schiller EJ, Latham BG (1987) A comparison of commonly used hydrologic design methods for rainwater collectors. J Water Resour Dev (3)
State of Indian Agriculture (2016) Ministry of Agriculture and Farmer Welfare, Government of India
Subrahmanyam VP (1967) Incidence and spread of continental drought. WHO/IHD Report 2
United Nations (2016) The world's cities in 2016, p 4. Retrieved 4 March 2017
Water policy for Delhi (2015) Department of Jal Board, Government of India
Wolff G, Gleick PH (2002) The soft path for water. In: Gleick PH (ed) The world's water: biennial report on freshwater resources. Island Press, Washington, DC, pp 2002–2003
Zerah M (2000) Water: unreliable supply in Delhi. Manohar, Delhi

Chapter 11
Future Urban Water Crisis in Mountain Regions: Example of Kathmandu Valley, Nepal

Basanta Raj Adhikari, Suresh Das Shrestha and Narendra Man Shakya

Abstract The civilization of Kathmandu Valley has started alongside the holy Bagmati River. However, rapid urban expansion and overpopulation have resulted not only in water shortage but ended up polluting the same water body based on which the settlements had grown, a classic example can be taken as that of Kathmandu and other surrounding cities in Kathmandu Valley. Due to resource availability and centralized government system, many people have migrated to the Kathmandu in recent decades. The annual population growth rate in the valley is 4.63% (CBS 2011) which has created haphazard urbanization resulting in water supply challenges. Geologically, Kathmandu Valley is an intermountain bowl-shaped basin comprised of both shallow and deep aquifers composed of fluvio-lacustrine sediments. Sandy gravel layers of northern side of the valley are considered the recharge zone due to the presence of unconsolidated coarse-grained deposits. Unfortunately, the area has the fastest urban growth and surface sealing, resulting in decrease in natural infiltration. Kathmandu Upatyaka Khanepani Limited (KUKL), the only one organization facilitated by the government, provides approximately 25–33% of the total demand of 350 MLD. Analyses have shown that there will not be water shortage between 2023 and 2025 if the Melamchi Water supply Project (MWSP) is completed within the allocated time. However, even that would not quench the thirst of the people as the demand is expected to rise in coming years. Proper planning, good governance, identification of water sources, and water treatment of wastewater are the long-term solutions for sufficient water supply. Alternate mitigation options, such as proper use of groundwater and surface water and spring water management with appropriate

B. R. Adhikari (✉)
Institute for Disaster Management and Reconstruction, Sichuan University-The Hong Kong Polytechnic University, Chengdu, China
e-mail: bradhikari@ioe.edu.np

B. R. Adhikari · N. M. Shakya
Pulchowk Campus, Institute of Engineering, Tribhuvan University, Kirtipur, Nepal

S. D. Shrestha
Central Department of Geology, Tribhuvan University, Kirtipur, Nepal

B. Ray and R. Shaw (eds.), *Urban Drought*, Disaster Risk Reduction,
https://doi.org/10.1007/978-981-10-8947-3_11

distribution system, would be useful. Unless the problem is tackled tactfully from different aspects and not only from demand supply and soon, the increasing social conflicts could be challenging for the planners in the future.

Keywords Urban · Groundwater · Crisis · Governance · Kathmandu

11.1 Introduction

11.1.1 Location

Kathmandu (27°32′13″ to 27°49′10″N, 85°11′31″ to 85°31′38″E, 1300 m asl), the capital of Nepal, is an intermountain basin lies the central development region of Nepal. Geologically, the basin is extensively covered with fluvio-lacustrine deposits of Plio-Pleistone and recent age over the rocks of the Kathmandu Complex. It is an oval-shaped basin with 25 km long in north–south direction and 30 km east–west direction. The altitude of Kathmandu Valley ranges from 1220 to 1500 m and occupies about (650 km^2) of area. The valley is surrounded by high hills; Shivapuri lekh (2732) in the north, Nagarkot (2166) in the east, Chandragiri (2550) in the southwest, and Phulchowki (2765) in the south. The centripetal type of drainage system of Bagmati River is the best example in the world (Holmes 1965), which drains water from this valley with its major tributaries like Bisnumati, Dhobi Khola, Hanumante, Manohara, Nakhu, Godavari, Balkhu, and Bosan Khola originating from the surrounding hills. All the water of Kathmandu Valley is drained out by the Bagmati River from only one exit along the southern edge of the valley at Katuwal daha (Fig. 11.1).

11.1.2 Climate and Population

The temperature of the Kathmandu Valley varies from place to place due to difference in altitude. The average mean maximum and mean minimum temperatures remain high from June to August and low from December to February. The temperature trend from 1971 to 2011 in the valley shows a continuous mean warming rate of 0.033 °C per year with the average maximum temperature increasing at 0.043 °C per year and minimum increasing at 0.02 °C per year (UNHABITAT 2015). Similarly, the precipitation in Kathmandu Valley generally occurs from June to September. Between June and September, 80% of the rainfall occurs where the valley receives 1500 m annual average rainfall. The Bagmati study team mentioned that the Kathmandu district seems to have three zones of precipitation zones, e.g., valley floor, lower mountain slopes, above Shivapuri hill (ICIMOD 1993). The total population of the Kathmandu Valley is 26,494504 (CBS 2012). The growth rate of the Kathmandu Valley is different in different decades having 1.65, 2.07, 2.66, 2.08, 2.25, and 1.35 in

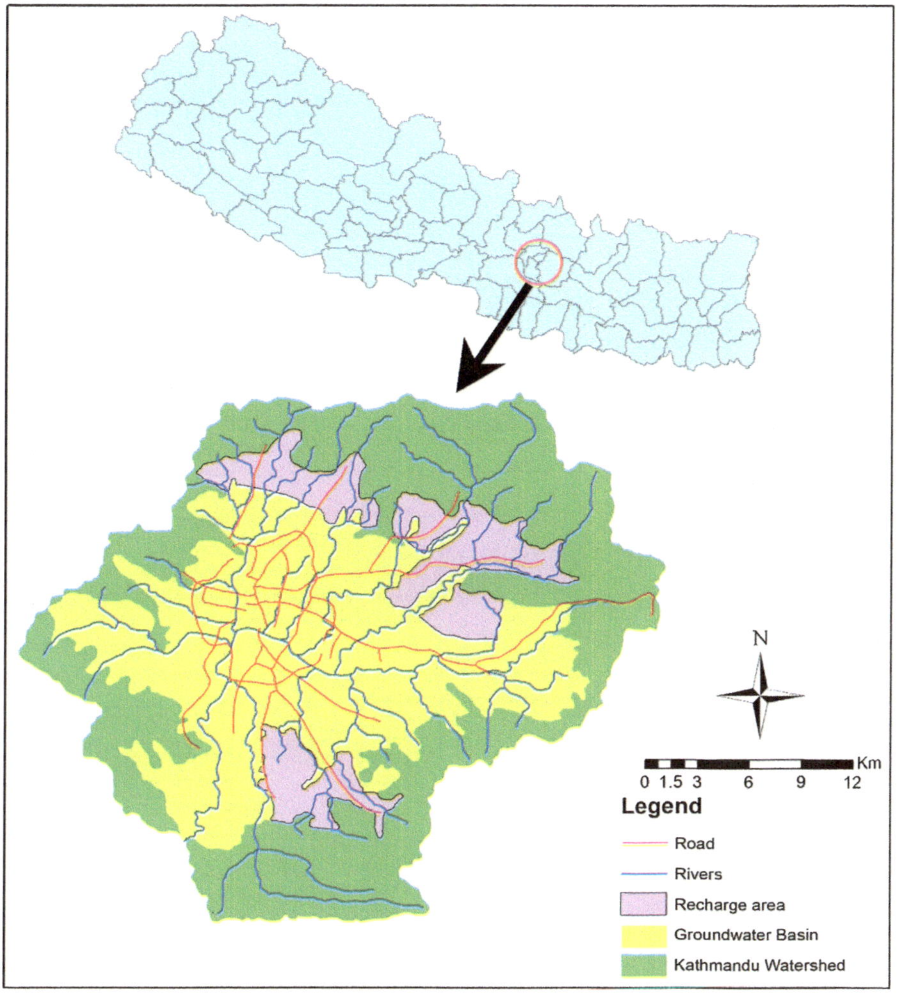

Fig. 11.1 Location map of the Kathmandu Valley. *Source* © Adhikari (2017)

the years of 1961, 1971, 1981, 1991, 2001, 2011, respectively (Fig. 11.2). Dramatic increase in population in the valley has been creating a big problem for sustainable groundwater use.

11.1.3 Land use

The land use change has occurred dramatically in past few decades due to geomorphological conditions, centralized administrative system, income generation oppor-

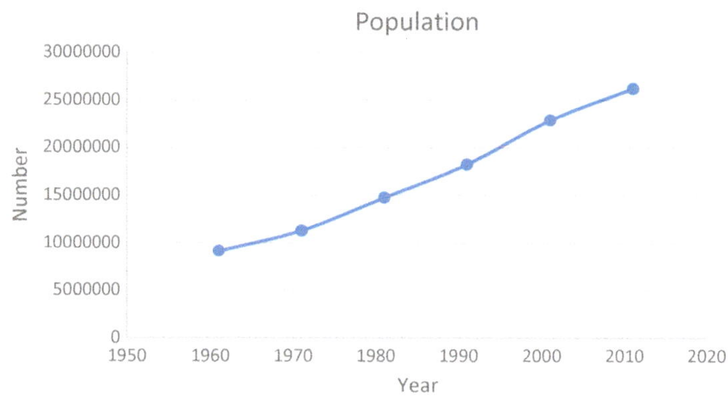

Fig. 11.2 Population growth of the Kathmandu Valley. *Source* CBS (2003, 2012)

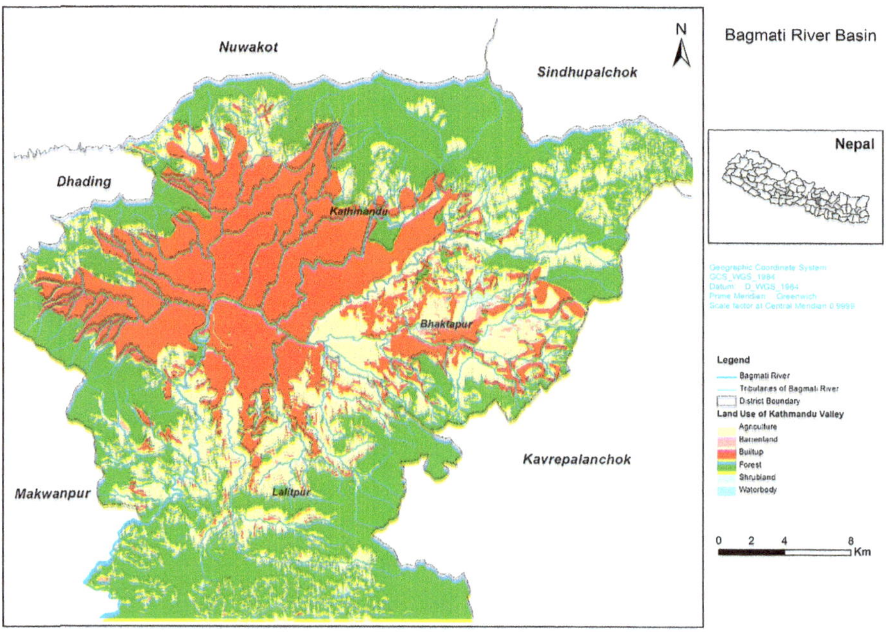

Fig. 11.3 Present land use map of the Kathmandu Valley. *Source* © Adhikari (2017)

tunities, real state industry, population growth, and existing plans and policies (Thapa 2009). Those activites are responsible for landuse pattern change, e.g., rural agricultural to urban settlement, which is truly reflected by the increasing population in different places (Fig. 11.4). The urban settlement has dramatically increased from 2003 to 2017. The central area of the valley is dominated by settlement and the agricultural land in the periphery (Fig. 11.3).

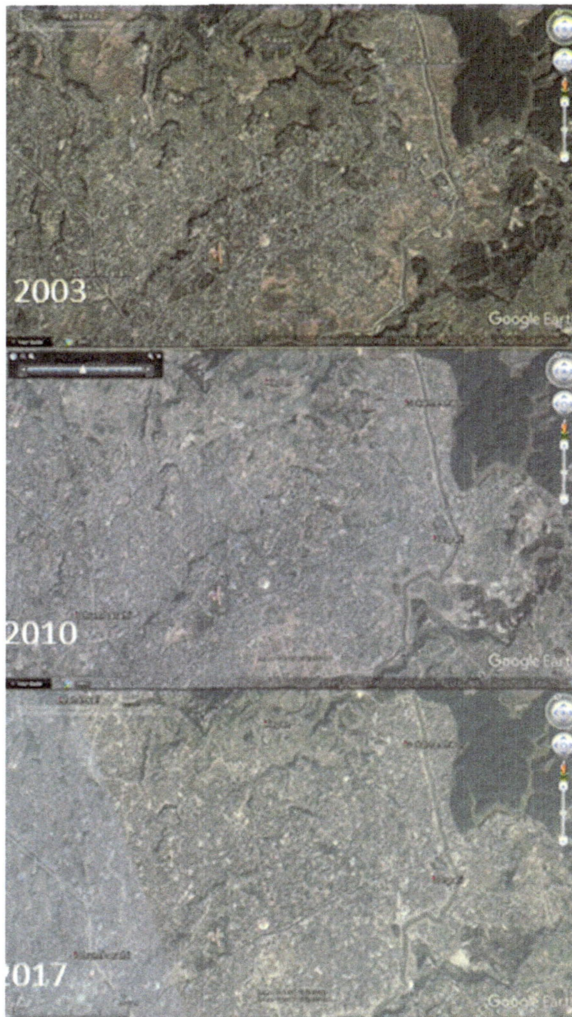

Fig. 11.4 Land use change in the northeast portion of the Kathmandu Valley. *Source* © Google Earth Image (2017)

11.2 Geology and Hydrogeology

11.2.1 Basement Geology and Quaternary Geology

The Kathmandu Valley is located in the center of a large syncline in the lesser Himalaya (Gansser 1964; Schelling 1992) surrounded by the Phulchowki and Chandragiri mountain ranges. The basement geology of the Kathmandu Valley consists of gneiss and schist of Precambrian age and limestone, meta-sandstone, slates and quartzite of Paleozoic age (Stöcklin and Bhattarai 1977). Kizaki (1994) has explained that Paleo Kathmandu Lake has gradually shifted from south to north due to the activity of Main Central Thrust (MCT) in the central Nepal. The basin-fill stratigraphy consists of lacustrine to fluvial sediments of Neogene–to-Quaternary sediment (Yoshida and Igarashi 1984). Sakai (2001) has divided into following six units; (1) Tarebhir Conglomerate; (2) Lukundol Formation; (3) Itaiti Formation; (4) Bagmati Formation; (5) Kalimati Formation; and (6) Patan Formation. Similarly, the northern and northeastern part of the Kathmandu Valley is divided into Gokarna and Thimi formations. According to Yoshida and Igarashi (1984) and Sakai (2001), Thimi formation is younger than Gokarna formation. These deposits consist of gravels, lignite beds, sand, silt, and clay, which are spread over the Kathmandu Valley. The grain size is decreasing from south to north. The clay layer thickness is high (~200 m) in the central part, which decreases toward the fringes.

11.2.2 Hydrogeology

Hydrological condition of the Kathmandu Valley is mostly controlled by the sediment distribution. The Well Log Data show that the geological units in the area are not uniform and are highly heterogeneous in lateral and vertical extent with interlayers of permeable and low-permeable geological units (Ganesh 2011). The groundwater aquifer is considered as a closed and isolated groundwater basin consisting of alluvial fans intercalated with clays and silts in the northern part; sand and gravel units overlain by thick lacustrine clay, and sand and gravel units are interbedded with silt and clay layers in the southern part (Shrestha MN 2012).

According to Binnie and Partners (1973), the aquifer system in the valley can be classified into seven types:

1. Interbedded: later extensive aquifers, more numerous toward the north
2. Liner: old river channel deposits
3. Bedrock: limestone (karst) of the southeast and southwest rims of the valley
4. Basal gravel: deep gravel overlying the bedrock in the southern part of the valley
5. River deposits: recent alluvial material
6. Gravel fans: from the hilly rim toward the valley
7. Gravel near surface: usually small thickness, locally widespread occurrences.

JICA (1990) has separated the subsurface of the valley into three groundwater districts: northern groundwater district, central groundwater district, and southern groundwater district. Northern groundwater district (NGD) lies in the northern part of valley from the foothills to Pashupati in south consisting of highly porous materials of micaceous sand and gravel. There are several impermeable fine layers in between. This zone is main recharge area for the Kathmandu Valley with water of low electrical conductivity (100–200 μs/cm) (Shrestha SD 2012).

Central groundwater district (CGD) is composed of thick impermeable black 'Kalimati' clay accompanied by some lignite and peat. Shallow aquifer is present up to 20 m which overlies the thick clay. The transmissivity (*T*) of the aquifer ranges from 32 to 960 m^2/day (JICA 1990; DMG/BGR/DOI 1998). Similarly, southern groundwater district (SGD) does not have well-exposed aquifer except along the Bagmati River. This district is characterized by a basal gravel of low transmissivity overlain by an impermeable clay. Pandey and Kazama (2011) have classified the aquifer system into three zones: the deep aquifer (0–85 m), clay aquitard (5–200 m), and deep aquifer (25–285 m) with volume of 7265 and 56813 million cubic meters (MCM) of shallow and deep aquifers, respectively.

11.3 Water Demand and Supply

Large population of the valley depends on groundwater as a main water source. Basnet et al. (2016) developed a robust groundwater model that can be used to understand surface water and groundwater interactions. The GSFLOW model has tried to provide some insights into groundwater extraction volume, especially that of deep aquifer system in the valley. A pumping sensitive analysis of groundwater system in Kathmandu Valley shows that the area near Dharmasthali, Dhapasi, Maharajganj, Sankhu, and Gokarna has a decline of 0.02–0.12 m, whereas the area near Balaju, Samakhusi, and Shywambu showed more decline of up to 0.12 to 0.23 m in head with per unit rise in pumping (m^3/s). The proposed extraction rate map prepared through this analysis (Fig. 11.5) also indicates that the northern part of the groundwater basin has more volume of water available per unit decline in head per year and the value of the extraction rate is decreasing as we move from northern part of groundwater basin to the southern part. Village development committee (VDC)-wise extraction rate map (Fig. 11.6) shows that Sangla, Baluwa, and Danchi VDCs have higher value of proposed extraction rate and Danchi VDC has highest extraction rate (6,273,967 m^3/yr) among them.

Similarly, Adhikari (2017) found using MODFLOW model that highest probable VDC-wise pumping rate of about 700 cubic meter per day is possible from 500 × 500 grid size area in the places Dhapasi, Mulpani, Manmaiju, Matatirtha, Daanchhi, Tikathali, Sirutar, Lubhu, etc. In these places, present-day pumping rate is about 200 cubic meter per day from a block of 500 × 500 m size. The present pumping rate and VDC-wise probable maximum pumping rate are as shown in Figs. 11.6 and 11.7, respectively.

KUKL is only the governmental organization in this valley which is supplying water by collecting drinking water using different water sources such as surface sources (35), deep tubewells (57), water treatment plants (20), service reservoirs (43) with about 1300 major operating valves (Shrestha SD 2012). Increasing population (1.65 million in 2001 to 2.53 million in 2011) and decreasing fertile land area in the valley from 62 to 42% between 1994 and 2000 (ICIMOD 2007) alter the dynamics of the hydrological environment (Thapa et al. 2017). The water demand has significantly increased from 35.1 in 1998 (Gyawali 1988) to 370 million liters per day (MLD) in 2015. The total groundwater extractions are increasing continuously, and the gap between recharge and the extraction is widening (Fig. 11.8).

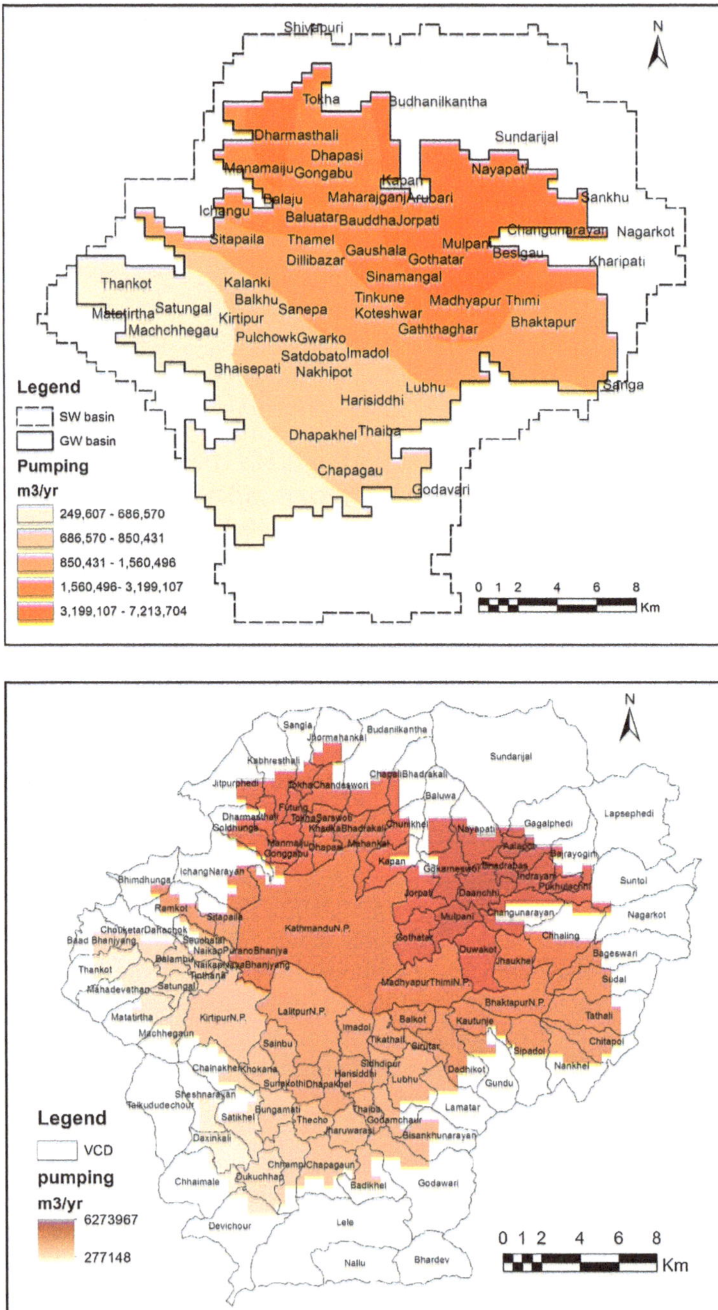

Fig. 11.5 Proposed groundwater extraction rate with VDC overlay of the valley. *Source* © Shakya (2017)

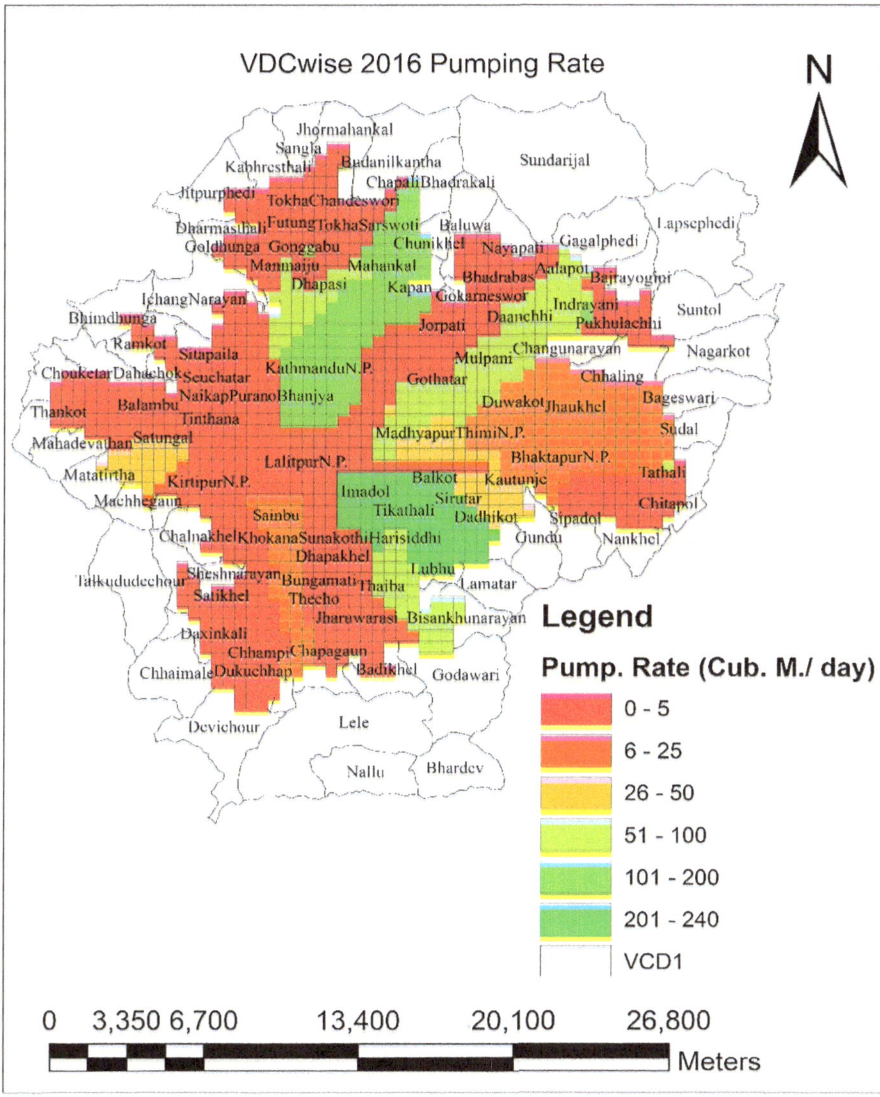

Fig. 11.6 Present (2016 AD) VDC-wise pumping rate from shallow aquifer. *Source* © Shakya (2017)

However, KUKL is only providing approximately 25–33% of the total demand of 370 MLD (KUKL 2015). Many private water supply companies have been providing the deficit water using deep aquifer pumping, Dhunge Dhara (traditional waterspouts), supply from private water vendors (~450 tankers in 2019; Shrestha and Shukla 2010), and bottled water companies (Thapa et al. 2017). Udmale et al. (2016)

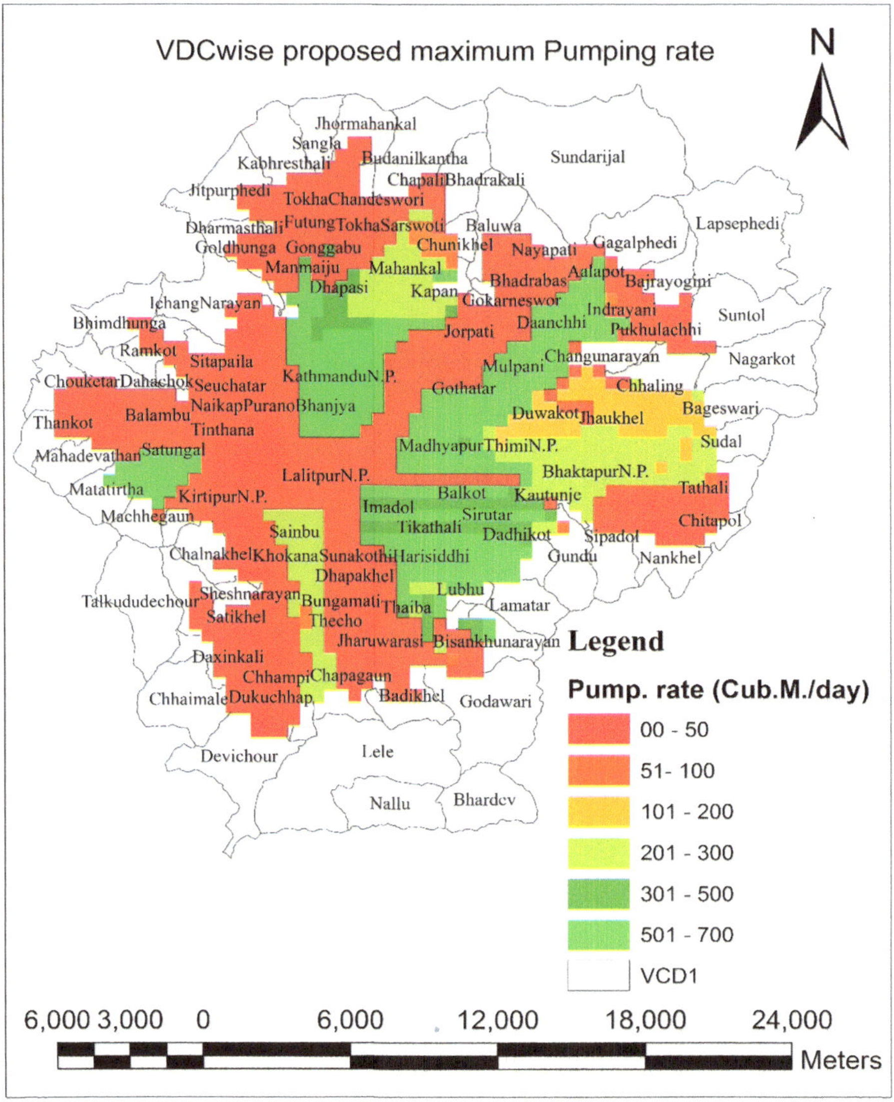

Fig. 11.7 VDC-wise probable maximum pumping rate from shallow aquifer. *Source* © Shakya (2017)

have calculated that the present supply deficit in 102 MLD which will increase 322 MLD in 2021.

Nepal Government started a MWSP in 1998 to maintain the water deficit demand diverting 170 MLD of freshwater to the valley through the tunnel in the first stage. However, the Sundarijal Water Treatment Plant has capacity of only 85 MLD (Udmale et al. 2016) which is still insufficient for the demand because the water

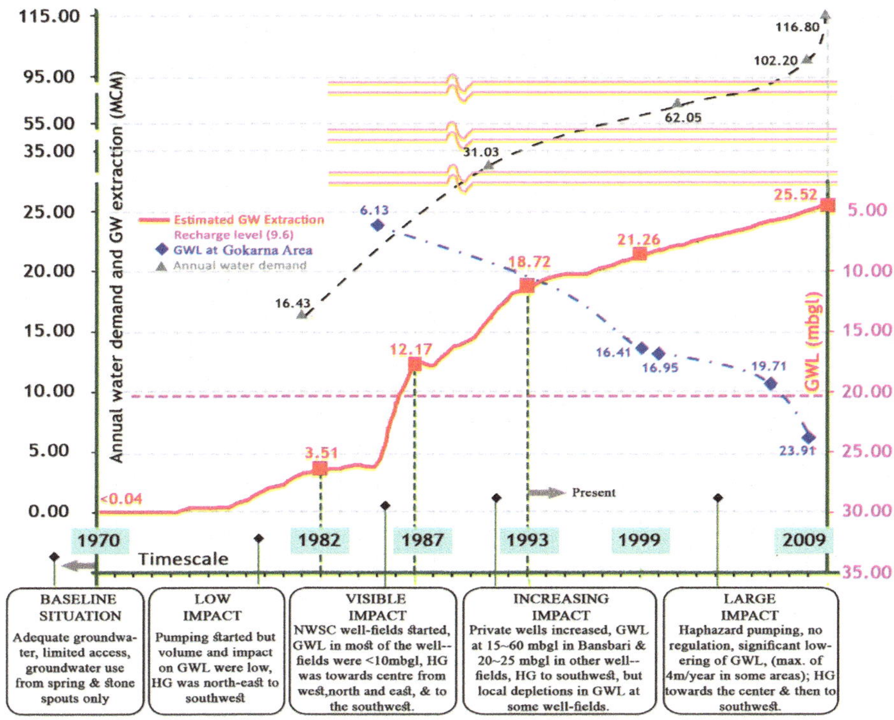

Note: 'mbg;' is meter below ground level; 'MCM' is Million Cubic meters; 'GWL' is Groundwater Level; 'HG' is hydraulic gradient; Annual groundwater extraction is calculated assuming 360days of well operation; GWL depletion is at Gokarna Area

Fig. 11.8 Impact of groundwater dynamics on groundwater level (modified after Pandey et al. 2012). *Source* © Pandey (2012)

deficit would be 102 MLD in 2016. Government has planned to supply 510 MLD addition water by 2023 in the second and third stages. The water demand will be sufficient in the valley after the treatment of all water from MWSP (445 MLD in 2021).

11.4 Water Management Strategies

Water management in the cities is always challenging and difficult in every part of the globe. Harvesting water from the sources and proper supply are hindered by many factors. Poor economic productivity, increased in water-related disease, contaminated water, long-distance travel to fetch water, uncontrolled rapid urbanization, political will/instability, weak institutional structure, and huge gap between demand and supply are the major problems in the valley (Chapagain 2013). Moreover, lack of effective human resource management, lack of regulation and monitoring body, improper maintenances, and lack of coordination between concern stakeholders are always playing a major role for the insufficient water supply. Khatri and Vairavamoorthy (2007) explained that population growth, emerging technology, increase in the costs, risk on critical infrastructure system, globalization and economic development, change in public behavior, climate change, governance, and privatization along with deteriorating infrastructure are major challenges to improve the water supply system in the Kathmandu Valley.

Similarly, CIUD (2003) has studied about the water consumption pattern in primary, secondary, renter's households and squatters and found that the water intake very much depends upon the type of house plumbing, types of toilet, kitchen, and washing facilities, i.e., complete plumbing (99 lcpd), yard type (41 lcpd).

The increase in population with changing lifestyle with modern buildings is putting stress on water resources in the valley, and the recharge area is decreasing due to unmanaged and haphazard built-up area. Therefore, direct storage of the water in shallow aquifer's storage space through artificial recharge needs to be initiated for the water balance. For example, the northern groundwater district and Patan area are considered for the good recharge zone which should be utilized for the artificial or natural recharge. Collection of spring water from the surrounding hills is very important with mandatory rainwater harvesting system. Government should actively organize for monitoring and maintenance of the shallow as well as deep aquifer, production, water quality, and subsidence and maintenance of water supply (Manandhar 2013). Awareness programs about water supply and sewer system and the ongoing water supply problem should organize in the local ward level to give them about the idea. Proper planning, good governance, identification of water sources, and water treatment of wastewater are the long-term solutions for sufficient water supply.

Besides governmental initiatives, many private organizations are working hard to solve this water crisis in core areas of city. A successful example can be taken from Lalitpur Metropolitan City where communities in Chysal, Pin Bahal, and other areas have taken measures like covering the wells, locking the well, and only distribute the water at certain allocated time. They collect water in the nighttime in the tank to distribute in daytime which is combination of rooftop rainwater collection also. Such kinds of measures are not restricted to core areas only but also in the outskirts too. Communities are managing in water supply in Godavari area for more than two decades. They are collecting water from the different sources and distribute water to the household and charge very nominal to maintain the system. Unless the problem is tackled insightfully from different aspects and not only from demand-supply aspect and soon, the increasing social conflicts could be challenging for the planners in the future.

11.5 Conclusions

Urban water crisis is very much pronounced in the Kathmandu Valley. The shallow and deep aquifer is controlled by valley sediments composed of gravel, sand, silt, and clay. Government has started MWSP to fulfill the increasing water demand of the Kathmandu Valley. The first and second phases of the MWSP will supply a total of 680 MLD. The water demand will be sufficient in the valley after the treatment of all water from MWSP (445 MLD in 2021). However, population growth, emerging technology, increase in the costs, and risk on critical infrastructure system, globalization and economic development, change in public behavior, climate change, governance, and privatization along with deteriorating infrastructure are major challenges to improve the water supply system in the Kathmandu Valley. Therefore, proper planning, good governance, identification of new water sources, and water treatment are the long-term solutions for sufficient water supply.

References

Adhikari B (2017) Study of Shallow groundwater hydrology of Kathmandu Valley, M.Sc. thesis, Water Resources Engineering Program, Institute of Engineering, Lalitpur, Nepal (unpublished)

Basnet S, Shakya NM, Ishidaira H, Thapa BR (2016) Simulation of Kathmandu Valley river basin hydrologic process using coupled ground and surface water model. In: International workshop on water security and climate change: challenges and opportunities in Asia, 29 November–01 December, 2016 Asian Institute of Technology, Bangkok, Thailand

Binnie and Partners (1973) Groundwater investigations, Kathmandu water supply and sewerage scheme. Appendix 4.5, World Health Organization-Project Nepal 0025 (unpublished)

CBS (2003, 2011, 2012) Central Bureau of Statistics, Government of Nepal

Chapagain DP (2013) Urban water supply sector reform in Kathmandu Valley. J Inst Eng 9(1):130–141

Centre for Integrated Urban Development (CIUD) (2003) Household water use survey and research in Urban Kathmandu Valley (Final Report), 43 p

DMG/BGR/DOI (1998) Hydrological conditions and potential barrier sediments in the Kathmandu Valley. Technical Cooperation Project-Environmental Geology, Final Report (Appendix 3), pp 17–59

Ganesh KC (2011) Numerical modelling of groundwater in Kathmandu Valley, Nepal, M.sc. thesis, Tribhuvan University

Gansser A (1964) Geology of the Himalayas. Wiley, London

Gyawali D (1988) Preliminary economic analysis of a water loss control program for Kathmandu/Patan system. East Consults, Kathmandu, Nepal

Holmes A (1965) Principles of physical geology. Thomas Nelson and Sons Ltd., New York, p 598

ICIMOD (1993) Economic and environmental development planning for Bagmati Zone: Kathmandu District: assessment current condition: V. 2, Part VIA, Bagmati Zone Study Team, International Centre for Integrated Mountain Development, Kathmandu

ICIMOD, MoEST and UNEP (2007) Kathmandu Valley environment outlook, Kathmandu, Nepal

JICA (1990) Groundwater management project in Kathmandu Valley, Final Report Supporting Report, p B1

Khatri KB, Vairavamoorthy K (2007) Challenges for urban water supply and sanitation in the developing countries, Discussion draft paper, UNESCO-IHE, The Netherlands

Kizaki K (1994) An outline of the Himalayan upheaval. Japan International Cooperation Agency (JICA), Kathmandu, p 127
KUKL (2015) Annual report, Kathmandu Khanepani Upatyaka Limited, Kathmandu, Nepal
Manandhar I (2013) Success to abundant water supply in Kathmandu Valley, M.Sc. thesis, University of Minnesota
Pandey VP, Kazama F (2011) Hydrologic characteristics of groundwater aquifer in Kathmandu Valley, Nepal. Environ Earth Sci 62(8):1723–1732
Pandey VP, Shrestha S, Kazama F (2012) Groundwater in the Kathmandu Valley: development dynamics, consequences and prospects for sustainable management. Eur Water 37:3–14
Sakai H (2001) Stratigraphic division and sedimentary facies of the Kathmandu Basin sediments. J Nepal Geol Soc 25:19–32
Schelling D (1992) The tectonostratigraphy and structure of the eastern Nepal Himalaya. Tectonics 11:925–943
Shrestha D, Shukla A (2010) Private water tankers operation in Kathmandu: analysis of water services and regulatory provisions, report submitted to Center for Research on energy, Environment and Water (CREEW), Kathmandu, Nepal
Shrestha MN (2012) Groundwater use in the Kathmandu Valley: an analysis of pre-and post-Melamchi scenerios. In: Shrestha S, Pradhananga D, Pandey VP (eds) Kathmandu Valley groundwater outlook. Asian Institute of Technology (AIT), The Small Earth Nepal (SEN), Center of Research for Environment Energy and Water (CREEW), International Research Center for River Basin Environment-University of Yamanashi (ICRE-UY), pp 90–97
Shrestha SD (2012) Geology and hydrogeology of groundwater aquifers in Kathmandu Valley. In: Shrestha S, Pradhananga D, Pandey VP (eds) Kathmandu Valley groundwater outlook. Asian Institute of Technology (AIT), The Small Earth Nepal (SEN), Center of Research for Environment Energy and Water (CREEW), International Research Center for River Basin Environment-University of Yamanashi (ICRE-UY), pp 21–30
Stöcklin J, Bhattarai KD (1977) Geology of the Kathmandu area and Central Mahabharat range, Nepal Himalaya. Report of Department of Mines and Geology/UNDP, 86 p (unpublished report)
Thapa RB (2009) Spatial process of urbanization in Kathmandu Valley, Nepal. Ph.D. dissertation, University of Tsukuba
Thapa BR, Ishidaira H, Pandey VP, Shakya NM (2017) A multi-mode approach for analyzing water balance dynamics in Kathmandu Valley, Nepal. J Hydrol Reg Stud 9:149–162
Udmale P, Ishidaira H, Thapa BR, Shakya NM (2016) The status of domestic water demand: supply deficit in the Kathmandu Valley, Nepal. Water 8(5)
UNHABITAT (2015) Cities and climate change initiative: abridged report, Kathmandu Valley, climate change vulnerability assessment, 30 p
Yoshida M, Igrashi Y (1984) Neogene to quaternary lacustrine sediments in the Kathmandu Valley, Nepal. J Nepal Geol Soc Sp. 4:73–100

Chapter 12
Drought and Urbanization: The Case of the Philippines

Emma Porio, Jessica Dator-Bercilla, Gemma Narisma, Faye Cruz and Antonia Yulo-Loyzaga

Abstract The Philippines is highly vulnerable to drought, resulting in severe impacts on crop productivity, water availability, and food security. This chapter explores water security in the country by examining the effects of drought on risk governance and its social impacts on the 1997–1998 and 2015–2016 El Niño episodes in Metro Manila, Iloilo City, and Cebu City. During these periods, widespread dryness occurred in both urban and rural areas, as rainfall was reduced by more than 50%. This decrease in rainfall affects most especially the urban poor as they experience acutely the dwindling supply of potable water, increasing costs of water, and compromised access to hygiene and sanitation services. Consequently, droughts have become a major concern for risk governance in major urban centers. Science-informed and contextually driven local climate adaptation plans (LCAP) seem to be the most appropriate response to mitigate and adapt to the effects of drought brought about by El Niño.

Keywords Drought · El Niño · Water security · Water governance
Risk governance

12.1 Introduction

In 1997–1998, the Philippines was hit with the effects of El Niño, the strongest of that century (Hilario et al. 2009). During that period, the severe drought affected over 70% of the country and resulted in damages such as water shortages and losses in rice and corn worth PhP3 billion (Jose 2002 in Table 2 of Hilario et al. 2009). The strong El

E. Porio (✉) · G. Narisma
Ateneo de Manila University, 1108 Quezon City, Philippines
e-mail: eporio@ateneo.edu

E. Porio · J. Dator-Bercilla · G. Narisma · F. Cruz · A. Yulo-Loyzaga
Manila Observatory, 1108 Quezon City, Philippines

J. Dator-Bercilla
Christian Aid, London, India

B. Ray and R. Shaw (eds.), *Urban Drought*, Disaster Risk Reduction,
https://doi.org/10.1007/978-981-10-8947-3_12

Niño of 2015–2016, which is often compared to the 1997–1998 event, also had severe impacts on the Philippines. On December 2015, the United Nations Economic and Social Commission for Asia and the Pacific (UNESCAP) and the Regional Integrated Multi-Hazard Early Warning System for Africa and Asia (RIMES) claimed through their advisory that the Philippines would be severely affected by El Niño (UNESCAP and RIMES 2015). History has shown that El Niño can reduce monthly rainfall in the Philippines by 50% during the peak periods of El Niño, which affected crop production and water availability for consumption.

In 2015, the Philippine Government responded to this dilemma by setting up an El Niño Task Force headed by the National Economic and Development Authority (NEDA). A Technical Working Group was convened by NEDA to formulate the Roadmap to Address the Impacts of El Niño (RAIN). The RAIN covered four dimensions, namely food security, energy security, health, and safety, and focused on 67 provinces that were projected to be most likely affected by El Niño, including Metro Manila (NEDA 2017). While El Niño is often associated with decrease in precipitation, water security was not part of the RAIN program of action.

This chapter explores water security in the Philippines by examining the influence of drought on risk governance and its social impacts on the 1997–1998 and 2015–2016 El Niño periods in the urban areas of the Philippines, specifically Metro Manila, and the highly urbanizing cities of Metro Iloilo and Metro Cebu in the Visayas region.

12.2 Influence of El Niño on Rainfall in the Philippines

In the Philippines, significant changes in rainfall, a main driver of climate variability, can have serious impacts on the country. One physical phenomenon that has a strong impact on rainfall in the Philippines is the El Niño Southern Oscillation (ENSO). During an El Niño event when ENSO is in its warm phase, the higher than average sea surface temperatures (SST) over central and eastern equatorial Pacific Ocean affect the spatial distribution of rainfall such that drier conditions are experienced over different parts of the Philippines, as well as, other areas in the western Pacific, particularly during its peak around December to February (Hilario et al. 2009; Salinger et al. 2014).

To illustrate, the rainfall anomalies during the strongest El Niños of 1997–1998 and 2015–2016 were obtained using the Tropical Rainfall Measuring Mission (TRMM) Multi-satellite Precipitation Analysis (3B42 version 7) daily precipitation data set produced at the Goddard Earth Sciences Data and Information Services Center (GES DISC) of the National Aeronautics and Space Administration (NASA) (Goddard Earth Sciences Data and Information Services Center 2016). Figure 12.1 shows widespread dryness occurring in both urban and rural areas during the season from January to May, with rainfall reduced by more than 50% in many parts of the country during the 1997–1998 El Niño including in three metropolitan cities of Manila, Iloilo, and Cebu. The decrease in rainfall in the 2015–2016 El Niño

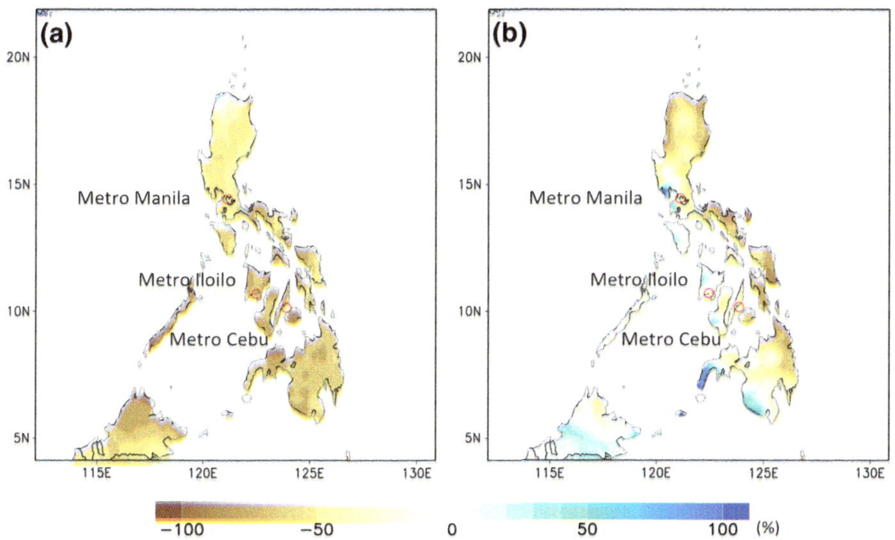

Fig. 12.1 Seasonal rainfall anomalies (%) (from January to May) in **a** 1998 and **b** 2016, relative to the average from 1998 to 2016 (baseline)

was less severe but nevertheless widespread. These dry spells of rainfall reduced by 21–60% for three consecutive months had strong impacts when coupled with exposure and vulnerabilities. The months of January to May are particularly important because agricultural planting often begins in the month of January. Furthermore, the lack of water due to drought not only threatens food production but can also lead to the competing use of scarce water resources in these periods.

Focusing on the key urban areas in this study, the monthly rainfall anomalies were obtained for Metro Manila, Metro Iloilo, and Metro Cebu (Fig. 12.2). Rainfall anomalies at Angat reservoir were also examined since this is the source of water supply for Metro Manila. Figure 12.2 also indicates the Oceanic Niño Index (ONI) centered in each month from the Climate Prediction Center (CPC) of the National Oceanic and Atmospheric Administration (NOAA)/National Weather Service (NWS). ONI measures the magnitude of an El Niño/La Niña event and is derived from a three-month running average of SST anomalies relative to a centered 30-year reference period over the Niño 3.4 region. Using this index, El Niño (La Niña) periods can be identified when the ONI is above +0.5 °C (below −0.5 °C) for at least five consecutive seasons (Climate Prediction Center 2016).

Figure 12.2 shows that there are comparable values of ONI between the 1997–1998 and 2015–2016 El Niños, implying that the two events have similar magnitudes. There are differing rainfall responses, however, in all of the study areas. In the case of Metro Manila, drier conditions were experienced in the metropolis in 1998 compared with Angat but not in 2016. In the case of Metro Iloilo, there was above normal rainfall in January 2016 before its decline in the succeeding months.

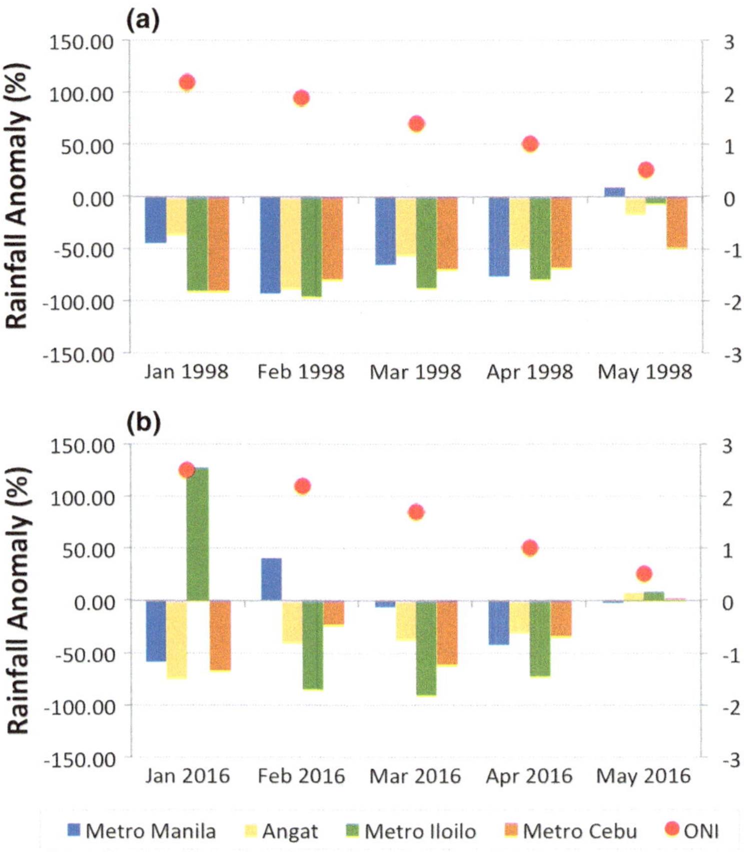

Fig. 12.2 Monthly rainfall anomalies (%) from January to May of **a** 1998 and **b** 2016, relative to the average from 1998 to 2016 (baseline). Red dots indicate the Oceanic Niño Index (ONI) centered in each month. Positive (negative) ONI values exceeding the +(–) 0.5 °C threshold for at least five consecutive seasons indicate El Niño (La Niña) events

Metro Iloilo was consistently dry from January to May 1998, with rainfall reduced by at least 79% from January to April. This was also true for Metro Cebu that had a rainfall decrease of at least 48% from January to May 1998.

The drought impacts brought about by El Niño can be compounded by urbanization and development associated with land use change and urban heat island effects. Studies have indicated changes in local climate (such as temperature, rainfall, circulation) due to modifications in the surface characteristics (e.g., Pielke et al. 2007, 2011; Mahmood et al. 2014). In Metro Manila, a warming rate of 0.8 °C per year from 1989 to 2002 was estimated due to its urbanization (Pereira and Lopez 2004). Furthermore, slow onset droughts due to future climate change can exacerbate the dry conditions during El Niño.

12.3 Social Impacts of Drought in the Metropolis

Rapid urbanization and continued economic expansion have increased the demand for water in Philippine cities (Albert 2001). As shown in Tables 12.1, 12.2, and Fig. 12.13 (see Appendix) showing urban–rural growth differentials (1950–2050), urbanization in the Philippines has been running at a fast pace and achieving its urban transition in the early 1990s (Porio 2009; World Bank Group 2017). Furthermore, urbanization, population growth, and land use changes have also led to excessive groundwater extraction and eventually land subsidence in the coastal areas of the metropolis (Rodolfo 2014). But access to basic services like potable water had not kept up with the growth of urban centers. Access to potable water and environmental sanitation services was most acute among urban poor communities in Metro Manila, Metro Cebu, and Metro Iloilo (Porio and Lao 2010).

Moreover, the supply of water has also been compromised by climate events like heavy rainfall and flooding, sea level rise (SLR), and drought. During the warm months from March to May, and especially during the El Niño years, access and affordability of water become a major challenge for city mayors and disaster risk reduction and management offices (DRRMOs). More significantly, the dwindling supply of potable water to the metropolitan areas hits most the vulnerable population like the urban poor. During El Niño periods, this often leads to increasing costs of water, further compromising the urban poor communities' access to sanitation and other services. Droughts, then, and the consequent dwindling water supply become a major concern for local officials and on the risk governance of cities. Thus, urban service deficits become acute during the drought periods as experienced in 1997–98 and 2015–2016, deemed the worst El Niño years.

The following section details the rising costs of water for the past years based on the 2015 Family Income and Expenditure Surveys (FIES) of the Philippine Statistical Authority. Across all income deciles, the cost of water rose greatly[1] during the El Niño year of 2015. Comparing the water costs in 1997, 2000, and 2015, it seemed to have risen tremendously over the years. But aside from being an El Niño year, the price increases of water were also due to the increasing development costs of infrastructure rehabilitation, generating new service delivery lines, and generating new sources of water.

In 2015, Metro Manila consumers paid more for their water (the lowest income decile paying Php3,644/month) compared to those in Metro Cebu (Php643/month) and in Metro Iloilo (Php278/month). Meanwhile, the fifth income decile in Metro Manila paid Php4,904 while those in Metro Cebu (Php1500) and Metro Iloilo (Php750/month) paid much less. The same pattern holds for those in the top income decile (tenth): Metro Manila's higher income families paid higher (Php8,389), compared to those in Metro Cebu (Php4,163/month) and Metro Iloilo (Php3,033/month).

[1] Increases of water costs across income deciles were computed from the Family Income and Expenditure Survey (FIES) of the Philippine Statistical Authority by Justin Charles See, Coastal Cities at Risk (CCAR), Social Sector, Manila Observatory and Ateneo de Manila University. The authors are also grateful to Justin Charles See for his technical assistance in preparing this chapter.

Interestingly, the Manila Observatory's Coastal at Risk (CCAR) project survey of urban poor communities shows that their average cost of water in 2016, ranged only from P420 to 650/month. Compared to the FIES sample population, the CCAR survey sample came from highly marginalized households located in riverine and coastline communities of the three flood basins in Metro Manila (Patankar et al. 2013).

The families' main sources of water or delivery lines also indicate the level of development of their communities. In Metro Manila, for example, almost three-fourths (75%) of the households have individual private connections or water piped into their homes, but only less than one-half (45%) have piped water in Metro Cebu and much less (21%) in Metro Iloilo. Some of them shared their faucets with other families/households (15% in Metro Manila, 10% in Metro Cebu but only 4% in Metro Iloilo). In Metro Iloilo and Metro Cebu, about 18 and 12%, respectively, obtained their water from rivers or streams. In the former, slightly more than a quarter (26%) shared their water access through a tube or piped well, while in the latter, only about 10% did. Interestingly, some families access their water supply by buying from peddlers or vendors (Metro Cebu, 5%, Metro Manila, 4%, and Metro Iloilo, 3.5%). The latter group pays the highest price for their water. Ironically, they also come from the lower-income decile groups, who cannot afford individual pipe connections to their abode.

Having an individual water connection to one's home costs much more in (about Php10,000) in Metro Cebu, compared to the cost in Metro Manila (Php3,000). In Cebu, the cost of water from the Metro Cebu Water District (MCWD) is Php20/m^3, while MCWD customers re-selling water per container is Php2–3 per 5 gallons or Php88/m^3 in urban poor communities with no or few water connections. Meanwhile, the cost for those buying water from water vending machines is Php1/250 ml or Php4,000/m^3. Ironically, the poor who can least afford it obtains their drinking water from these expensive sources.

12.3.1 Security of Tenure and Drought Impacts

Aside from the hefty costs of water connection, the urban poor household must show proof of ownership of land/house or contract/lease from the structure/landowners. For most of the informal settlers, this can prove to be a challenge because of those in the informal sector, written contracts are not really the norm. Verbal agreements tend to be the norm between landowner and/or homeowner. Often the meter is in the owner's name, and the renter is often charged higher than what is recorded in the meter.

As illustrated in this case (Fig. 12.3), the effects of El Niño on the urban poor households are very much related to family structure, gender, health, and occupation of the household head (Albert 2001). The case above also highlights that the poor needs to be supported by the city's waterworks systems with its networked infrastructure of piped water, drains, roads, electricity; its services that include public transport, health care, emergency services; and its protected and managed ecosystems

Illustrative Case. Lolit, married, living in an informal settlement with her unemployed husband, with 5 children, 2 grandchildren in Quezon City, Metro Manila.

Lolit, age 48 years old, lives with her unemployed husband, 3 children and 2 grandchildren in Commonwealth Market, Quezon City, in a house lot of 24 square meters, which she acquired with the help of their Iglesia ni Kristo church leader 10 years ago. Two of their daughters got married without finishing their schooling and lived in nearly informal settlements as well. Twenty years ago, they moved to Manila from Surigao del Norte when they only had 2 children and her husband earned a living through construction work. Over the years, the latter however had several health complaints (poor eyesight, respiratory ailments, etc.) that he could only obtain intermittent work. Thus, the burden of supporting the family falls on Lolit as a home service masseuse. But at the height of summer season, she suffered a minor stroke so she had to stop working. After eight weeks, she had no choice, but to resume servicing her clients. She felt that the extreme heat and tiredness from her 12-15 hours daily work brought about her stroke. She also observed that since the El Nino last year, the flow of water in their home had lessened but their water costs have not. Before the dry spell, she was paying about Php3200/month. With the drought and dwindling supply of water, she was paying P3500/month. Aside from her health, Lolit also feel strongly about her family's hygiene and sanitation needs heavily compromised (less bathing, less cleaning, washing after ablution, etc.) by the dry spell during El Niño and the irregular water flow in their faucet.

Fig. 12.3 Illustrative case

that deliver a sustained supply of ecosystem services (Satterthwaite and Doodman 2013: 291). The case above also underscores the need for a comprehensive proactive response system to drought and episodes of El Niño (Porio 2017a).

The challenges toward water security among the poor in Metro Manila, therefore, have to be located within a larger set of social, political, and economic challenges that besiege urban poor families in general. Water insecurity also becomes acute during extreme events like typhoons during the southwest monsoons and flooding as illustrated during Typhoon Ketsana. Data from the surveys of Manila Observatory shows that the poor's water costs almost doubled during and after Ketsana (Patankar et al. 2013).

As seen in the distribution of water costs by income decile, the higher the income decile, the higher they consume as well as pay. But as seen from the costs of water above, those without individual faucets or whose water is delivered by trucks or who have to buy it from water vendors, end up paying much more than those with individual faucets. This cost structure does not contribute to the building of resilience and reducing the vulnerability of low-income households, especially during drought periods. Nor does it really build the capacity of the poor in coping with the scarcity of water (e.g., reduced baths for both adults and children, compromising family hygiene and environmental sanitation). More significantly, the poor depending on the water delivered by the city fire trucks makes them more beholden to their local politicians and landlords (Porio 2017c).

12.4 Water Governance in the Philippines

To understand how the Philippines responds to drought, one must understand the country's governance system as well as its water governance approach. The Philippines has a decentralized and devolved mode of governance with decision-making powers specifically resting on the local government units (LGUs). Despite this, water regulation and consumption are not one of the functions brought down to the local level. Thus, local governments can only respond to the impacts of a water crisis but do not wield the main decision-making powers over water resources in the country.

The Philippines has a Water Code (PD 1067) developed in 1976. The policy appointed a National Water Resources Council mandated to control and regulate the utilization, exploitation, development, conservation, and protection of water resources (Tabios 2010). In the Philippines, laws should have accompanying Implementing Rules and Regulations (IRR); the IRR of the country's Water Code were developed in June 1979. These include multiple decision makers: among them, but not limited to, the National Water Resources Board (NWRB), Department of Environment and Natural Resources (DENR), Department of Public Works and Highways (DPWH), Department of Health (DOH), Department of Interior and Local Government (DILG), Local Government Units (LGU), Local Waterworks Utilities Administration, Water Districts (LWUA), Metropolitan Water Works and Sewage (MWSS), Philippine Atmospheric, Geophysical and Astronomical Services Administration (PAGASA), National Irrigation Administration (NIA), and the National Power Corporation (NPC) (Tabios 2010).

Tabios (2010) concluded that in the Philippines, there are overlapping and fragmented roles and responsibilities of major water-related agencies, with the various institutions operating in hierarchies of authority. These agencies also come from various sectoral perspectives operating and acting on multiple thematic concerns (Malayang 2004; Tabios 2010). For elaboration, please see Fig. 12.4.

On top of these, Tabios (2010) notes that the regulation of surface water differs from that of groundwater in the Philippines. For instance, groundwater utilization, exploitation, and development are normally managed by institutions such as the DPWH, DOE, and LWUA, while surface water is regulated by institutions such as the DENR, MWSS, NIA, and the LLDA. Institutions like PAGASA, which monitor climate parameters related to water resources, are engaged in policy planning, data monitoring, scientific modeling, capacity development, and information and education. But they are not, in any way, engaged in issues of operations of water facilities, regulatory functions, and financing. Tabios (2010) proposed that an integrated water resource management must encourage water governance that does not distinguish between surface water and groundwater.

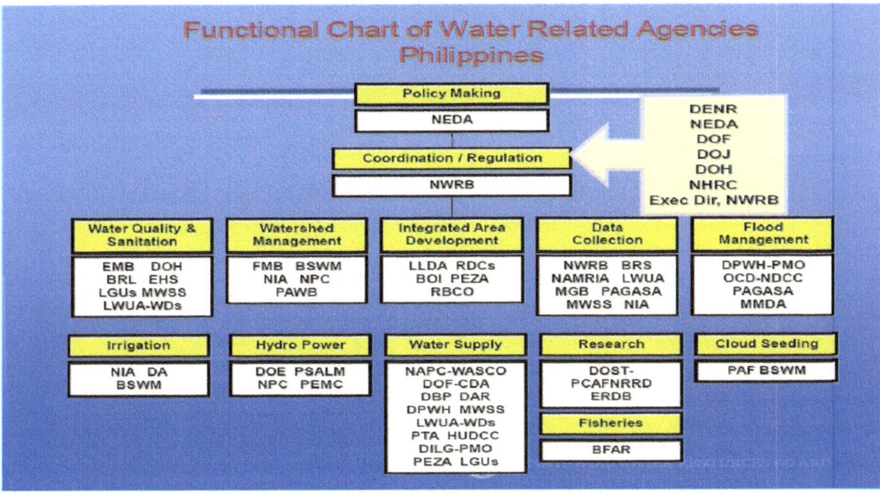

Fig. 12.4 Water-related agencies in the Philippines (Tabios 2010)

12.5 Drought Governance in Metro Manila

Metro Manila is an urban center based in the National Capital Region (NCR) made up of 16 cities and one municipality, which each having their respective local governments. However, when it comes to risk reduction in major thoroughfares, the functions shift to the regional level, leaving the responsibilities with the Metro Manila Development Authority (MMDA).

Two concessionaires, namely the Manila Water Company, Inc. (MWCI) and the Maynilad Water Services (MWSI), provide water and sewerage services in Metro Manila. Both these private concessionaires invested heavily for their capital expenditure programs (e.g., pipe rehabilitation and extension projects, refurbishment, and construction of water reservoirs and pumping stations, development of new water sources). According to the concessionaires, there are no new water sources to meet the increasing demand for water in the coming years. Thus, they have concentrated their investments on non-revenue water reduction initiatives to meet the increasing demand for water (https://www.adb.org/sites/default/files/institutional-document/33810/files/philippines-water-supply-sector-assessment.pdf).

Because 97–98% of the water used in the metropolis comes from Angat Dam, Metro Manila's long-term water security is questionable. Located about 40 km northeast of Manila, this multi-purpose dam completed in 1967 also provides water for irrigation and hydropower generation. Although the MWSS has been planning to build a new dam since the mid-1990s to augment the current water supply, this plan has been repeatedly postponed. Thus, MWSS plans to upgrade the older aqueducts to respond to the projected water demand increases in Metro

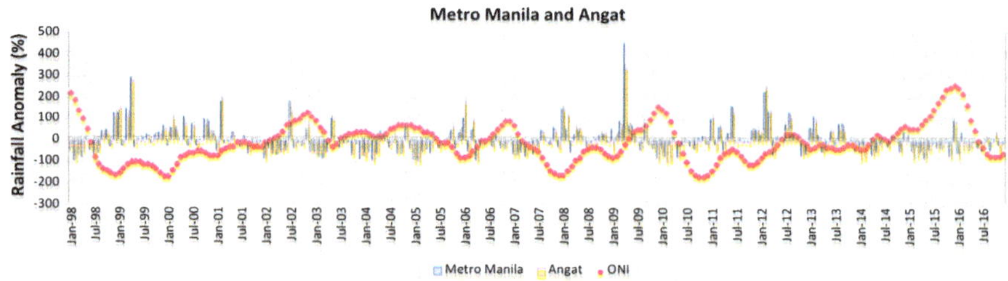

Fig. 12.5 Monthly rainfall anomalies (%) and ONI from 1998 to 2016 for Metro Manila and Angat show reductions in rainfall during El Niño events. Shaded areas indicate El Niño episodes in 1997–1998, 2002–2003, 2004–2005, 2006–2007, 2009–2010, 2014–2015, and 2015–2016

Manila (https://www.adb.org/sites/default/files/institutional-document/33810/files/philippines-water-supply-sector-assessment.pdf).

Metro Manila and Angat have experienced dry conditions in the past El Niño events (Fig. 12.5). During the El Niño years, MMDA did not respond with very concrete policy actions contrary to its response in 2010 when Angat Dam reached critical water levels. Manila, Caloocan, Malabon, Pasay City, Parañaque, Quezon City, and Navotas areas, served by water concessionaire Maynilad Water Services, Inc., were severely affected. According to Maynilad, 45 barangays were without water supply. Water rationing became the preferred immediate response by LGUs. Apart from this, Caloocan City encouraged water conservation while Quezon City also emphasized on the prevention of water leakages (Padua and Laude 2010).

There were very few documentation on LGU actions during the El Niño years in the Philippines. Makati City issued a resolution on March 1998, which was City Resolution 98-018, declaring that Makati City was under a State of Calamity as a result of El Niño. They needed to issue the resolution to allow them to mobilize the then Calamity Fund (now the Local Disaster Risk Reduction and Management Fund following the passage of Republic Act 10121 of the Philippine Disaster Risk Reduction and Management Law) to respond to the crisis.

Following the drought forecasts in 2015–2016 and the projected critical levels of Angat Dam that supplies most of the potable water needs of Metro Manila, the NWRB established a Technical Working Group on Angat Reservoir Operation and Management (TWGAROM). It allowed the engagement of and collaboration between the NWRB, MWSS and its concessionaires Manila Water and Maynilad, NIA, PAGASA, NAPOCOR, and the Angat Hydropower Corporation (AHC) (National Water Resources Board 2015). MWSS also adjusted the water allocation for Metro Manila. The lower water allocation resulted in the reduced water supply to some 355,000 households in the metropolis (Ranada 2015). The NWRB then issued a proposed water allocation scheme below (Fig. 12.6).

In response, MWSS concessionaire Manila Waters, which serves some 125,000 households, developed a mitigation plan that included the intensification of the leak repair programs, the reactivation of deep wells for possible supply augmentation,

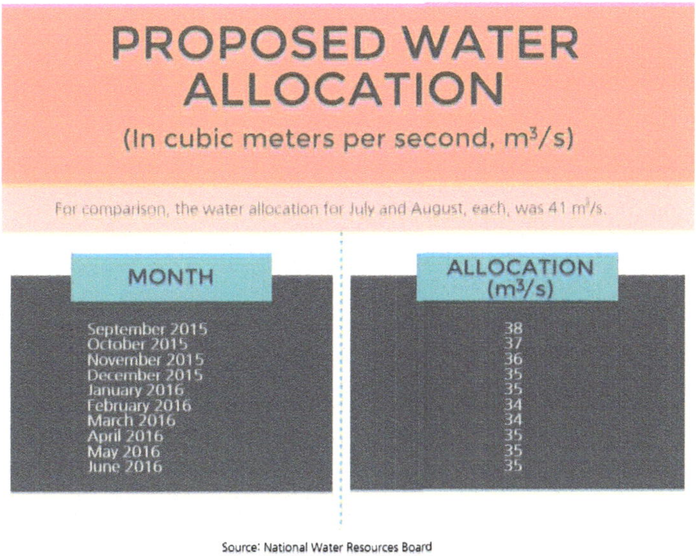
PROPOSED WATER ALLOCATION

(In cubic meters per second, m^3/s)

For comparison, the water allocation for July and August, each, was 41 m^3/s.

MONTH	ALLOCATION (m^3/s)
September 2015	38
October 2015	37
November 2015	36
December 2015	35
January 2016	35
February 2016	34
March 2016	34
April 2016	35
May 2016	35
June 2016	35

Source: National Water Resources Board

Fig. 12.6 Proposed water allocation structure during El Niño months (http://www.gmanetwork.com.news/scitech/science535165/rotating-water-shortages-implemented-as-el-nino-strengthens/story/)

and the implementation of supply and pressure management schemes, among others (Manila Water Company Inc. 2015).

The DILG, on the other hand, reminded the LGUs that under Section 21 of RA 10121, a Local Disaster Risk Reduction and Management Fund (LDRRMF) had been set up to address issues like El Niño. According to the DILG, a Php19-billion national calamity fund has also been set up to augment LGU resources to address impacts of hazards like the El Niño. However, to access the funds, the LGU must fulfill the following requirements: (Department of Interior and Local Government 2016) Local DRRMC Damage Report/Calamity Impact Assessment Report/Work and Financial Plan. The requirements include, among others, a project description and justification; a local legislative council resolution declaring a State of Calamity or Imminent Danger which a clear budget allocation for necessary action; a certificate to complete projects funded by the Office of the President duly issued by the Local Chief Executive (LCE) and affirmed by the local legislative council through a council resolution; an endorsement from the chairperson of the Regional Disaster Risk Reduction and Management Council (RDRRMC); certification from the LCE on the emergency nature of the activities charged to the Calamity Fund; certification from the Local Accountant or Finance Officer on the non-availability of other funding sources except for the Calamity Fund; and a certification validating that infrastructure project, for which funds have been requested, do not have insurance coverage.

Out of concern for the 15 million residents of Metro Manila, MWSS established a Water Security Legacy Project. It is a collaborative initiative between MWSS, its concessionaires Maynilad and Manila Waters, and other stakeholders. The focus of the project, relevant to drought, includes the following (http://mwss.gov.ph/learn/water-security-legacy-roadmap/):

(1) Water infrastructure development and resource management protection—which includes water management and protection;
(2) Water distribution efficiency;
(3) Tariff rationalization and business plan review;
(4) Partnership development;
(5) Communications and knowledge management;
(6) Organizational excellence.

12.6 Drought Governance in Iloilo

Iloilo City is one of those areas severely affected by the dwindling supply of water during droughts. Similar to Metro Manila, the city experiences significant reductions in rainfall during El Niño periods (Fig. 12.7). During the 2015–2016 El Niño, Iloilo province declared a state of calamity and mobilized Php33.7 million of their provincial fund to mitigate El Niño impacts. Iloilo City has declared water-related crises over the recent years: a State of Imminent Calamity in 2007 and a State of Water Crisis in 2009. However, in both cases, there was barely any interest in the impacts of drought brought by El Niño to highly urbanizing cities (HUCs) like Iloilo. It appears that droughts and regular episodes of El Niño remain a disaster hidden from the consciousness of local officials and planners in Iloilo and those in the rest of the country.

The objective of the following section is to examine how meteorological drought during the El Niño years affected Iloilo City. Specifically, there is a need to look into how the development sectors of the city were affected by El Niño and whether programs and policies that were set up to respond to the event were effective.

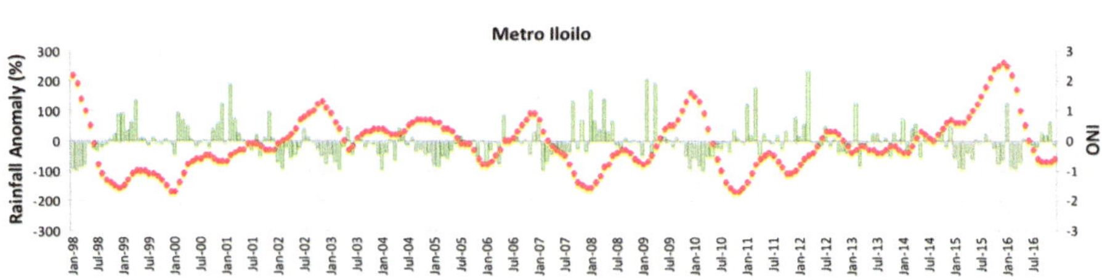

Fig. 12.7 Monthly rainfall anomalies (%) and ONI from 1998 to 2016 for Metro Iloilo show severe reductions in rainfall during El Niño events. Shaded areas indicate El Niño episodes in 1997–1998, 2002–2003, 2004–2005, 2006–2007, 2009–2010, 2014–2015, and 2015–2016

12.6.1 Profile of Iloilo City

To know more about the risk and impacts of drought in Iloilo City, it is important to understand its context. Iloilo City, as it is known now, has six districts namely Jaro, La Paz, Mandurriao, Molo, Arevalo, and Iloilo City proper; it also has 180 barangays or villages. As of 2015, its population was estimated to be at 447,992. The city's population density as of 2010 is 5,423 per square kilometers (Source: Ecological Profile of Iloilo City 2015). Iloilo is classified as a highly urbanized city by the Department of Interior and Local Government of the Philippines. However, it is a city that has *rurban* characteristics. Its growth as a city was founded on the dynamic local agricultural, fisheries, textile industry, and trade. The city was already a preferred trade hub even before its colonization by the Spaniards—proof is the large number of Chinese settlers in the city. Its irregular coastline and safe harbor attracted trading ships. However, the old part of the city, as it is known now, was a reclaimed swampland. The origin of the economic activities that fueled the city's growth started from the old towns of Jaro, Molo, Arevalo, and the municipality of Oton.

The main production centers that created the wealth of Iloilo were all outside the city center. It only became a trading hub (see Fig. 12.8). Across time, the wealth of a few grew, creating a middle class in the towns of the province of Iloilo. These were families who were able to access education for themselves and for their children. Over the years, the value given to education turned Iloilo City into a center of education for the region. It has also become a regional hub for governance. Its progress attracted the region's population to look for more opportunities there given the economic status of the city. Along with this progress and flow of migration for jobs, also came the influx of informal settlers. In 2015, the Iloilo City Urban Poor Association Office estimated 6,954 households in informal settlements located in nearby creeks, rivers, shorelines, road easements, and private lots. These are areas with greater risk and higher geophysical and social vulnerability.

The Metro Iloilo Water District (MIWD), which has a capacity to deliver 20,043 m^3/day in contrast to the water demand of the entire Iloilo which is 83,296 m^3/day, provides for the water needs of Iloilo City. In 2015, the number of active water connections in the city was 18,634, with only 87,580 of the residents having water access via pipelines from the MIWD. But MIWD claimed to have served 170 barangays (see Fig. 12.8). In Iloilo City, the average monthly consumption per connection is 26.80 m^3 and the average per capita per day consumption is 26.80 m^3. The water service hours are also limited. MIWD is only able to provide 18–24 h/day of water service in Jaro district, 12–18 h/day in La Paz, 6–12 h/day in Mandurriao and city proper, and 1–6 h/day in Arevalo. Hence, other water needs of the city residents are provided by domestic private wells, water pumps, and other commercial service providers who extract the groundwater resource of Iloilo City, and public faucets (Source: Ecological Profile of Iloilo 2015).

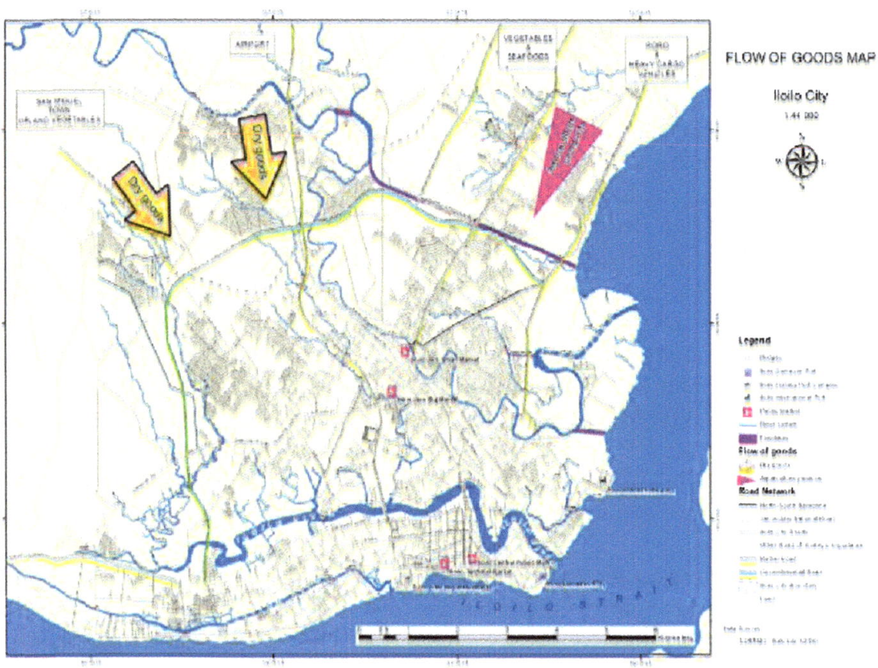

Fig. 12.8 Flow of goods to Iloilo City from neighboring towns. *Source* Ecological Profile of Iloilo 2015

12.6.2 El Niño and Governance in Iloilo

As noted in the previous section, Iloilo City was hit by a meteorological drought in 1998 and in 2015–2016. However, there was far less literature on the impacts on the city than there is in the municipalities of the province of Iloilo where the food supply of Iloilo City comes from.

In both years of study, the water from Tigum River of Maasin, Iloilo dropped at critical levels. While Iloilo's watershed includes the Tigum and Aganan Rivers and the critical role of Iloilo and Jalaur River Watersheds, Tigum–Aganan remains to be the primary source for 73% of MIWD's water supply (Engility Corporation and Strata Consulting and Environmental Law Institute 2013). MIWD reports low pressure, thus, inability to deliver enough water to the consumers. However, it has been established that MIWD cannot distribute sufficient water because of its narrow pipes and has yet to resolve the issue of its choice of a bulk water supplier for Iloilo City. These factors, accompanied by the reported drying of wells in 2015 and the severe water levels in wells in 2016, led to acute shortage of tap water resources. Severe water shortages were reported in the district of Jaro namely the villages of Lanit, Buntatala, Bitoon, Camalig, Cubay, Tabuc Suba, and San Isidro and in villages

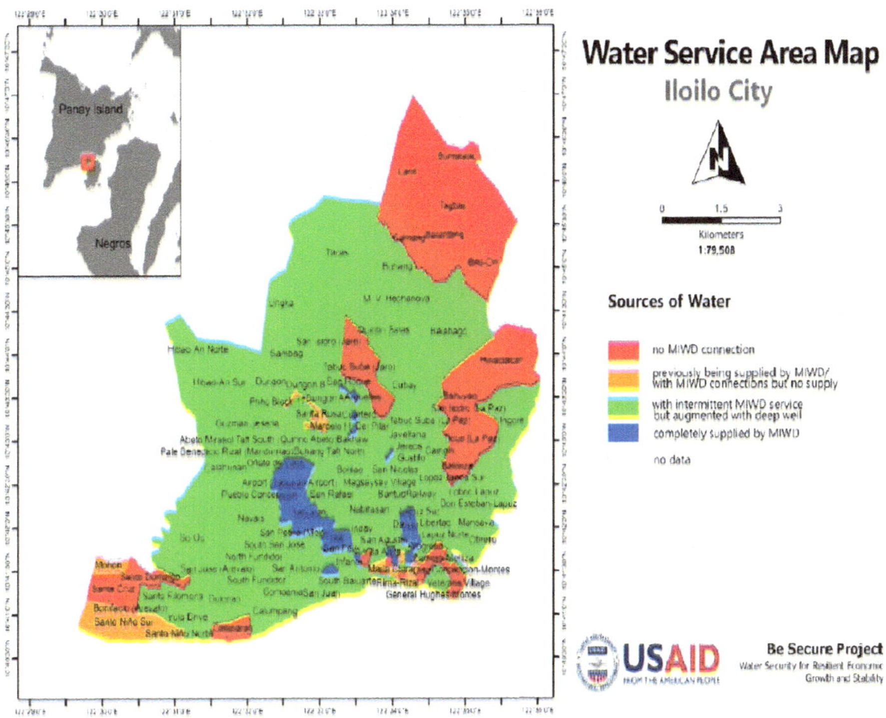

Fig. 12.9 Service areas and water sources of Metro Iloilo. *Source* Ecological Profile of Iloilo City 2015

of San Juan, Boulevard, and Calumpang in the district of Molo. These complaints were largely coming from areas where informal settlements were located.

Iloilo City declared a "state of imminent water crisis" in 2015 via a City Disaster Risk Reduction and Management Council Resolution No. 001-2015 and a state of calamity Board Resolution No. 001-2016. From there, the city identified possible alternative water distribution mechanisms and alternative water sources and allocated Php3 million from the Local Disaster Risk Reduction and Management Fund in 2015–2016 (see Fig. 12.9).

Water rationing and water delivery were the preferred coping mechanisms of the city government in 2015. It utilized the LDRRMF allocation for the purchase of containers for water distribution and fuel expenses for transporting water. They also purchased a water tank that doubled as a fire truck for water distribution. To ensure the efficient distribution of water, Iloilo City partnered with the private sector—the Iloilo City Action Group (ICAG) and the Iloilo Federation of Filipino Chinese Chamber of Commerce (Fig. 12.10).

Because Iloilo City has suffered drought impacts during the El Niño years and dry spells, the government of Iloilo City planned and pursued actions that were identified in their Local Climate Change Action Plan. These plans and actions tried to address

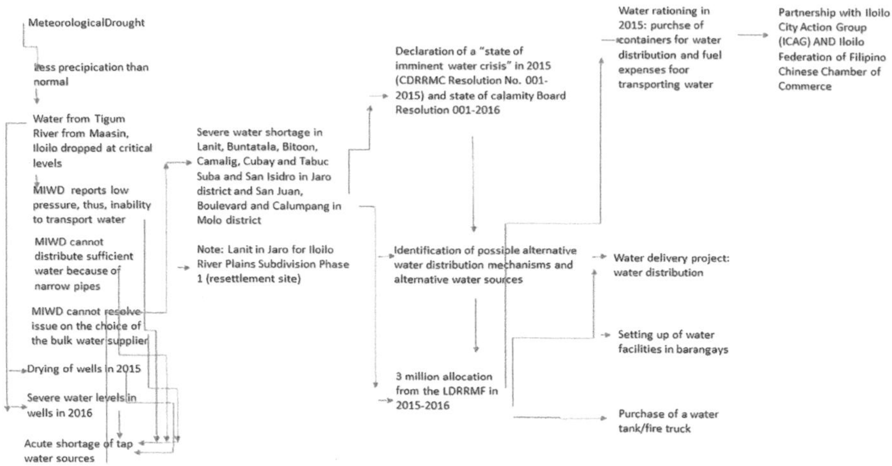

Fig. 12.10 El Niño and Iloilo City government response. *Source* Ecological Profile of Iloilo City 2015

the issue of drought and dry spells seen as effects of the growing business district of Iloilo City. As water demand increases, the Local Climate Change and Action Plan of Iloilo City for 2014–2028 includes the following programs to address issues of irrigation for agricultural areas, food security, water supply development, and tourism development (Source: Iloilo City Local Climate Change Action Plan 2014–2018):

(1) Implementation of the Jalaur River Dam Project to secure water supply development;
(2) Information, Education, and Campaign (IEC) and advocacy on household water conservation;
(3) Implementation of a Department of Public Works and Highways (DPWH)-proposed Water Impounding Dam Project at the confluence of Tigum–Aganan River near the Jaro Floodway Channel;
(4) Retention of fishponds at the Strategic Agriculture and Fisheries Development Zones (SAFDZ) areas for possible water impounding;
(5) Retrofitting of water supply facilities;
(6) Deputation of the City Environment and Natural Resources Office (ENRO) by the National Water Resources Board (NWRB) for water permit regulation;
(7) Regulation of commercial water dealers to monitor the volume of groundwater extraction and related business permits
(8) Preparation of education, information, communication, and advocacy materials on the design and construction of private dug wells;
(9) Replication of rainwater harvesting and barangay water resource mapping initiatives;
(10) Inclusion of drought in the City Disaster Risk Reduction and Management (CDRRM) and other development plans, and

(11) Barangay (or village)-level emergency protocols for drought, dry spells, El Niño.

As an adaptive mechanism, Iloilo City started setting up water facilities in barangays. The Department of Interior and Local Government, Province of Iloilo, Department of Environment and Natural Resources, and the Metro Iloilo Economic and Development Council, which Iloilo City is a part of, undertook a joint study, in partnership with the Canadian Urban Institute, to assess the state Tigum–Aganan Watershed. The study found that El Niño and other climate events will enhance drought in Iloilo. Thus, joint action will be needed.

12.7 Urbanization, Economic Expansion, and Demand for Water in Metro Cebu

Water-related risks and vulnerabilities have increased over the past decade in Metro Cebu. These have been intensified by extreme weather events such as heavy rainfall, floods, and drought brought about by El Niño (Winsemius 2015). Figure 12.11 shows the impacts of El Niño events on Metro Cebu, with prolonged dry periods especially in 1998 and in the 2015–2016 strong El Niño episodes. The rapid urbanization of the region and its economic expansion has also exacerbated the effects of these hazards.

Metro Cebu is composed of seven rapidly urbanizing cities and six municipalities, spread equally to the north and south side of the city center. Cebu City, the capital of Cebu province, forms the core of Metro Cebu as well as the regional capital of the Central Visayas region. Thus, the central business district, major educational institutions, and the international port are mostly located in Cebu City. Most of the territories compromising Metro Cebu, however, are in the uplands. This is exemplified by Cebu City where 70% of the population reside in the 30% lowland plains; meanwhile 70% of its area are settled by 30% of the population in the upland barangays. In the latter, access to water, infrastructure, and other basic services pose a big challenge to most

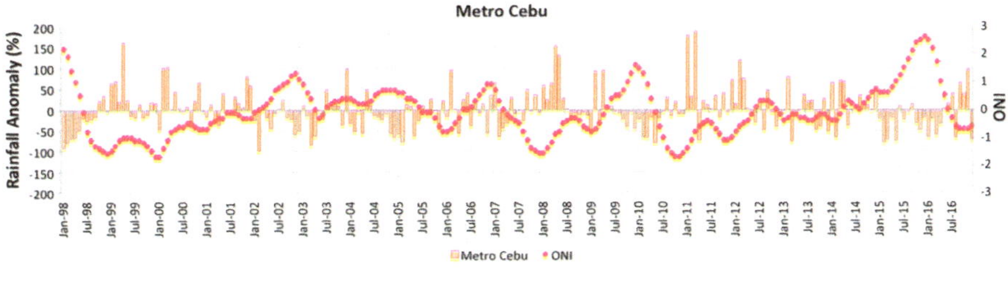

Fig. 12.11 Monthly rainfall anomalies (%) and ONI from 1998 to 2016 for Metro Cebu show drier conditions during El Niño periods, similar to Metro Manila and Metro Iloilo. Shaded areas indicate El Niño episodes in 1997–1998, 2002–2003, 2004–2005, 2006–2007, 2009–2010, 2014–2015, and 2015–2016

residents, whose incomes fall below the poverty threshold. It seems that the social geographies of water access and distribution in Metro Cebu are highly unfavorable for those in the highlands and the coastal urban poor barangays.

Consultation meetings with Cebu's urban poor organizations and urban poor federation showed that aside from the scarcity of water and increasing water costs, they also reported that the price of rice and seafoods went up from the January—May period due to El Niño. In upland communities, they had to resort to eating *camote* (sweet potato) and *tuge* (a kind of root crop), to supplement their food needs. Meanwhile, among the island communities in Palawan inhabited by Tagbanua tribe, they eat *kurot* (a kind of tuber), during the dry season. Just like in other parts of the country, these food items are associated with poverty and eaten during lean months. The increasing costs of water also affected their sanitation habits, depriving children of water for bath and related hygienic practices (Porio 2017b; Porio and Lao 2010). In fact, the 2015–16 drought resulted in political mobilizations in some parts of Mindanao (NFHM 2016, Drought/Final%20%20-20NFHM%20report.%20Mindanao%20drought%20report.pdf).

Meanwhile, major industrial companies, factories, and commercial establishments are concentrated in the urban centers of Cebu and Mandaue. Talisay City hosts quite a number of small- and medium-sized businesses alongside residential subdivisions and informal settlements, just like other cities in the metropolis. In terms of economic expansion and infrastructure development, Metro Cebu continues to be one of the fastest growing metropolises in the region. Thus, it has also become the top tourist destination in the country. With the influx of tourists and their demand for services, the demand for water is heightened especially during the summer months.

12.7.1 Water Governance in Metro Cebu

The Metropolitan Cebu Development Council (MCDC) is mandated to formulate development plans and coordinate their implementation across 15 cities and municipalities. But unlike the Metro Manila Development Authority (MMDA), it does not have legal and institutional powers. Thus, it has little influence on the water governance of the metropolis.

Since 1974, the Metropolitan Cebu Water District (MWCD) has been responsible for distributing potable water from their reservoirs located in various areas of Metro Cebu. Deep wells supply 70% of the water supply in the metropolis. Water shortage has always posed a serious challenge to the city, in part, due to deforestation, degradation of its watershed and seawater intrusion into the underground water supply. MCWD data showed that Cebu's demand for water rose from 265,000 cubic meters in 2007 to 350,000 cubic meters in 2017 and projected to be 490,000 cubic meters in 2028 (see Fig. 12.14 in the Appendix). Thus, the major problem facing MCWD is where to get water to support a rapidly growing urban center to address the water demand of 1.5 million residents in addition to those of the commercial, industrial, and service sectors (Berondo 2005) (http://www.philstar.com:8080/cebu-news/291728/water-crisis-cebu-part-i).

Because of the prolonged El Niño in 2015–2016, the wells in both the city and province have been drying up and seawater has been seeping into the deep wells. The persistent shortage, especially during the summer months, has led officials to propose buying water from nearby Bohol Island. But this has been opposed by some local officials both from Cebu and Bohol.

A compromised environmental sanitation contributes to the scarcity of water supply. Many households do not have toilets or septic tanks, so they dump their wastes on the ground or into rivers, creeks, and other water channels. In addition to domestic wastes, commercial and industrial establishments also unload their chemical wastes and other pollutants into water channels and drainage systems. MCWD officials have reported that some commercial and industrial establishments (e.g., malls, hotels, restaurants, factories) even dump their sewage or liquid waste into drainage channels around the city. Moreover, leakage from septic vaults and runoffs from farm fertilizers, sewage and open dumps add to the water contamination, especially those from the ground sources.

But water concessionaires have been assured by MCWD that these water sources are constantly monitored to prevent contamination. But the NGO Cebu Uniting for Sustainable Water (CUSW) estimated that about 95,320 cubic meters of liquid containing bacteria/viruses and chemicals are drained into the city's aquifers and groundwater supply. This is the reason why MCWD would assert that only their water supply is safe to drink and use. Former DENR Sec. Elisea Gozum asserted (confirmed by Cebu City's Health Officer, Dr. Stella Ygonia) that 31% of illnesses are due to water contamination (e.g., amoebiasis, *E-coli*, cholera) causing the national government to lose about Php67 billion annually. Meanwhile, over-extraction of water from ground sources had also resulted to water contamination due to saltwater intrusion and depletion of underground aquifers (Berondo 2005) (http://www.philstar.com:8080/cebu-news/291728/water-crisis-cebu-part-i).

About 80% of MCWD's water supply is underground water. To prevent further depletion of the water supply, MCWD implemented a self-moratorium in underground water extraction. But this does not solve the increasing demand for water from the ever-expanding commercial and industrial sectors of the city. Meanwhile, the Water Resource Center of the University of San Carlos reported that over-extraction of groundwater has caused saltwater to intrude coastal aquifers. Cebu's denuded mountains are critically linked to water depletion in the dams/reservoirs because rainwater goes directly to rivers and seas. Deforestation, environment degradation, and pollution alongside a rapidly growing population and migration have exacerbated Cebu's water-related woes (Berondo 20050) (http://www.philstar.com:8080/cebu-news/291728/water-crisis-cebu-part-i). At the height of the 2015–2016 El Niño, former Presidential Advisor for Environmental Protection Nereus Acosta announced that Cebu was one of the most ill-prepared provinces for climate change. This announcement is very significant because Cebu is one of the fastest growing regions and demand for water is in fact higher than that of Mindanao. It was also one of the highest at-risk areas and among the most vulnerable during the dry spell (Codilla 2015) (http://cebudailynews.inquirer.net/56501/cebu-is-high-risk-area-for-el-nino-acosta).

12.7.2 Drought Governance in Metro Cebu

In Metro Cebu, the threats to the dwindling water supply are much more felt during the warm months from March to April. The prolonged El Niño has led to the drying up of some wells while seawater has also been seeping into these wells. The following section details the largely reactive management by local DRRM officials of the water crisis brought about by the 2015–2016 El Niño.

On April 6, 2016, the MCWD announced that the water deficit had reached 19,000 cubic meters per day from the 15,000 cubic meters per day two days before that. Two days later, then Mayor Michael Rama declared a state of emergency due to the increasing water deficit in the metropolis. Accordingly, the mayor was advised by PAGASA (Mactan) that the dry spell will last up to June 2016; thus, he needed to make the declaration. Two days later, Cebu's Provincial Disaster Risk Reduction and Management Council (PDRRMC) also placed the province under a state of calamity due to the lack of water in different areas caused by the El Niño. Both the Jaclupan and Buhisan dam water supplies were both at critically low levels, prompting the MCWD to purchase close to 10,000 cubic meters per day from bulk water suppliers (https://www.rappler.com/nation/128781-cebu-city-state-calamity-water-shortage).

About 44 areas or communities in the cities of Cebu, Lapu-Lapu City, and the town of Liloan experienced low pressure to no water supply for 15 h from 6 am to 9 pm from late March to early April 2016 (Rappler, April 8, 2016). Out of the 44 areas affected by the water shortage, 35 areas were from Cebu City. Of the latter, 80% are from the urban poor communities along the coastal lines and in the upland mountain barangays. About 21,000 households were affected by the dry spell brought about by the El Niño. This led the Cebu City Disaster Risk Reduction and Management Council (CCDRRMC) to distribute water to the urban poor barangays in the coast and in the mountain barangays. Fire trucks of the city rationed water to the consumers, for example, to the 12 urban poor barangays fronting the South Reclamation Properties, the latest commercial and residential development to sprout in southern Cebu City. The CCDRMC also established hotlines in their Command Control Center (or C3), so the barangay captains and other officials can call to request water augmentation in their respective areas of jurisdiction (see Fig. 12.12). CCDRMC also organized the task force "Water is Life" to attend to the requests for water augmentation, headed by the city administrator (the little mayor), the Association of Barangay Councils (ABC) president and the head of the city's Quick Response Team (QRT) (http://cebudailynews.inquirer.net/91598/water-crisis-worsens).

Meanwhile, the city government reported the following agricultural damage caused by the drought: Php58 million loss of livestock and Php22 million of poul-

Fig. 12.12 Residents of Barangay Mambaling, Cebu City lineup with their plastic pails to get water from the barangay fire truck. *Photograph Credit* Cebu City Councilor and DRRM Officer Dave Tumulak

try losses (http://www.gmanetwork.com/news/news/regions/562346/cebu-province-under-state-of-calamity-due-to-water-shortage/story/).

This reactive type of managing the water crisis brought about by the drought is a common scenario during the summer months. Declaring a state of emergency and water rationing by the fire trucks to areas without water seem to be the time-honored response of the mayor, city council, and the DRRM office every time a water crisis emerges. Aside from being reactive, local governance units and their task forces last only for the duration of the emergency period. After El Niño and the monsoon rains arrive, earlier strategies of conserving water are now focused on preventing excess water in canals, creeks, and other water channels from flooding the city. This Sisyphean cycle seems to characterize the risk reduction strategies of local government units, despite the continued advocacies by other branches of the government, private sector, and civil society for the city to craft a responsive water governance in Metro Cebu, through an integrated climate change adaptation and mitigation and disaster risk reduction management (CCAM-DRRM) policies and programs (Porio 2017b).

To conclude, water-related risks and vulnerabilities have increased over the past decade in the Central Visayas region of the Philippines, especially in Metro Cebu, but the response of the government, private sector, and civil society has been quite reactive and uneven. Despite several declarations of a state of emergency or calamity every summer, a comprehensive, systematic, and proactive water governance regime is yet to be established.

12.8 Summary and Conclusion

Urbanization, population growth, and the rapid economic expansion of the metropolitan areas of the National Capital Region (NCR or Metro Manila), Metro Iloilo, and Metro Cebu have also increased the demand for water. During El Niño periods, the demand for water hits crisis proportions and usually leads to a declaration of a state of emergency or calamity by the mayor, the head of the DRRMO.

While El Niño may hit a country, its characteristics may vary according to ecosystem and scale. The availability of options, i.e., surface vs. groundwater, or both, influences responses to the phenomenon. Socio-demographic patterns of in-migration and population growth in areas experiencing rapid urban expansion impacts the water security of formal and informal sectors differently. The dwindling water supply during El Niño episodes shows that majority of the poor, who do not have access to formal channels of water distribution, suffers mostly from the water price hikes and compromised hygiene and sanitation. While urban residents have options to access potable water, i.e., dug wells in Iloilo City or surface water in Metro Cebu versus water access through alternative water distributors such as peddlers who also rely on a single water source, i.e., Angat, water costs will be cheaper. This, however, will sacrifice future water security because of the use of groundwater resources. In Cebu City, some coastal barangays are already experiencing saltwater intrusion into their wells because of excessive groundwater extraction.

In this study, it appears that cities in the South like Iloilo and Cebu experienced greater reduction in rainfall compared to Metro Manila during El Niño years. But the availability of water sources created a sense of water security among stakeholders, as presented in the case of Metro Manila. The infrastructure necessary for the distribution of water supply plays a critical role in addressing El Niño risks. This was made apparent in the concern over narrow pipes and limited water flow, both in Iloilo City and Metro Cebu. The capacity of various stakeholders to reduce risks for a water crisis will mitigate drought problems. However, while coping or reactive mechanisms such as water rationing and anticipatory actions like water allocation help, there is a greater need for adaptive actions. These include the preparation of a Water Security Legacy Plan led by MWSS and their partners in Metro Manila, as well as a prepared Local Climate Change Action Plan that attempts to address drought in Iloilo City.

The Local Government Code and the Urban Development and Housing Act, both passed in 1992, devolved the delivery of basic services, but water governance was not included in the devolved functions. Water generation, supply, and distribution have therefore been largely under different modes of public–private partnership schemes in different cities and towns. In Metro Manila, the two water concessionaires have made headways in developing their sustainable water strategic policies and plans for the metropolis, but this has yet to be developed in Metro Cebu. Ironically, Metro Cebu is highly at risk to the impacts of drought, excessive groundwater extraction, drying up of wells and saltwater intrusion, and its water governance deficit is especially very high.

Preparing a DILG-mandated Local Climate Change Action Plan (LCAP) that is science-informed and contextually driven seems to be the most appropriate response to mitigate and adapt to the effects of drought brought about by El Niño. But most local government units are still struggling with producing integrated risk assessments necessary for a well-crafted LCAP. Moreover, the implementation of the plans that need massive capital investments in infrastructure development and capability building poses great challenges for both national and local governance systems. But the Department of Interior Local and Government (DILG), which mandates all LGUs to have their LCAP formulated and implemented alongside the Philippine Disaster Resilience Foundation initiative of the private sector, provides some hope for more responsive water governance.

Appendix

See Tables 12.1, 12.2 and Figs. 12.13, 12.14.

Table 12.1 Percentage of urban–rural population in the Philippines

	Urban	Rural	Total
Population (in millions)	41.9	50.5	92.3
Percentage (%)	45.3	54.7	100

Source 2010 Census of Population and Housing, Philippine Statistics Authority 2010

Table 12.2 Urban growth in the Philippines

Population size category (millions)	Philippines—urban population (millions)			
	Number of urban areas	2000	2010	Ave. annual expansion rate (%)
10 or more	1	12.20	16.52	3.1
5–10	0	–	–	–
1–5	1	1.01	1.53	4.1
0.5–1	3	1.37	2.05	4.1
0.1–0.5	16	2.24	3.18	3.6
Total	21	16.83	23.28	3.3

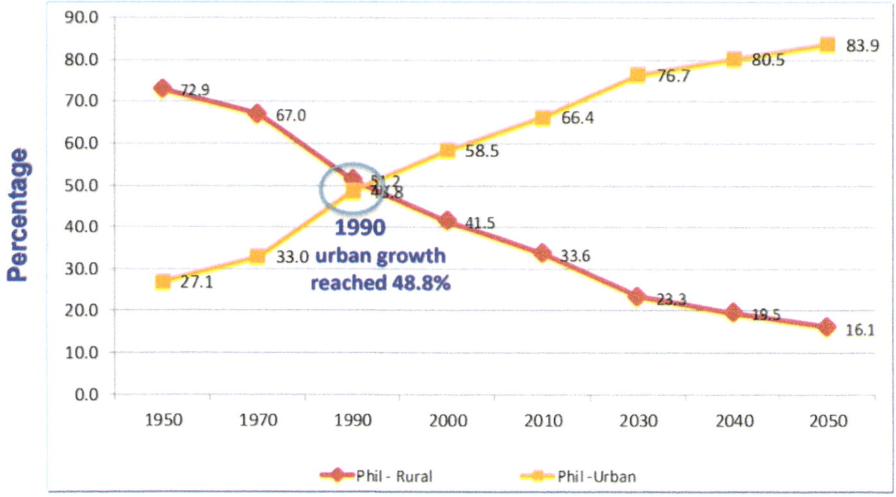

Fig. 12.13 Percentage of urban–rural population in the Philippines (1950–2050). *Source* 2010 Census of Population and Housing, Philippine Statistics Authority 2010

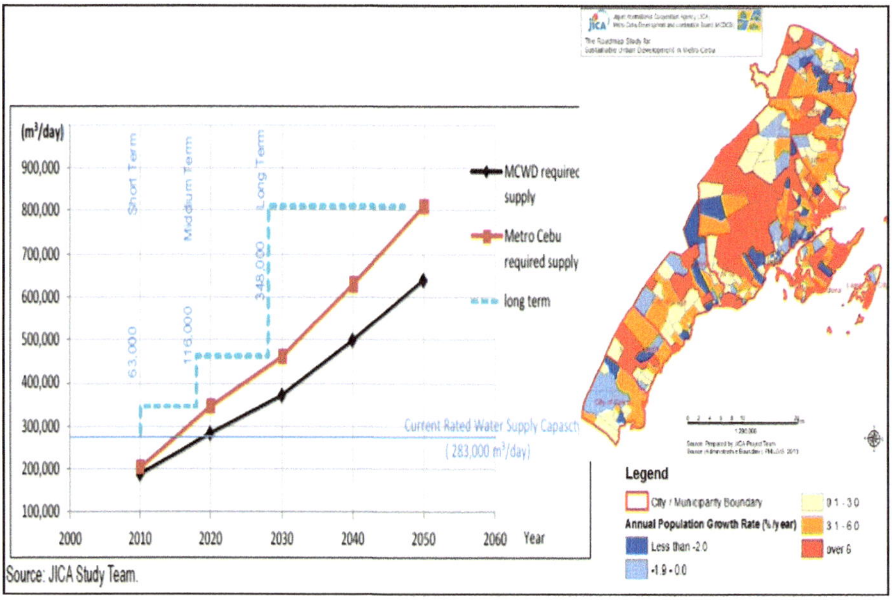

Fig. 12.14 Current rated water supply capacity in Metro Cebu. *Source* Platform for Smart, Inclusive, and Sustainable City Region: Report, 2016

References

Albert RG (2001) Assessing the impact of the Asian Financial Crisis and El Niño on Poverty in the Philippines by Jose Ramon G. Albert. http://dirp4.pids.gov.ph/ris/ris/wbi/Country%20Report_Philippines.PDF

Berondo W (2005) Water crisis in Cebu. http://www.philstar.com:8080/cebu-news/291728/water-crisis-cebu-part-i

Codilla M (2015) Cebu is high risk area for El Niño: Acosta. http://cebudailynews.inquirer.net/56501/cebu-is-high-risk-area-for-el-nino-acosta

Climate Prediction Center (2016) Cold & warm episodes by season. Retrieved 14 Oct 2017. http://origin.cpc.ncep.noaa.gov/products/analysis_monitoring/ensostuff/ONI_v5.php

Department of Interior and Local Government (2016) LGUs can tap calamity funds, other available funds to combat El Nino. http://www.dilg.gov.ph/news/LGUs-can-tap-calamity-funds-other-available-funds-to-combat-El-Nino-DILG/NC-2016-1057

Engility Corporation and Strata Consulting and Environmental Law Institute (2013) An assessment of water security Iloilo Tigum-Aganan, Watershed, Washington, USAID, DC

Goddard Earth Sciences Data and Information Services Center (2016) TRMM (TMPA) Precipitation L3 1 day 0.25 degree x 0.25 degree V7, Edited by Andrey Savtchenko, Goddard Earth Sciences Data and Information Services Center (GES DISC). Accessed 30 June 2017, 10.5067/TRMM/TMPA/DAY/7. https://disc.gsfc.nasa.gov/datacollection/TRMM_3B42_Daily_7.html

Hilario F, de Guzman R, Ortega D, Hayman P, Alexander B (2009) El Niño Southern Oscillation in the Philippines: impacts, forecasts, and risk management. Philippine J Dev 36(1):9–34

Iloilo City Climate Change Technical Working Group (2014) Local Climate Change Action Plan 2014-2018, Iloilo City, April 2014

Iloilo City Government (2016) Iloilo City Ecological Profile. http://www.iloilocity.gov.ph/images2013/BlueProfile.PDF

Iloilo City Urban Poor Association Office (2015) Updated numbers of informal settlers colonies

Israel D (2016) Cebu City in state of emergency due to water crisis.cnn. Retrieved 10 May 2017, from http://cnnphilippines.com/regional/2016/04/09/cebu-coty-state-of-emergency-water-crisis.html

Jose AM (2002) ENSO impacts in the Philippines: examples of ENSO-society interactions [online] in Hilario et al (2009). http://iri.columbia.edu/climate/ENSO/societal/example/Jose.html

Macasero R (2016) Cebu City declares state of calamity due to water shortage. Rappler. Retrieved 10 May 2017, from http://www.rappler.com/nation/128781-cebu-city-state-calamity-water-shortage

Mahmood R et al (2014) Land cover changes and their biogeophysical effects on climate. Int J Climatol 34(4):929–953. https://doi.org/10.1002/joc.3736

Malayang B III (2004) A model of water governance in the Philippines. In Rola AC, Francisco HA, Liguton JPT (eds) Winning the water war: watersheds, water policies and water Institutions. Philippine Institute for Development Studies, Makati City, Philippines, pp 59–83

Manila Water Company Inc. (2015) Manila water sets plans in motion to mitigate El Nino Effects. http://www.manilawater.com/Pages/ReadNews.aspx?MBID=196

Maynilad, Manila Water in JV to Address Cebu Water Crisis—The Manila Times Online (2016) The Manila Times Online. Retrieved 10 May 2017, from http://www.manilatimes.net/maynilad-manila-water-in-jv-to-address-cebu-water-crisis/256869/

National Water Resources Board (2015) NRWB holds press conference on preparation for El Nino impacts in water supply, NWRB, Manila, July 2015. http://www.nwrb.gov.ph/index.php/news-articles/611-nwrb-holds-press-conference-on-preparation-for-el-nino-impacts-in-water-supply2

NEDA (2017) Statement of socio-economic planning secretary Emmanuel Esguerra on the status of the roadmap to address the impact of El Nino (RAIN). http://www.neda.gov.ph/2016/04/05/statement-of-socioeconomic-planning-secretary-emmanuel-f-esguerra-on-the-status-of-the-roadmap-to-address-the-impact-of-el-nino-rain/

NFHM (2016) Final report of the national fact finding and humanitarian mission to Kidapawan City North Cotabato, Philippines, April 4–6, 2016. Drought/Final%20%20-20NFHM%20report.%20Mindanao%20drought%20report.pdf
Padua R, Laude P (2010) Quezon City, Caloocan Move to Conserve Water. http://www.philstar.com/metro/551375/quezon-city-caloocan-move-conserve-water
Patankar A et al (2013) Impacts of extreme weather events and implications for adaptation planning for coastal cities. APN Sci Bull 3:16–75
Pereira RA, Lopez ED (2004) Characterizing the spatial pattern changes of urban heat islands in Metro Manila using remote sensing techniques. Philippine Eng J 25(1):15–34
Pielke RA Sr et al (2007) An overview of regional land-use and land-cover impacts on rainfall. Tellus B 59(3):587–601. https://doi.org/10.1111/j.1600-0889.2007.00251.x
Pielke RA Sr et al (2011) Land use/land cover changes and climate: modeling analysis and observational evidence. Wiley Interdisc Rev Clim Change 2(6):828–850. https://doi.org/10.1002/wcc.144
Porio E (2009) Urban transition, poverty and development in the Philippines. http://pubs.iied.org/pdfs/G02570.pdf
Porio E (2017a) Risk and resilience: interrogating prosperity, inequality and climate-disaster impacts in Metro Manila. Conference Proceedings, Philippine Institute of Development Studies
Porio E (2017b) Citizen participation, environmental security and risk governance: integrating climate change adaptation and mitigation and disaster risk reduction and management (CCAM-DRRM) in Metro Cebu. Assessment report submitted to Misereor Foundation (Aachen, Germany) and the PagTambayayong, Foundation for Mutual Aid, Inc
Porio E (2017c) Prosperity and inequality in Metro Manila: reflections on housing the poor, climate risk and governance of cities. In: Wungaeo C, Rehbein B, Wungaeo S (eds) Globalization and democracy in Southeast Asia: challenges, responses and alternative futures. Palgrave-Macmillan, London
Porio E, Lao M (2010) A qualitative assessment of the effects of an IFC project on the reduction of poverty through a water concessionaire in Metro Manila. A report submitted by the Institute of Philippine Culture to the International Finance Corporation (IFC)
Ranada P (2015) 355,500 homes in NCR face water disruption starting September, Rappler, Metro Manila. http://www.rappler.com/nation/104371-metro-manila-homes-water-disruption-el-nino
Rodolfo K (2014) On the geological hazards that threaten existing and proposed reclamations of Manila Bay. Philippine Sci Lett 7(1):2014
Salinger MJ, Shrestha ML, Ailikun, Dong W, McGregor JL, Wang S (2014) Climate in Asia and the Pacific: climate variability and change. In: Manton M, Stevenson LA (eds) Climate in Asia and the Pacific, pp 17–57. Springer, Netherlands. https://doi.org/10.1007/978-94-007-7338-7_2
Satterthwaite D, Doodman D (2013). Towards resilience and transformation of cities. In: Environment and urbanization, vol 25, issue 1
Tabios G III (2010) Water governance in the Philippines and some thoughts on governance. Retrieved from http://www.fao.org/fileadmin/user_upload/groundwatergovernance/docs/Shijiazhuang/Presentations-PDFs/Day3/PS8_TABIOS_Philippines.pdf
UNESCAP and RIMES (2015) El Nino 2015/2016 impact outlook and policy implications. UNESCAP, Advisory Note, Dec 2015
Winsemius HB et al (2015) Disaster risk, climate change, and poverty: assessing the global exposure of poor people to floods and droughts. Policy Research Working Paper 7480, World Bank, Washington, DC
World Bank Group (2017) Philippines urbanization review: fostering competitive, sustainable and inclusive cities. The World Bank Group, Washington, DC

Chapter 13
Coping with Urban Water Insecurity in the Colonial City of Kolkata and Implications on Sustainable Development Goals

Bhaswati Ray and Rajib Shaw

Abstract The Sustainable Development Goals adopted in September 2015 aim at ensuring clean water and sanitation to all, building resilience in infrastructure through innovations, and making cities inclusive, resilient and sustainable, at a time when global water systems are challenges by variability in climatic conditions, rapid growth of population, and environmental degradation. Moving from a tank-based surface water supply system, the colonial capital since 1773 had its first pumping station in 1820 to distribute water from River Hugli to the main city by gravity. Though having sufficient supply of freshwater, the current water supply system in the megacity of Kolkata is crippled with intermittent supply, old worn-out zonal mains, high leakage loss, inadequate coverage, dependence on groundwater, poor water quality, and low-cost recovery. In this paper, an attempt has been made to study the existing water supply system and its implications on the Sustainable Development Goals 2015 and to also explore coping strategies for a resilient and sustainable water future. It is evident that the formal water supply system is far from adequate and communities are seeking resilience in an informal parallel water supply system.

Keywords Water systems · Communities · Sustainable

13.1 Introduction

Providing clean water to all is the greatest challenge facing urban water systems in the twenty-first century, under the impacts of exponential growth of population, rapid and mass urbanization, higher levels of consumption, changes in land use pattern, and climate change. Over one billion people worldwide lack access to clean and safe

---

B. Ray (✉)
Sivanath Sastri College, University of Calcutta, Kolkata, India
e-mail: bhaswati173@gmail.com

R. Shaw
Graduate School of Media and Governance, Keio University, Fujisawa, Japan

B. Ray and R. Shaw (eds.), *Urban Drought*, Disaster Risk Reduction,
https://doi.org/10.1007/978-981-10-8947-3_13

water, and 2.4 billion people live without adequate sanitation (Vo 2007; Cain and Gleick 2005; Cosgrove and Rjisberman 2000; Gleick 1999, 2003). The importance of water in the global sustainable development agenda beyond 2015 is evident from the fact that water has been recognized as a key to a sustainable future at the 2012 United Nations Conference on Sustainable Development. The key global challenges of poverty, hunger, illiteracy, inadequate health services, gender inequality, degraded ecosystems, climate change, and disasters as also their impacts are governed by the availability of safe drinking water. The Sustainable Development Goals adopted in September 2015 thus aim at ensuring clean water and sanitation to all, building resilience in infrastructure through innovations, and making cities resilient and hence sustainable.

Sustainable management of water resources and capacity building are essential components for a resilient and sustainable water supply system in most urban areas across the globe. The Indian cities are no exception. Most Indian cities are faced with water insecurity issues under the twin impacts of mass urbanization and climate variability. The urban population of India is 377 million and accounts for 31% of the country's total population (Census of India 2011). The average annual increase of urban population is estimated at eight million (Census of India 2011). The decadal urban growth is recorded at 3.4% between 2001 and 2011, an increase from 2.1% between 1991 and 2001. Nearly, 43% of this urban population resides in the metropolitan or million-plus cities. The urban population is also projected to increase rapidly from 377 million in 2011 to 600 million by 2030, an increase of more than 200 million in 20 years (Planning Commission 2012). The increase is likely to be concentrated in the large cities. The massive size of the mega cities of India is well documented in the ranking of world cities by the United Nations based on population size. Mumbai and Delhi are placed among the top 10 cities, while Kolkata is ranked among the top 15 (United Nations Human Settlements Programme 2011).

The Indian cities, particularly the million-plus cities, are affected by severe water insecurity, evident from the inadequate availability of safe drinking water as well as from the frequent urban floods during monsoon months. In a study by Mathur et al. (2000), it was seen that only 2.7% of the municipalities under study supplied more than 100 l of water per capita per day. The 100 l benchmark is in accordance with the norm for urban water supply by the National Institute of Urban Affairs (NIUA), fixed at 95–125 l per capita per day, though it is higher for the medium-sized towns at 150 l per capita per day (National Institute of Urban Affairs 1997). It may be noted that the minimum desirable supply of drinking water per person per day is higher for the Class I towns and is 220 l according to the National Commission of Integrated Water Resource Management (NCIWRD) (1999). The hours of supply are also limited in all urban centers, except for Thiruvananthapuram and Kota, where the supply of water is for 24 h. The average supply of water to the urban centers is for 4.3 h only, and 30% of the water is lost during transmission (Asian Development Bank 2007). The National Water Policy 2002 encourages private participation and recommends private involvement in building, owning and operating, water resources facilities for increased resilience and capacity building (Ministry of Water Resources 2002).

Like most Indian cities, the megacity of Kolkata also faces severe water stress under the impacts of rapid urbanization and expanding city size. The water supply system, one of the oldest in the country built during colonial period, has failed to keep pace with the rapid urban growth. The study focuses on the water insecurity issues in the megacity of Kolkata and its implications on Sustainable Development Goals. The scope of the paper is defined by the appraisal of drinking water insecurity in the colonial capital of Kolkata and the associated unsustainable withdrawal of groundwater, community participation for improved water supply and wastewater management. It thus addresses some of the targets specified in SDGs particularly those in goals 6, 9, 11, and 15. Goal 6 on water supply and sanitation aim at ensuring availability of water and sanitation for all, sustainable management of water resources, and capacity building through innovations and community participation. Goal 9 emphasizes innovations for improved infrastructure, Goal 11 on building inclusive and resilient cities, and goal 15 recommends ecosystem management.

13.2 The Colonial City of Kolkata

Kolkata, referred to as Calcutta till 2001, is one of the biggest cities of eastern India sprawling over an area of 187.33 km^2 and with a population 4.5 million according to the Census of India, 2011. With the addition of Joka I and Joka II Gram panchayats on September 1, 2012, the total area under the jurisdiction of Kolkata Municipal Corporation now stands at 200.71 km^2. Kolkata is the third largest megacity in India. The population of the urban agglomeration was 14.1 million in 2011 (Census of India 2011). The city, evolved from the merging of three rural hamlets of Kolkata, Sutanuti, and Gobindapur, was the hub of commercial and business activities for the British since 1690 and was nominated to the capital of India in 1773. Located in the Ganga Delta region only 100 km from the Bay of Bengal, it is at an elevation of only 1.5–9.0 m. Much of the city was a vast marshy wetland, remnants of which still remain as the East Calcutta wetlands and occupies the eastern peripheral parts of the city, reclaimed over the decades to accommodate the city's increasing population. To protect the wetlands from being reclaimed and converted into urban use to accommodate the expansion of Kolkata, the East Calcutta wetlands are designated as Ramsar site with restrictions on alternate use under the Ramsar Convention held on August 19, 2002.

Flanked by the River Hugli on the west and the River Kulti to the east, the city is hydrologically endowed with enough freshwater. Kolkata also has rich reserves of groundwater available at a depth of only 3 m from the surface. Yet the city is water insecure.

13.3 Evolution of the Urban Water System in the Colonial City of Kolkata

The water supply system of Kolkata built by the British during colonial period is one of the oldest in the country. In the early days, the drinking water supply in Kolkata was from River Hugli, though the banks and the river were prone to polluting activities like bathing, washing, and idol immersion. Water pollution was thus a major concern for the European residents in this trade capital of the British East India Company. Having revulsion for the polluted water of River Hugli, the Europeans living in Kolkata preferred to use stored rainwater, even for the purpose of drinking. Improved and extensive storage facilities were needed as the city grew in terms of population and importance, particularly since 1773 when the city was nominated to become the capital city of British India. Kolkata continued to serve as the capital of India till 1912. It was during this period that more tanks were dug to conserve rainwater to enhance drinking water supply in the city. The oldest of them popularly known as Lal Dighi at Dalhousie Square, the heritage hub of the city, was deepened further in 1709 to ensure improved water supply to the garrison city at Fort William, around which the colonial capital of Kolkata later expanded. Between 1805 and 1836, large water tanks were excavated at College Square, Wellington Square, and Wellesley Square that ensured increased water supply in the city.

Later, in 1820, a small pumping station and a system of open masonry aqueducts were constructed (Ray and Shaw 2016; Dasgupta 1991). The completion of a pumping station at Chandpal Ghat in 1822 allowed the lifting of water from River Hugli into masonry aqueduct for distribution by gravity to a population of 200,000 residing in parts of present-day central Kolkata. Apart from the supply of drinking water, the distribution system also provided water for the washing of streets and for fire hydrants. Pumping stations operated for seven hours daily, between 6 a.m. and 10.00 a.m. in the morning and between 4 p.m. and 7 p.m. in the evening. By 1870, Kolkata had piped water supply in all major streets through stand posts that were 500 in number according to the Kolkata Metropolitan Water and Sanitation Authority. There was no purification except for the sand and charcoal filters in individual houses.

The first water treatment plant (WTP) was constructed at Palta in 1870. The total capacity of the treatment plant was six million gallons per day (MGD). A pumping station and a reservoir of capacity one million gallon were also in use. A second pumping station was also constructed at Wellington Square. The site of the treatment plant being 30 km upstream at Palta helped reduce the risk of contamination and diminished the presence of tidal seawater. The residents thus had access to filtered water through the distribution network at a per capita availability of 60 l per day. In 1888, three underground reservoirs constructed at Wellington Square, Halliday Street (Md. Ali Park), and Auckland Park (Bhowanipur) helped augment supply. The supply of filtered water, however, did not exceed seven million gallons daily and continued to be intermittent. Hence, as a measure of economy, unfiltered water supply was extended for street watering and fire fighting. Gradually, an additional engine was installed at Tallah, the pumping plant at Chandpal Ghat was strengthened, and unfiltered water supply was further extended. The Chandpal Ghat Pumping Station was transferred to the Port Commissioners following an extension of the unfiltered water pipes in 1871. This pumping station supplied all the unfiltered water required for the town area. In 1897, another pumping station for unfiltered water supply was constructed at Watgunge. The supply of unfiltered water grew from a little over a million gallons (4.546 million l) per day at Chandpal Ghat to 33 million gallons (150 million l) per day from Mallick Ghat and Watgunge Pumping Station in 1921.

It soon became evident that the works should be enlarged. The capacity of the water treatment plant at Palta was increased in 1914, and an elevated reservoir was also constructed at Tallah to ensure an increased supply of water to the city. After independence, production capacity of the water treatment plant at Palta, renamed as the Indira Gandhi Water Treatment Plant in 2002, was increased to 120 MGD in 1952 and further to 160 MGD in the year 1968. For a total population of 2.70 million, the supply of water was 130 l per person per day. For more efficient distribution of filtered water further into the expanding city, the existing underground reservoirs at Wellington Square, Halliday Street, and Auckland Square were converted into booster pumping stations. Continuous supply could not, however, be maintained due to wastage in the water supply system (Ray and Shaw 2016) which amounted to 3% in the Palta network (Calcutta Environmental Management Strategy and Action Plan 1995). The supply was further declined toward South Kolkata, the outer limits of area served by the Palta waterworks, because of leakages in the zonal mains and in the distribution lines. Hence, the supply was being supplemented by groundwater sources.

According to the Kolkata Municipal Corporation, the city is now served by five water treatment plants. River Hugli remains the predominant source of the water supplied in Kolkata, after being treated at the three water treatment plants currently located at Palta, Garden Reach, and Dhapa. The 260 MGD capacity Indira Gandhi Water Treatment Plant at Palta is the oldest treatment plant in city. Water from Palta is distributed in the city through four zonal mains and a distribution network of 3,800 km, after getting lifted at the pumping station at Tallah. The other important water treatment plants are at Garden Reach and Dhapa, with respective capacities of 135 and 120 MGD. The Dhapa Water Treatment Plant, however, runs at a capacity of 30 MGD and supplies water in the eastern peripheral wards. Water supply to these wards is hence insufficient, and many parts continue to remain outside the supply network. Municipal supply is supplemented by groundwater in eastern fringe areas, known to be affected by arsenic above permissible limit. There are two other water treatment plants though with smaller capacities of 8 and 5 MGD at Jorabagan and Watgunge, respectively (Fig. 13.1).

The combined capacity of the five water treatment plants is 438 MGD though supply to the city is only 315 MGD (Table 13.1). Groundwater supplies another 30 MGD in the peripheral wards inadequately served by the municipal supply from treatment plants, lifted with the help of power-driven tube wells. The average per capita availability of water is 130 l per day, against an NIUA recommended 95–125 l per capita per day.

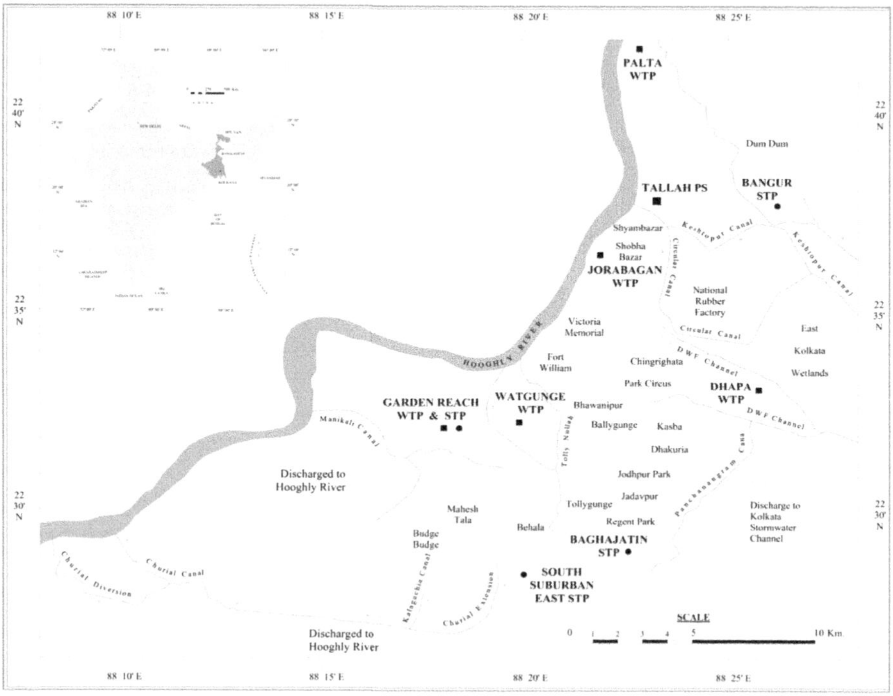

Fig. 13.1 Water and sewage treatment plants in Kolkata

13.4 Urban Water Insecurity in Kolkata and Its Spatial Dimension

In spite of being equipped with enough freshwater, Kolkata suffers from water insecurity. Even in the best water supply wards in the city, nearly 35% of the total population continues to remain uncovered by municipal supply. Another 20–25% is covered by single tap connections within their premises. 8000 stand posts provide water to more than half of the slum and squatter population who do not have access to piped water supply. The population living in slums or squatter colonies account for 32.4% of the total urban population of Kolkata. Because of the uncontrolled continuous flow during hours of supply, 60% of the water from the stand posts is never utilized, leading to immense wastage of supply. Backflow from the stand posts during hours of no supply also adds to the wastage. The duration of municipal water supply is also far from adequate and varies markedly in different parts of the city (Table 13.2). According to the Kolkata Municipal Corporation, daily hours of municipal water supply in different parts of the city vary between 9 h in the main city or central Kolkata to 6 h in the peripheral wards (Table 13.2). The supplied water is also not safe always, as the intermittent supply and the low pressure in supply mains increase the risk of contamination. It may be mentioned that pressure in the supply lines is

Table 13.1 Water supply in the city of Kolkata

Water treatment plants	Capacity in MGD	Supply in MGD	Area covered	Remark
Indira Gandhi Water Treatment Plant	260	200	Central or main city	Treated surface water
Garden Reach Water Treatment Plant	140	135	Southern peripheral	
Jorabagan Water Treatment Plant	8	5	Central city	
Watgunge Water Treatment Plant	5	3	Western peripheral	
Dhapa Water Treatment Plant	30	10	Eastern peripheral	
Total	438	315	315	
Groundwater	30	30	30	Untreated
Total	468	345	345	

Source By authors (based on data provided by Kolkata Municipal Corporation)

kept low in an attempt to reduce the high transmission and distribution loss to the tune of 30% because of the old dilapidated networks. The municipal water supply in the city also suffers from microbiological contamination as is evident from the 3000 water samples analyzed by the Federation of Consumers Association West Bengal and Better Business Bureau. Eighty percentage of samples were found to contain fecal bacteria or *Escherichia coli*, typhoid-causing *Salmonella,* dysentery-causing *Shigella,* and *Vibrio* that causes cholera. Though chlorinating the water kills these microbes, not the slightest trace of chlorine was found in even a single sample (The Telegraph, April 14, 2003).

The residents particularly those in the peripheral wards have their own groundwater sources to supplement municipal supply, making the population vulnerable to arsenic contamination and salinity. The water supply system of the city, one of the oldest in India, is crippled with inadequate and intermittent supply, low per capita availability, inadequate coverage, low-cost recovery, and excessive dependence on groundwater, often arsenic contaminated. Major deficiencies in the water delivery system include intermittent and irregular supply, low pressure, high leakage loss, and over-exploitation of groundwater (Majumdar and Gupta 2007). The city also suffers from inadequate zonal mains and old supply networks in an advanced state of dilapidation (Ray and Shaw 2016; Sivaramakrishnan 1993). Kolkata Municipal Corporation's recovery of operational costs, which is merely 15%, is one of the lowest among all Indian cities (McKenzie and Ray 2009). The policy of not pricing the water used for domestic consumption has earned the authorities criticism for promoting wastage (ADB 2007; McKenzie and Ray 2009) and lack of capital for network improvement. Only central Kolkata, served by the Indira Gandhi Water Treatment

Table 13.2 Urban service delivery status in Kolkata and its spatial dimensions

Area	Population %	Coverage %	Liters/capita/day	Hours of supply	Metered connections %
National average		81	123	4	25
All KMC area		92	134	8	0.1
Central	69	100	146	9	None
Eastern periphery	6	76	109	6	None
Western periphery	7	45	120	6	None
Southern periphery	18	83	120	6	None

Source Kolkata Municipal Corporation, 2012

Plant, is above the national average in terms of all parameters of water supply including coverage in terms of population and area as well as duration of supply per day. The percentage of population covered by municipal supply is particularly poor in the peripheral areas, especially the eastern and western peripheries.

13.5 Community Response to Urban Water Insecurity in Kolkata

It is evident that municipal supply of drinking water is insufficient in the city of Kolkata. The inability to ascertain the provision of universal and equitable access to treated and safe drinking water may be attributed more to delivery inefficiencies, bureaucratic inertia, lack of investment, poor cost recovery, as well as institutional and governance issues, rather than actual limitations on production capacity. To cope with the insecurities in the formal water supply system, urban communities are seeking resilience in an informal parallel supply system. The study thus aims to assess the sustainability and resilience in the urban water system in the city, based on questionnaire survey conducted in 300 households in the city of Kolkata. For the selection of households, stratified random sampling was used to ensure participation of households from the central or main city as well as from the peripheral wards located toward the outer limits of the city proper. Supply in central Kolkata is above the national average and mostly from Indira Gandhi Water Treatment Plant located at Palta. The southern and eastern peripheral wards are served by the water treatment plants located at Garden Reach and Dhapa, respectively (Fig. 13.2). Here the supply is poor in terms of per capita availability, coverage, and also on water quality issues. Survey was also conducted in the low-income slum pockets with perpetual scarcity

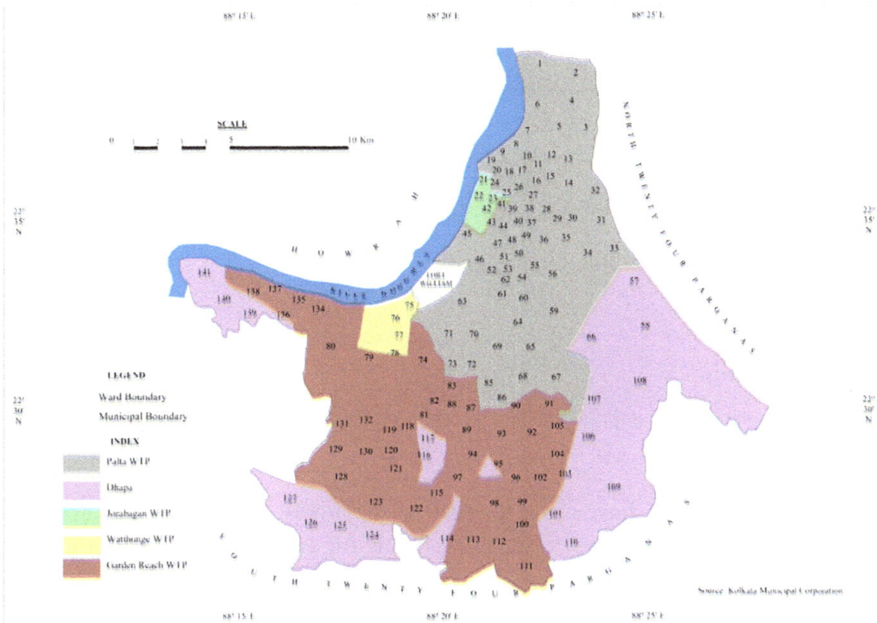

Fig. 13.2 Wards served by the water treatment plants in Kolkata

of drinking water and inadequate sanitation facilities. The aim was to assess the availability of water in terms of hours of supply and the adverse impacts of reduced pressure in the supply mains. Coping strategies adopted by the urban communities were also evaluated for an inclusive, resilient, and sustainable urban water system. The survey thus addressed the goals 6, 9, and 11 of the SDGs. The wastewater management system in the East Kolkata Wetlands was also explored.

With respect to the hours of supply, major parts of the city had fewer hours of supply than claimed by the civic authorities. While in the main city, the hours of municipal water supply ranges between 6 and 8 h; majority of the surveyed population complained that the duration of municipal supply was much less (Fig. 13.3).

Central Kolkata gets water supply three times a day, in the morning from 6.00 a.m. to 9.00 a.m., from 11.30 a.m. to 12.30 p.m. in the afternoon, and from 4.30 p.m. to 6.00 p.m. in the evening. In the peripheral wards, supply is only twice a day, once from 6.00 a.m. to 9.00 a.m., and again from 4.30 p.m. to 6.00 p.m. The coverage percentage is also quite low as is evident from Table 13.2, except for the core city where 100% of the area is covered by municipal supply. The supply is for 6–8 h for majority of the households even in the core area. However, the respondents highlighted that earlier pressure in the supply lines was adequate to enable lifting of water to personal overhead tanks during hours of supply so that tap water was available for 24 h. In an attempt to reduce leakage loss, pressure in the supply lines is now kept low, making it difficult to lift water to the overhead tanks during municipal supply. The peripheral

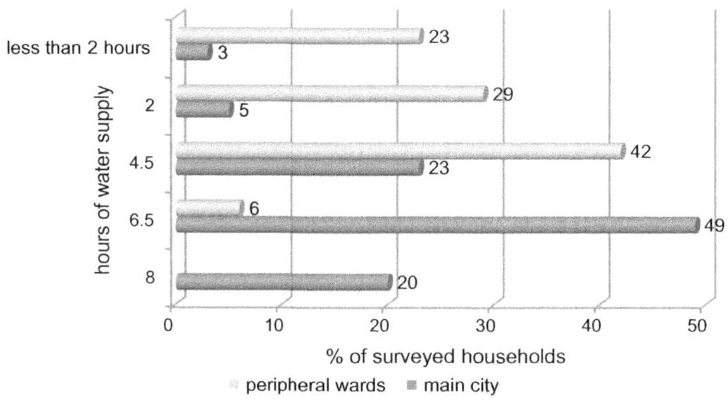

Fig. 13.3 Duration of water supply in main city and peripheral areas

wards, on the other hand, have fewer hours of water supply. For 42% of the surveyed households, the municipal supply is for 4.5 h while for another 29% the supply is for 2 h only. Hence, the residents are compelled to supplement municipal supply with informal sources like the unregistered watermen.

Urban communities have thus evolved their own water management strategy to supplement the inadequate municipal supply. Many of the households, particularly in the peripheral wards, have their own groundwater sources. The percentage of surveyed households dependent only on municipal supply is 89% in the main city or central Kolkata while it is only 4% in the peripheral wards. Even the Municipal Corporation supplies groundwater to households in the peripheral wards through a network of tube wells, reservoirs, and pipelines. It is, however, already noted that groundwater in parts of Kolkata has arsenic above the permissible limit. Hence, dependence on groundwater exposes the urban population to various health hazards. It was revealed during primary survey that 92% of the respondent households have never tested the quality of the groundwater they use for drinking. Many of the households are also compelled to buy water from the waterman who often sells the same municipal water, available at community taps and stand posts, at a price though municipal supply within the city is not priced (Fig. 13.4). In fact, 37% of the surveyed households supplement the municipal water supply with water bought from the waterman in the peripheral wards while it is only 5% in the main city. Hence, the civic authorities could explore the possibilities of water metering for resource mobilization. Forty-one percentage of the surveyed households in the peripheral wards supplement the municipal water supply with groundwater sources while another 18% of the households depend only on groundwater sources, in the absence of municipal supply (Fig. 13.4).

Community participation in water management is also evident among the slum population without household connection or without any access to piped water supply. The stand posts are an integral part of the urban water system in Kolkata and provide drinking water for the poor households and the slum population. Apart from

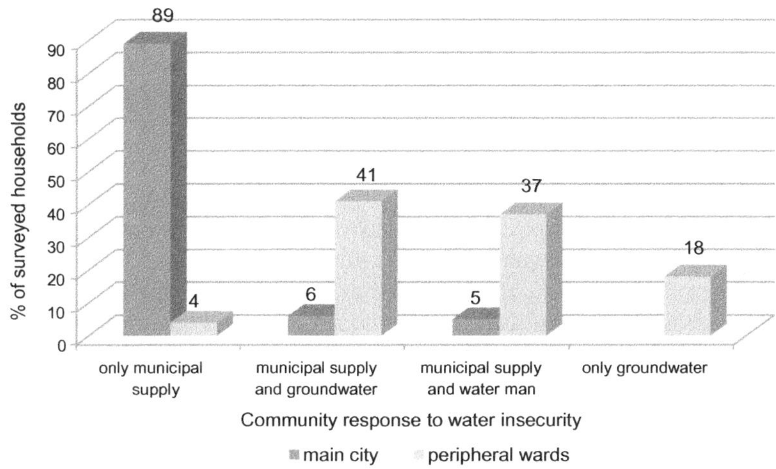

Fig. 13.4 Sources of municipal water supply in central and peripheral areas of Kolkata

supplying drinking water to the poor households in the city, they are also a source of economic sustenance for the poor slum population. Shelters near the stand posts double as vending spaces during the daytime. Tea stalls and street kitchens cluster around the stand posts and supply food to the daily commuters. Hence, these stand posts provide income opportunities to the poor households in the city. In spite of the high underutilization of the available water, immense wastage, and contamination of water due to backflow during hours of no supply, the stand posts are an indispensable component in the urban water system of Kolkata. Water is also often supplied to the poor slum households by the municipal corporation through mobile water tanks during hours of acute crisis.

It is thus evident that urban water supply is inadequate in the city of Kolkata and needs to be supplemented from alternate sources including groundwater sources and the waterman, particularly in areas not covered or inadequately covered by the municipal supply. The dual water system, with the municipal supply supplemented by groundwater or the waterman, also raises doubts about the quality of drinking water particularly in areas not covered or inadequately covered by municipal supply. No treatment facilities are available except for domestic filters, used by only 63% of the surveyed households. The old dilapidated network results in leakage loss, and the low pressure in the supply mains exposes the drinking water to contamination, including fecal contamination, during hours of no supply. Stand posts that are an integral part of the water supply system in Kolkata result in wastage as does the unmetered domestic supply. The absence of water meters and the consequent financial burden on the municipal corporation prevent innovative adaptations for a resilient urban water system.

13.6 Kolkata's Wastewater Management

Building on the experience of the MDGs, the Sustainable Development Goals 2015 emphasize improving the quality of water with reduced pollution, eliminating the dumping, and release of hazardous chemicals into natural water systems, reducing the proportion of untreated wastewater by 50% and promoting safe reuse and recycling. In spite of such efforts and global initiatives, the extent of wastewater treatment is extremely poor in low-income and lower-middle-income countries, suggesting an urgent need for implementing affordable solutions and promoting safe water reuse (United Nations Economic and Social Council 2016). Ecosystems are also effective in providing economical wastewater treatment services, at least as long as these ecosystems are healthy and the ecosystem's pollution carrying capacity is not exceeded by pollutant load. There are natural limits to the assimilative capacity of ecosystems, beyond which they are threatened, and they fail to perform a purifying role (United Nations Economic and Social Council 2016). Treated and partially treated wastewater when used for ecosystem management enables improved resource efficiency and benefits the environment through nutrient recycling, reduced water pollution, development of fisheries, reduced freshwater abstraction, and aquifer recharge (United Nations Economic and Social Council 2016).

Kolkata is fortunate in having a twin river system, with the River Hugli to bring in freshwater and the River Kulti to flush out its wastewater. The wastewater drains into River Kulti through a system of channels constructed in 1934 to carry dry weather and stormwater flow as well as sewage into the river. These channels continue to drain nearly 75% of the city's rainwater and sewage. The system was developed in the wetlands of East Kolkata. These wetlands continue to function as the city's wastewater treatment plant, treating almost one-third of its sewage. The East Kolkata Wetlands can receive and effectively treat 810 MLD of city sewage every day that act as nutrient for the fish and the farmlands. They also purify the water through oxidation and natural aeration in *bheris* or fish ponds and in the channels, nearly 2,000 km in length. The city thus got its wastewater treatment plant in the pisciculture system in local *bheris* that use wastewater of the city brought in by the network of channels. It is estimated that these *bheris* produce 8,000 t of fish annually. The naturally treated wastewater then flows into River Kulti, 28 km to the east of the city. To stop the filling up of the wetlands under the impact of rapid urbanization, the East Kolkata Wetlands were declared a Ramsar Site in 2002. The area is since designated as a no development zone.

The city's sewage system was first engineered by the colonial rulers in the late 1800s for a small city to discharge its waste into a tidal creek east of the city. The system now stands choked with sewage and garbage and hence is unable to handle very high-intensity rainfall exceeding 6 mm per h. Four small capacity sewage treatment plants (STP), namely Bangur STP (45 MLD), Garden Reach STP (45 MLD), Bagha Jatin STP (2 MLD), and South Suburban East STP (45 MLD), carry treated water into River Hugli but continue to be largely unconnected to the city sewage it has to treat.

The city thus has its unique natural system of wastewater treatment in the East Kolkata Wetlands. It is already proven that the planned use of wastewater can be beneficial to ecosystem functioning and benefits the wetlands as fisheries and aquatic ecosystems thrive better. The wetlands also help in recharging the fast depleting aquifers.

The wetlands in Kolkata have, however, been built upon, and silting continues to be a major problem in many wastewater conduits such as stormwater drains, sewers, and canals. The channels are choked with increased amount of garbage, and the sewer lines built during colonial period are in a state of decay. The Bagjola canal along Kolkata's northeast and the Krishnapur canal in the north are already choked. The blockage is up to 50–70%, while Tolly's Nallah in south Kolkata shows a blockage up to 40% (Biswas 1999). Besides, the gully pits are blocked because of poor solid waste management system resulting in a time lag for water to reach the pumping stations. The closing of the *bheris* under urban expansion and the destruction of the natural system of recycling of the sewage in them caused a chocking of the city with wastewater particularly in the monsoon season. Hence, both the channels and sewer lines need to be repaired and upgraded if necessary to carry the increasing amount of sewage as the city grows. The wetlands need to be preserved to ensure a sustainable wastewater management system for the colonial city of Kolkata.

13.7 Urban Water Management and Implications on Sustainable Development Goals

The city of Kolkata suffers from inequitable and inadequate supply of drinking water. Only 69% of the population in central Kolkata has access to municipal supply. In other parts of the city, the percentage is much lower. For the peripheral wards, it is below 20%, the lowest being 6% in the eastern peripheral wards. The hours of supply vary mostly between 4.5 and 6 h, while in many pockets, the daily supply is for 2 h only. The urban water system suffers from old dilapidated network, high leakage loss, risk of contamination during hours of no supply, and poor cost recovery. The peripheral wards continue to depend on groundwater with arsenic above the permissible limit. The slum population accounting for 32.4% of the city's total population is without access to individual connections and depends mostly on the stand posts. The supply in the stand posts along with unmetered supply in the city adds to wastage in the already limited municipal supply. Under the impacts of inadequate maintenance of the canals, lack of preparedness of the civic authorities and the high-intensity rainfall in short spells, the low-lying parts of Kolkata are under constant threat of waterlogging in the monsoon season.

From the above analysis, it is quite evident that water supply needs to be improved throughout the city of Kolkata to facilitate universal and equitable access to safe drinking water by 2030 in accordance with the SDGs. In an attempt to ensure availability and uniform access to water for all, the Kolkata Municipal Corporation is already working on increasing the number of booster pumping stations to improve supply.

It is also indicated in the SDGs that community participation needs to be supported and strengthened for improved water management and capacity building. The urban communities in Kolkata have already developed their own system of drinking water supply to supplement the inadequate municipal supply. These measures may be strengthened through certain adaptations involving the local communities. One such measure could be to incorporate the waterman in the formal water supply system and ensure that the supply by them is from the municipal stand posts and not from other sources. This could help in revenue generation through the introduction of proper licensing policies. It would, at the same time, ensure that the quality of water is not compromised with. Gradual phasing out of the individual or institutional groundwater sources may also be thought of to ensure sustainable withdrawals, in keeping with the targets specified in the SDGs. Such measures along with network improvement to reduce wastage and water metering for resource mobilization would ensure sustainable and resilient urban water management.

Protection of ecosystems and wastewater treatment, other major SGDs targets, are well taken care of by the unique natural system of wastewater treatment in the East Kolkata Wetlands. However, the sewer lines that carry the wastewater are in a state of decay and need to be repaired and upgraded to carry the increasing amount of sewage as the city grows and make the wastewater management system sustainable. Flow in the channels also needs to be restored. Pisciculture in the bheris of the East Kolkata Wetlands is a source of sustenance for the urban poor, particularly those residing next to the canals or bheris, thereby contributing to enhanced economic viability of the wetlands. Thus, the colonial city of Kolkata is naturally equipped with a wastewater treatment plant that not only takes care of the refuse from the city but also engages in economic upliftment of the urban poor through pisciculture.

For wastewater management, the Kolkata Municipal Corporation has upgraded its sewerage and stormwater drainage system by building new brick sewer systems, installing mechanical sewer machines, constructing new pumping stations, cleaning and dredging of canals, and protecting the wetlands. Garbage collection at household level would also help reduce the dumping of garbage at gully pits. Plastics are often responsible for the choking of the gully pits and sewer lines. The controlled use of plastic would help keep the gully pits clean.

13.8 Conclusion

The existing urban water system in the city of Kolkata is far from adequate. It is thus important to channelize efforts to strengthen and improve the existing water system to make it sustainable and resilient. Community participation is already evident in improving the availability of drinking water, particularly in the peripheral wards where a parallel informal water supply network exists. The informal supply network if included within the formal water supply system could help in removing the gaps in municipal supply, both spatial as well as across different socioeconomic groups and ensure equitable access. Since the parallel water supply system comes with a cost, water meters may also be introduced at household level. Charges could be made proportional to the quantity of water consumed above a minimum of 100 l. It would not only help in resource mobilization for the civic authorities and better maintenance of the dilapidated supply network, but would also help in reducing wastage substantially. Replacement of old worn-out pipelines is highly recommended to reduce leakage in the supply system. Demand management options also need to be explored. The stand posts may be fitted with a knob to regulate the supply of water and minimize the wastage of water. The ownership right on the stand posts may be bestowed upon the local communities against a nominal charge so as to ensure proper functioning of the regulators. While the phasing out of individual groundwater sources is recommended, the existing system of municipal supply using groundwater may be maintained for emergency use. To improve resilience in the urban water system, rainwater harvesting and recharging of groundwater are recommended. Such measures may be made mandatory, at least at the institutional level. Innovations in urban water system would also involve wastewater treatment and reuse, use of gray water, rainwater harvesting, and community participation. The possibility of public–private partnership in the water sector also needs to be explored in accordance with the National Water Policy, 2002. Dredging and desilting the wastewater channels as well as the protection of the East Kolkata Wetlands would, on the other hand, improve efficiency in the system of wastewater treatment in the city. The system of treating sewage through the older fish farming techniques also needs to be enhanced. Coordination between the various departments responsible for urban water management needs to be strengthened for better implementation of project initiatives.

The city administration (Kolkata Municipal Corporation) has already adopted many of the suggested interventions to cope with water security in the megacity of Kolkata. Seven newly constructed booster pumping stations, when fully operational, are expected to augment supply and coverage. Two others are proposed. In the strategic plan for 2025, Kolkata Municipal Corporation envisages to phase out the use of groundwater for consumption and provide surface water supply for the entire city by 2025. For better resource mobilization, the possibility of introducing water meters is also under consideration.

References

Asian Development Bank (2007) Benchmarking and data book of water utilities in India. Asian Development Bank, Metro Manila, Philippines

Biswas AK (1999) Water crisis—current perceptions and future realities. Water Int 24:363–367

Cain NL, Gleick PH (2005) The global water crisis. Issues Sci Technol 21:79–81

Calcutta Environmental Management Strategy and Action Plan (1995) Working paper 1. Situation report (revised version November 1995). Department of Environment, Government of West Bengal, Kolkata

Census of India (2011) Final population tables. Government of India, New Delhi

Cosgrove WJ, Rjisberman FR (2000) World water vision: making water everybody's business. Earthscan Publications, London
Dasgupta B (1991) Calcutta's urban future: agonies from the past and prospects for the future. South Asia Books, New Delhi
Gleick PH (1999) The human right to water. Water Policy 1:487–503
Gleick PH (2003) Global freshwater resources: soft-path solutions for the 21st century. Science 302:1524–1528
Majumdar C, Gupta G (2007) The debate over municipal water pricing: evidence from Kolkata, India. Water Resour Dev 23(4):571–582
Mathur OP, Sengupta P, Bhaduri A (2000) Option for closing the Revenue Gap of Municipalities, 2000/01 to 2004/05. National Institute of Public Finance and Policy, New Delhi
McKenzie D, Ray L (2009) Urban water supply in India: status, reform options and possible lessons. Water Policy 11(4):442–460
Ministry of Water Resources (2002) National Water Policy. Ministry of Water Resources, Government of India, New Delhi
National Commission for Integrated Water Resource Development (1999) Integrated Water Resource Development: A Plan for Action, Volume-I. Ministry of Water Resources, Government of India
National Institute of Urban Affairs (1997) Financing urban infrastructure in India. National Institute of Urban Affairs, New Delhi
Planning Commission (2012) Twelfth Five-Year Plan (2012–2017) Volume II: sectoral policies and programmes. Government of India, New Delhi
Sivaramakrishnan KC (1993) Managing urban environment in India. Times Research Foundation, Delhi
Ray B, Shaw R (2016) Water stress in the megacity of Kolkata, India and its implications for urban resilience. In: Shaw R, Rahman A, Surjan A, Parvin GA (eds) Urban disasters and resilience in Asia, 1st edn. Elsevier, United Kingdom
The Telegraph, April 14, 2003, Kolkata
United Nations Human Settlements Programme (2011) Cities and climate change: global report on human settlements. Earthscan, London
United Nations Economic and Social Council (2016) Progress towards the Sustainable Development Goals: 2016 session. New York, 24–27 July 2016
Vo PL (2007) Urbanization and water management in Ho Chi Minh City, Vietnam-issues, challenges and perspectives. GeoJournal 70(1):75–89

Chapter 14
Urban Drought Management Through Water Conservation: Issues, Challenges and Solutions

Kapil Gupta and Vinay Nikam

Abstract Urban water conservation efforts have begun in times of water shortage due to drought or other calamities. Increasingly many cities in India are now drawing water through large diameter water pipelines by pumping from water sources further and further away from the city. The rate of augmentation works has been overtaken by the rapid increase in population in urban areas due to migration. Moreover, new supplies are now becoming increasingly difficult to find. Thus, to cater to the increasing urban population, demand reduction through water conservation as a long-term measure offers a way out of the present crisis. The poor living in slums are already managing with limited water and so the population having access to piped water and bathrooms are the target group for urban water conservation measures. The various water conservation measures, viz: water conservation devices, leak detection and repair, water reuse, metering and incremental rates are briefly described and discussed. Two applications of rainwater harvesting and wastewater reuse in Thane city in Mumbai Metropolitan Region have been described as an outcome of drought/water shortage.

Keywords Urban · Drought · Conservation · Reuse · Rainwater harvesting

14.1 Introduction

Rapid urbanization in India has resulted in increased demand for water in urban areas. However, the very process of urbanization has resulted in rapid paving of urban surfaces, thereby impeding groundwater recharge and depleting the groundwater table. The urban water managers can no longer depend on the groundwater sources, and they are forced to look for alternate water sources. Increasingly many cities in

K. Gupta (✉)
Indian Institute of Technology Bombay, Mumbai, India
e-mail: kgupta@iitb.ac.in

V. Nikam
EnviroTech Consultants, Thane, India

B. Ray and R. Shaw (eds.), *Urban Drought*, Disaster Risk Reduction,
https://doi.org/10.1007/978-981-10-8947-3_14

India are now drawing water through large diameter water pipelines by pumping from water sources further and further away from the city. This process has its own challenges and drawbacks—as rainfall occurs in India mostly from June to September with very little rainfall occurring in the remaining eight months, most cities are struggling for water from March to June. The cities now have to conserve water during the four months preceding the monsoon. This chapter looks at the existing and potential water conservation measures to alleviate urban drought with examples from the Mumbai Metropolitan Region.

14.2 Need for Water Conservation as a Long-Term Strategy

Most urban water conservation efforts have begun in times of water shortage due to drought or other calamities. These efforts were ad hoc measures aimed at saving water on a short-term basis. While these were successful to varying degrees in reducing water use substantially, water use increased again after the crisis was over.

The rate of augmentation works has been overtaken by the rapid increase in population in urban areas due to migration. Moreover, new supplies are now becoming increasingly difficult to find. Thus, to cater to the increasing urban population, demand reduction through water conservation as a long-term measure offers a way out of the present crisis.

14.3 Water Consumption Per Capita

In the 1990s, the Central Public Health and Environmental Engineering Organization (CPHEEO) under the Ministry of Urban Development had recommended 135 litres per capita per day (lpcd) with the breakup as shown in Table 14.1. However, this figure dates back to the time when toilet flushing tanks were designed using 12.5 litres per flush. Many of the old buildings still use the older flush tanks which use as much as 12.5 litres per flush. Nowadays, low-volume flush tanks (3.5 litres per flush or less) with the same flushing efficiency are available. Many utilities are now even shifting towards waterless toilets. Retrofitting the buildings with such devices and installing the latest water conservation devices in the new buildings offer the safest and surest way of reducing water consumption without any loss of flushing efficiency or causing any inconvenience or change in the lifestyle of the individual. This has also been recognized in a study carried out by the Ministry of Environment, Forest and Climate Change in which they reworked the water demand for individual use as shown in column 2 of Table 14.1 and identified the total as 86 lpcd. However, during the water crisis in Mumbai during April 2016, a water consumption survey was carried out by the authors' amongst 150 students at IIT Bombay, and the conservative water use for various purposes is shown in column 5 of Table 14.1. It is seen that even after including an additional 10% for contingency, the requirement works out to 62 lpcd.

Table 14.1 Per capita water consumption

S. No.	Category	Consumption (in lpcd)		
		CPHEEO	MoEFCC	IIT Bombay
(1)	(2)	(3)	(4)	(5)
1	Human consumption[a]	7	7	5
2	Bathing	20	20	20
3	Flushing	45	21	10
4	Washing	40	15	10
5	Misc[b]	23	23	11 + 10%
	Total	135	86	62

[a]Drinking and cooking; [b]contingency

Thus, a saving of just over 50% from the estimates prepared by CPHEEO can be achieved through improved practices without any inconvenience to the individual.

However, in emergencies, research indicates that 20 lpcd is the minimum quantity of safe water required to realize the minimum essential levels of health and hygiene (WHO-WEDC 2013).

14.4 Water Conservation Measures

The poor living in slums are already managing with limited water and so the population having access to piped water and bathrooms are the target group for urban water conservation measures. The various water conservation measures are briefly described and discussed.

14.4.1 Water Conservation Devices

The maximum saving in urban water use would be achieved by retrofitting the toilets with low-volume toilet flushing cisterns as already discussed. Other measures would include pressure-reducing valves and faucet aerators so that less amount of water is consumed per wash.

14.4.2 Leak Detection and Repair

The most common leakages in buildings are from partially closed taps, toilet cisterns or float operated self-closing valves malfunctioning due to a defective float or the float coming loose in the water storage tanks. A slow steady drip from a tap can waste around 45 lpcd of water while a leaking cistern can waste 45 litres per hour. A preventive annual replacement will go a long way in reducing leakages and prevent wastage of valuable water.

14.4.3 Water Reuse

Water recycling and reuse are now being increasingly adopted by large utilities for non-potable purposes. For example, many airports now treat the wastewater and reuse the treated water for non-potable purposes. Similar measures are being adopted by railways and roadways for washing the coaches.

14.4.4 Metering and Incremental Rates

Demand reduction can be adopted through the installation of water metres and appropriate pricing policy, incentives and penalties. It should be mentioned that in most cities, water rates are highly subsidized and water rates should be increased rationally towards a price that reflects the real cost of water and to make consumers lower their water use to the desired level. Changing from a flat rate schedule to an increasing block rate requires metering if it does not already exist. Metering can be expensive in terms of the cost of the metre and the cost of reading and billing. Even though it takes time for metres to save enough water to offset their high initial cost, metering with incremental pricing is a cost-effective water conservation measure.

14.5 Modifications in Bye-laws for Reuse of Wastewater

Application of bye-laws for recycling and reuse of wastewater can have a significant role in reducing dependence on municipal water supply.

For example, to meet flushing and gardening water requirements, the Thane Municipal Corporation has approved the following modifications in bye-laws for recycling and reuse of wastewater from 30 July 2016:

Types of wastewater are defined as

1. Black wastewater: wastewater from toilets,

2. Grey wastewater: wastewater from sinks, bathrooms,
3. Non-residential wastewater: wastewater from commercial, medical and industrial.

14.5.1 Applicability on Case-by-Case Basis

Modified bye-laws in the Mumbai Metropolitan Region mandate that in each of the following cases, a wastewater treatment plant should be provided and the treated water should be reused.

(i) Construction sites having an area exceeding 4000 m^2;
(ii) Group housing with area exceeding 2000 m^2 or building having more than 50 tenements or with water consumption exceeding 20,000 lit/day;
(iii) Government, semi-government organization, lodging, hotels, commercial, educational, industrial buildings having built-up area beyond 1500 m^2 and if water consumption over and above 20,000 lit/day;
(iv) All hospitals with more than 40 beds; and
(v) Garages servicing vehicles.

Suitable fines and penalties would be levied at the prevailing rates in case of violations.

14.6 Implementation

Establishing a programme for long-term urban water conservation measures requires a different, more carefully planning approach than setting up a temporary emergency conservation programme. Urban bodies can choose from a variety of water conservation options including water saving devices, water restrictions, water reuse, landscaping charges and incremental water rates. Implementation can be successful only when there is a firm commitment by the people themselves, the urban bodies and the elected representatives to conserve this valuable and limited natural resource.

14.7 Case Studies

Two case studies from Thane city incorporating urban water conservation measures are presented (Fig. 14.1). Case I demonstrates micro-level planning to optimize the use of rainwater and reduce the peak flow while Case II describes the measures initiated by the municipal authorities to recycle treated wastewater for gardening and flushing.

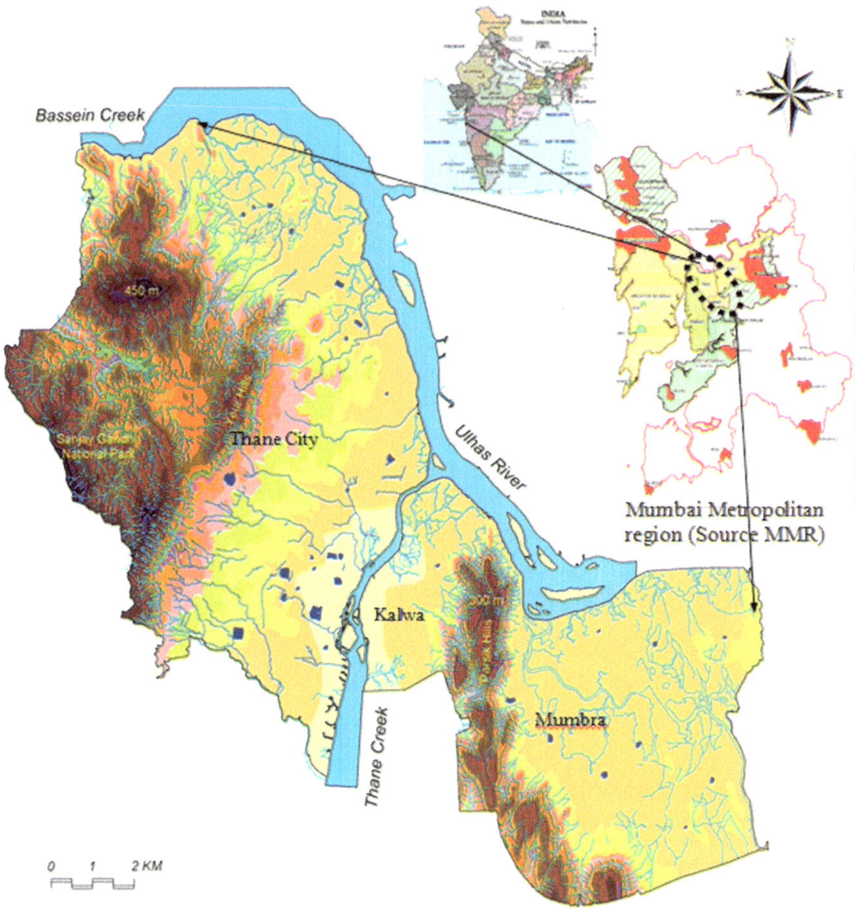

Fig. 14.1 Location of Thane (CDP 2006)

14.7.1 Case I: Micro-Level Planning to Optimize the Use of Rainwater and Reduce the Peak Flow

14.7.1.1 Rainwater Harvesting

The responsibility of developing and implementing urban runoff and flood control strategies is on the local body according to the BPMC Act (Bombay State Provincial Municipal Corporation 1949). The Thane Municipal Corporation (TMC) has taken up integrated urban design with the goal of developing an all-inclusive rainwater harvesting programme aimed at providing water, reducing contributions from building to drains and thereby reducing water logging and flooding in Thane. The development plans involve a proactive process which recognizes the opportunities for urban water conservation, landscape architecture and stormwater management infrastructure are intrinsically linked. Rainwater harvesting has been made mandatory.

14.7.1.2 Rainwater Harvesting System Flow Path

This section describes the flow path of rainwater falling on a building within the catchment area and the processes involved. The rainwater schematic has been shown in Figs. 14.2 and 14.3.

The rainwater reuse and recharge system has been designed for daily rainfall for 100 mm. The area breakup and runoff generated is shown in Fig. 14.4 while the mass balance of stormwater has been shown in Fig. 14.5. Calculations show that 185.75 cum/day runoff is generated and can be conserved from an area of 2200 m^2.

14.7.1.3 Rainwater Harvesting Components

The following are the rainwater harvesting components proposed to be installed in the premises:

Flush Valve

Flush valve has been proposed to flush first rainwater to prevent silt and other impurities coming into rainwater harvesting system.

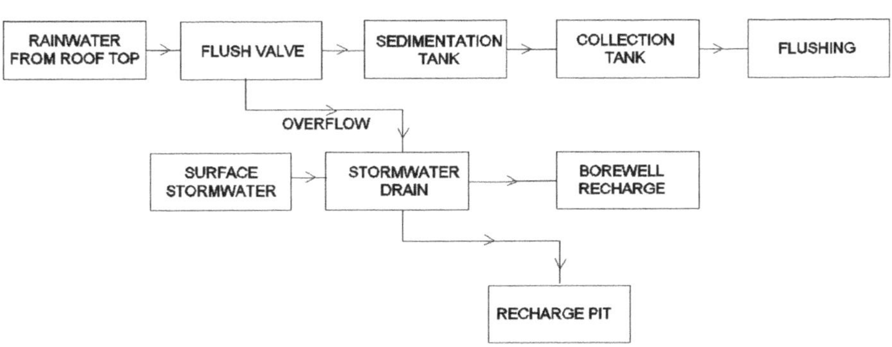

Fig. 14.2 Rainwater harvesting system schematic

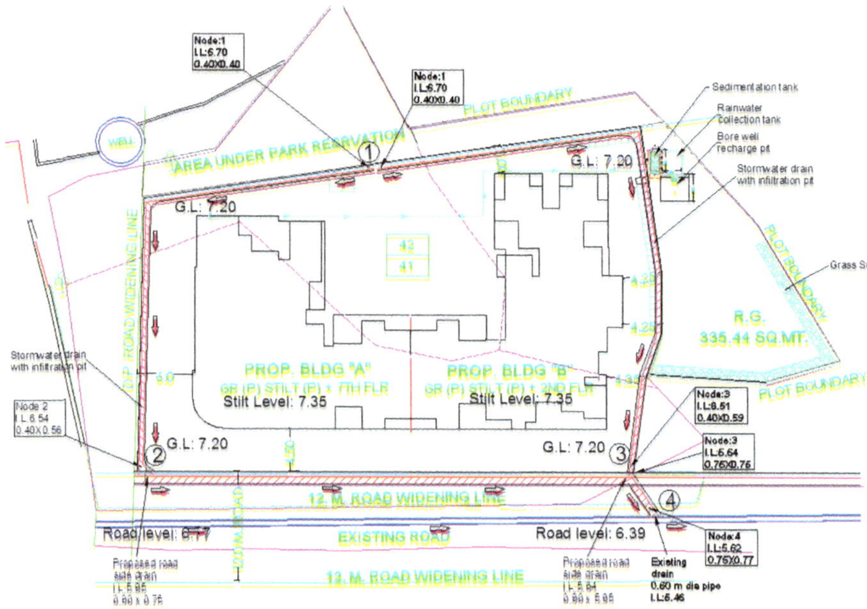

Fig. 14.3 Representative schematic of sedimentation tank and collection tank provided at a residential premise in Thane

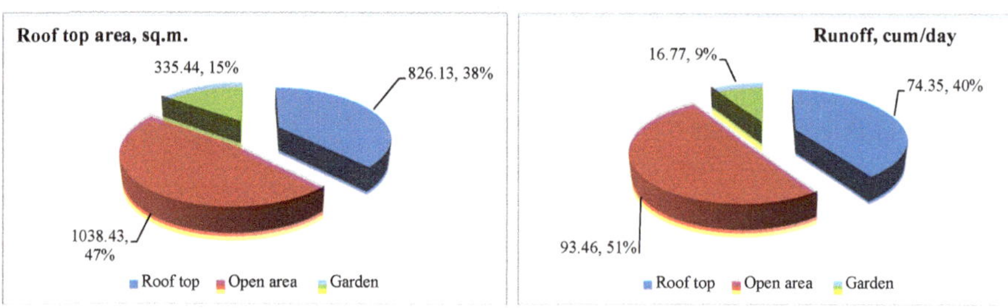

Fig. 14.4 Area breakup and runoff generated

Sedimentation Tank

Rainfall falling within the rooftop catchment area is directed towards a single downspout and collected through drain. The water is collected in a sedimentation tank (Fig. 14.6). Fines are trapped in first two compartments of sedimentation tank.

Collection Tank

Rainwater from sedimentation tank is collected in collection tank to reuse for flushing.

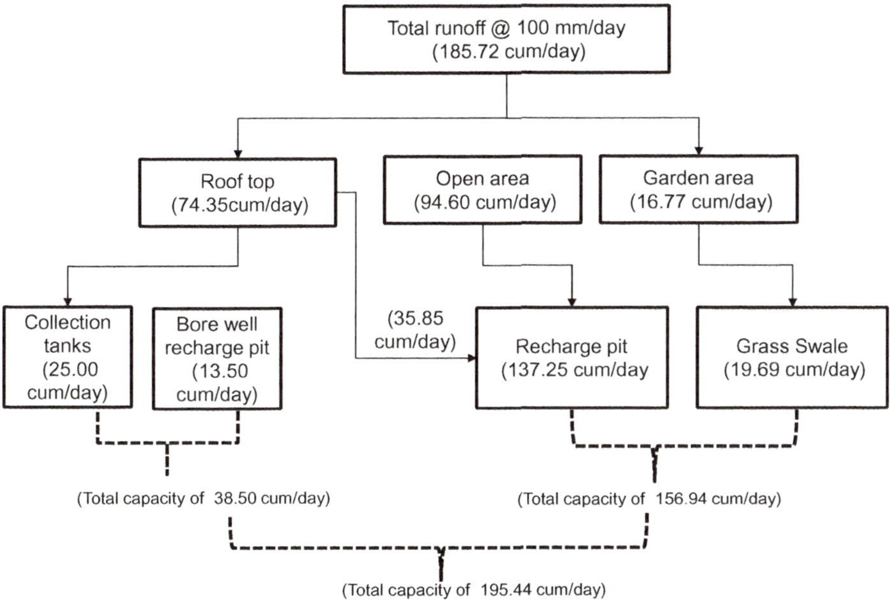

Fig. 14.5 Stormwater mass balance

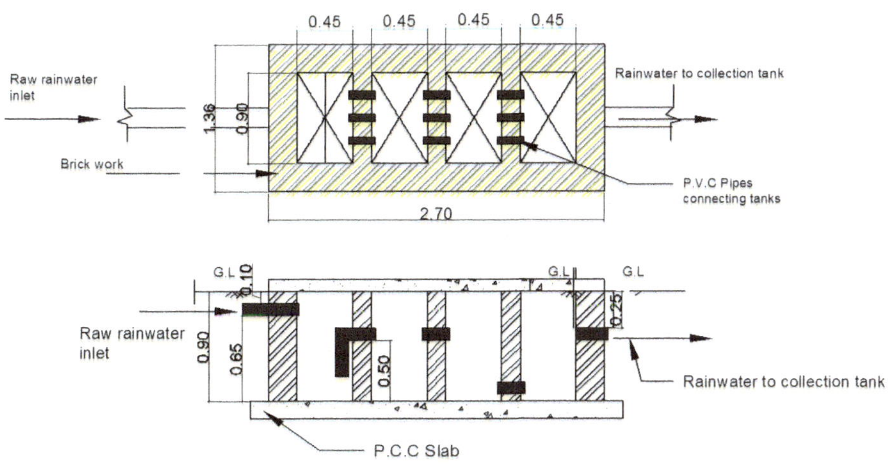

Fig. 14.6 Sedimentation tank (all dimensions in m)

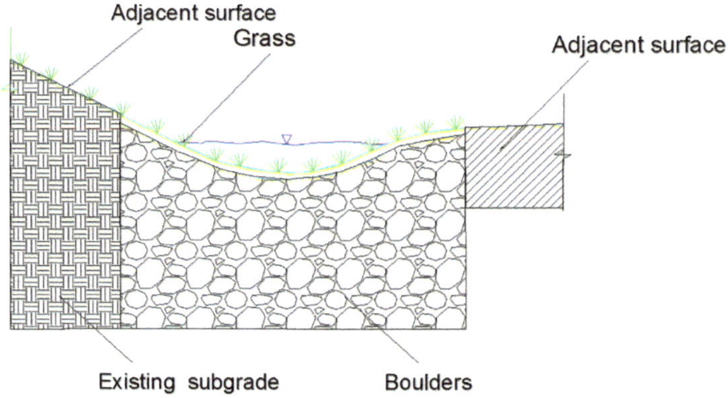

Fig. 14.7 Typical details gravel swale

Fig. 14.8 Stormwater drain cum infiltration pit

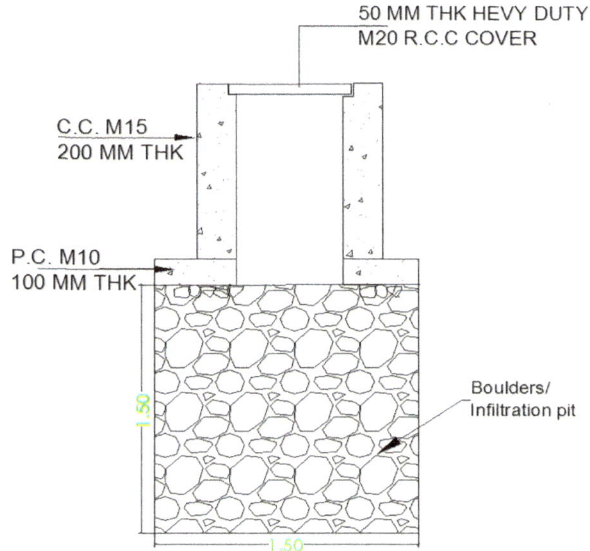

Surface Rainwater Recharge Though Grass Swale and Infiltration Pit Within Stormwater Drain

Surface rainwater from garden area is used for recharging groundwater table though grass swale and infiltration pit within stormwater as shown in Figs. 14.7 and 14.8, respectively.

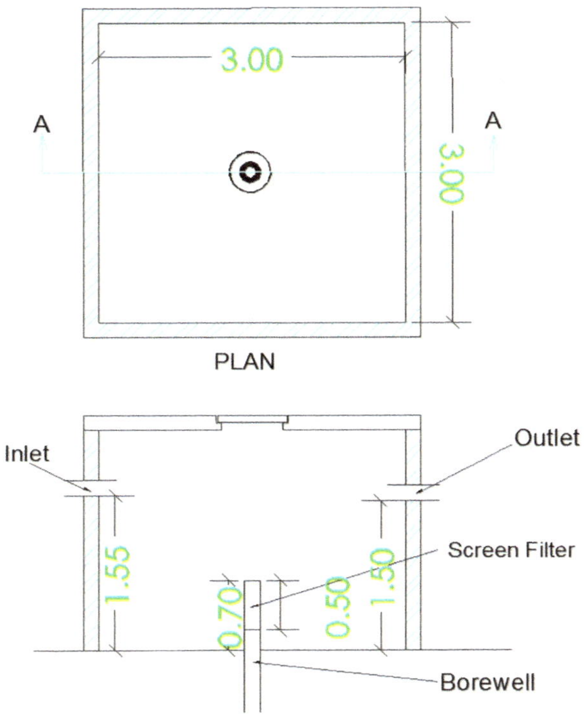

Fig. 14.9 Details of borewell recharge pit

Borewell Recharge Pit

Excess rainwater from collection tank is collected in borewell recharge pit for recharging groundwater table as shown in Fig. 14.9 and the details are summarised in Table 14.2.

14.7.2 Case II: Recycling Treated Wastewater for Gardening and Flushing

Assessment of the modifications in the Development Control Plan (DCR) on water demand and wastewater disposal has been carried out for sector 6, which is one of the rapidly developing sectors of Thane. The index map of sector 6 is shown in Fig. 14.10. The area under residential zone is 477.34 ha which is 17.90% of total residential zone of Thane. The broad categorization shows that 40.18 ha (8%) is rural, 160.14 ha (34%) is developed, ongoing development is of 78.15 ha (16%), whereas 198.89 ha (42%) is yet to be developed as depicted in Fig. 14.11.

Table 14.2 Design of rainwater harvesting

Rooftop		
Rooftop area	826.13	m^2
Coefficient of runoff	0.9	
Rainfall per day	100.0	mm
Rainwater collection	74.35	m^3/day
Open area		
Area	1051.10	m^2
Coefficient of runoff	0.90	
Rainfall per day	100.0	mm
Rainwater collection	94.60	m^3/day
Garden area		
Area	335.44	m^2
Coefficient of runoff	0.50	
Rainfall per day	100.0	mm
Rainwater collection	16.77	m^3/day
Water conservation from rooftop in a collection tank per year		
Rainwater collection tank	10	m^3/day
Flushing tank	15	m^3/day
Total capacity of tank proposed	25	m^3/day
Design of borewell recharge pit		
Number of pits	1	
Length	3.00	m
Width	3.00	m
Depth	1.50	m
Total collection capacity	13.5	m^3/day
Design of recharge pit for surface stormwater		
Number of pits	1	
Length	122.00	m
Width	1.50	m
Depth	1.50	m
Total collection capacity	137.25	m^3/day
Design of grass swale		
Number of pits	1	
Length	35.00	m
Width	1.50	m
Depth	0.75	m
Total collection capacity	19.69	m^3/day
Total rainwater generated	185.72	m^3/day
Total rainwater conservation capacity	195.44	m^3/day

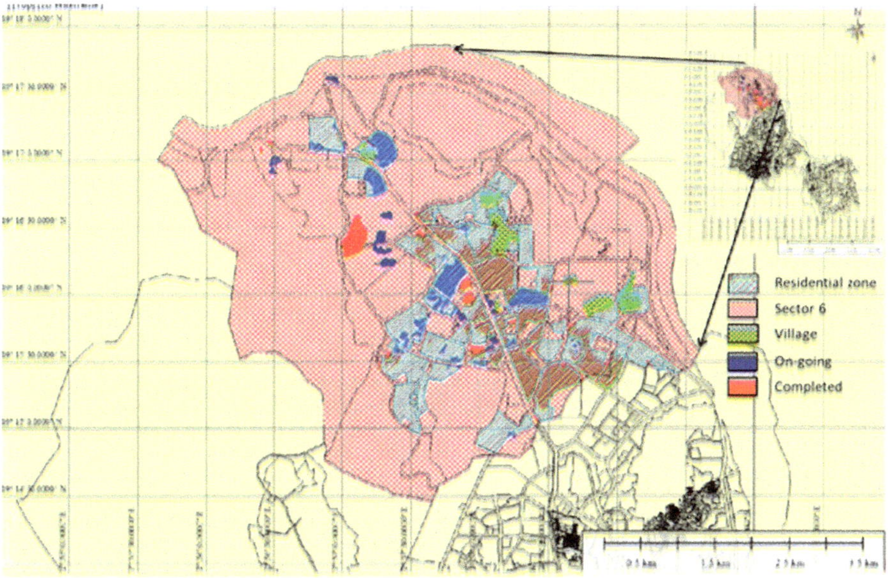

Fig. 14.10 Planning sector 6 of Thane Municipal Corporation

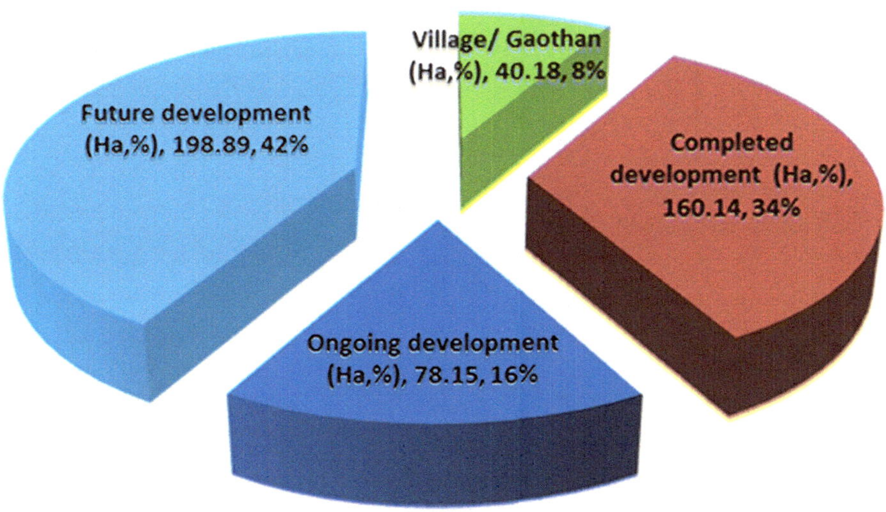

Fig. 14.11 Breakup of development in sector 6 of Thane Municipal Corporation

The wastewater generation (grey and black) has been computed for different categories of development and given in Table 14.3. After implementation of modified bye-laws, 13.96 Mld grey water will be treated and reused for gardening and flushing.

Table 14.3 Wastewater generation (grey and black) for different categories of development

Category	Wastewater (Mld)		
	Black	Grey	Total
Rural	1.23	2.02	3.25
Completed development	4.90	8.07	12.97
Ongoing development	2.39	3.94	6.33
Future development	6.09	10.02	16.11
Total ongoing and future development	8.48	13.96	22.44

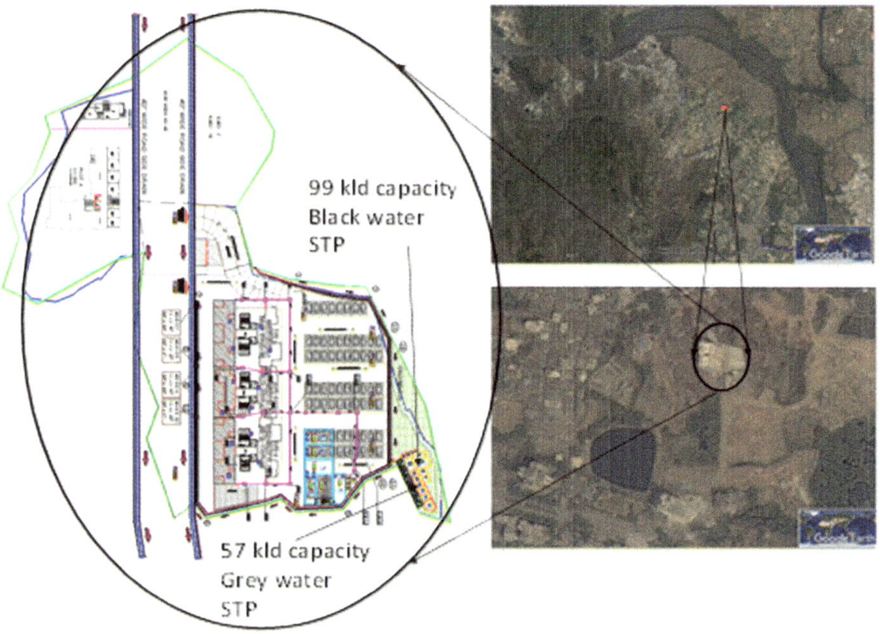

Fig. 14.12 Separate grey and black wastewater treatment systems for residential project in Thane (*Aerial view source* Google Earth)

14.7.2.1 Application of Amended DCR for Residential Project

Separate grey and black wastewater treatment systems have been implemented for one project at Kasarvadavali located in sector 6 in Thane. The entire project is proposed to be developed in two phases. Phase I comprises of three buildings. A separate grey water treatment system has been proposed for each phase. Separate grey and black wastewater treatment systems of 57 and 99 kld capacity, respectively, for Phase

I as shown in Fig. 14.12 have been completed. Treated wastewater is proposed to be used for gardening and flushing.

14.8 Conclusions

This chapter has discussed various methods of water conservation which may be implemented without affecting the present lifestyle, mainly by retrofitting efficient plumbing fixtures in old buildings and installing better ones in the new buildings. Two applications of rainwater harvesting and wastewater reuse in Thane city in Mumbai Metropolitan Region have been described.

References

City Development Plan (CDP) (2006) Thane Municipal Corporation, Thane

WHO-WEDC (2013) Technical note on drinking water, sanitation and hygiene in emergencies—how much water is needed in emergencies

Chapter 15
Urban Water and Health Issues in Hong Kong

Emily Ying Yang Chan and Janice Ying-en Ho

Abstract A basic necessity of health is the provision of a safe, harmless, reliable, accessible and affordable water supply. In cities, community well-being and public health protection rely heavily on the integrity and reliability of lifeline infrastructure that supports water supply and sewage systems. Hong Kong is a city in south-eastern China, which hosts seven million urban inhabitants. It is characterized by being one of the most densely populated vertical cities globally with the highest number of high-rise buildings that exceed 150 m. Public health and community well-being are extremely sensitive to water availability in this city. Although the city has its own rainfall catchment and reservoir system, 80% of its lifeline water supply is imported from Dongjiang (the East River) of Guangdong Province of mainland China via a dedicated aqueduct. Any industrial accidents that might lead to river pollution in its neighbourhood community and breakdowns of lifeline water-related infrastructure (water, sewage pipes, and electricity which drives all the water pumps in Hong Kong's 300 plus ultra-high-rise buildings) would bring major urban public health crisis. In addition, as a dengue fever-prone coastal metropolis, stagnant water management and water quality of its beaches, marine and rivers are all significant to human health. This chapter discusses and examines the health risks associated with water resources management (drinking water supply, sewage management and environmental water issues). Three cases (water pipe lead pollution crisis in 2015, the potential impact of climate change on rainfall and vector-borne disease, and the disaster risks asso-

E. Y. Y. Chan (✉) · J. Y. Ho
Division of Global Health and Humanitarian Medicine, Jockey Club School of Public Health and Primary Care, Chinese University of Hong Kong, Hong Kong, China
e-mail: emily.chan@cuhk.edu.hk

E. Y. Y. Chan
International Centre of Excellence for Collaborating Centre for Oxford University and CUHK for Disaster and Medical Humanitarian Response (ICoE-CCOUC), Integrated Research on Disaster Risk (IRDR), Hong Kong, China

E. Y. Y. Chan
WHO Thematic Platform for Health Emergency & Disaster Risk Management Research Network, World Health Organization, Geneva, Switzerland

B. Ray and R. Shaw (eds.), *Urban Drought*, Disaster Risk Reduction,
https://doi.org/10.1007/978-981-10-8947-3_15

ciated with the current water drainage infrastructure) are included to illustrate the vulnerability and urban resilience towards water issues in this urban context.

Keywords Urban health · Health risks and impact · Water supply
Sewage management · Environmental water issues · Hong Kong · China

15.1 Introduction

Water is essential for sustaining life and providing good health in many aspects. Ranging from drinking water to water for personal hygiene and food preparation, and the stagnant puddles of water outdoors; all of these water-related issues can have an effect on human health. Globally, water stress and scarcity are already a reality for more than 2 billion people, and this number will continue to increase due to growing populations and increasing water usage (UN-Water, n.d.). It will be a challenge to conserve and manage water resources in urban areas to maintain the health of its residents.

The earliest civilizations were built in locations where there was access to water bodies. The most long-lasting cities flourished with a stable access to basic needs to meet the basic requirements for health, namely clean water and sanitation, food and nutrition, shelter and clothing, health services, and information (Bolton 2006). Although many major cities are still located along rivers, people have changed from direct usage of the water body to a system-based dependency (Shaw and Thaitakoo 2010) as urban populations have increased and technology has improved.

Asia, in particular, has seen a rapid increase in urbanization in the last fifty years. About 53% of global urban population and 16 mega-cities were located in Asia in 2014 (United Nations, Department of Economic and Social Affairs, Population Division (UNDESA) 2014). By 2050, the urban population in Asia is projected to see a 61% increase, with 404 million and 292 million urban residents added to India and China, respectively (UNDESA 2014). These urban areas in Asia require the same stable access to basic needs as did ancient cities, while the demand is in much larger quantities. However, many of these areas are already water stressed. A 2016 study by MIT projected that by 2050, there is a high chance that densely populated watersheds in Asia will be experiencing *severe* water stress (Fant et al. 2016). Thus, water is now a more critical issue for this region than ever.

"WASH", the acronym for water sanitation and hygiene, is a public health approach to disease prevention that emphasizes the access to safe water supply, adequate sanitation services and better hygiene practices (World Health Organization (WHO), n.d.). In urban areas, the water issues of main concern include not only the drinking water supply, but also the drainage and sewage systems, and water prevalent in the environment. Water sources can come from rainfall, rivers, lakes, man-made reservoirs, groundwater, seawater or even importation from neighbouring regions. This water is treated by centralised infrastructure to meet established health standards and then delivered into people's homes. This chapter will take a look at water from the public health framework and address the health issues related to drinking water supply, wastewater in sewage systems and water in the environment through the lens of the city of Hong Kong.

15.2 Water Supply Through the Public Health Framework

A water supply must be harmless, adequate, reliable, accessible and affordable. As adapted from World Health Organization Guidelines for Drinking-Water Quality (2011a, p. 83), the public health framework for assessing drinking water supply includes the following aspects and definitions:

Quality: the presence of regularly verified quality of water supply that is free of hazardous substances harmful to human health;
Quantity: the amount of water used to ensure hygiene practice is not compromised, measured by service level;
Reliability: the percentage of time in a given period during which drinking water is available;
Accessibility: the percentage of population with ease of acquiring quality drinking water supply; and
Affordability: the cost of the water measured by the tariff paid by domestic consumers.

Using Hong Kong as an example of urban water context, these drinking water supply parameters will be described and assessed.

15.2.1 Hong Kong and Its Water Supply History

Hong Kong is a coastal city located in south-eastern China with a population of over seven million people. By 2043, the city population is projected to increase to 8.22 million (Census and Statistics Department 2015). It is one of the most densely populated places globally, with an average density of 6,777 persons/km^2 as of 2016, and up to 57,530 persons/km^2 in the Kwun Tong District (Census and Statistics Department 2017). Hong Kong has also been ranked top in the world for the highest number of high-rise buildings exceeding 150 metres (The Skyscraper Center 2017). The dense population and vertical built environment produce a unique condensed environment demanding substantial amount of water resources.

Up until the early 1900s, the colonial city of Hong Kong faced chronic water scarcity problems. Reservoirs and catchment zones were developed in the late 1800s to 1970s for rainfall collection, creating up to a total storage capacity of 586 million cubic metres from 17 reservoirs, several of which were ground-breaking engineering projects (Water Supplies Department (WSD) 2016a). The city also adopted the use of seawater for toilet flushing during this time, becoming the first and the only city to comprehensively use seawater in this way (Liu 2013). However, these were not enough water to sustain the population growth and industrial development, especially during the successive drought periods in the 1920s and 1960s. A stable long-term water source was needed.

In 1960, the Dongjiang-Shenzhen Water Supply Project began as a longer-term solution for Hong Kong's water supply. The Project pumps water from Dongjiang River, 80 km north of Hong Kong, to the Hong Kong border, where it is received by the Muk Wu Pumping Station, before being treated and distributed to the rest of the territory (WSD 2016c). Successive agreements with Guangdong between 1960 and 1990 financed the Project's expansion to the current full operational capacity. During that time, reliance on the Dongjiang water quickly rose to the current 70–80% of Hong Kong's consumption.

The agreements with Guangdong were initially made on a fixed quantity supply of water and were often found failing to meet water quality standards. Overestimation of annual water consumption led to excessive imports of Dongjiang water every year, which were stored in the local reservoirs. The unused water coupled with large amounts of rainfall led to the overflow of reservoirs, great water wastage and huge financial losses of up to HKD1.718 billion in the years 1994–1998 alone (Liu 2013). Following the handover of Hong Kong to China in 1997, agreements on the Dongjiang-Shenzhen Water Supply Project have continued, with negotiations over the water quality and oversupply of water. It was not until 2003, when the building of sealed pipelines under the DongShen Renovation Project was completed, that the water quality improved drastically (WSD 2017). Later, in 2006, negotiations successfully enabled a "flexible" quantity of supply agreement. The provisions of this agreement established the annual water supply amount and price, but enable adjustment of daily water supply according to the rainfall and reservoir conditions. Thus, between 2006 and 2015, the annual supply quantity allotted in the Dongjiang water supply agreement remained at a constant 820 million cubic metres (Director of Bureau: Secretary for Development 2017, p 224). However, the actual quantity imported averaged 702 million cubic metres annually, only reaching close to the allotted amount in the 2011 drought year (Director of Bureau: Secretary for Development 2017, p. 224).

15.2.2 Water Quality

Setting water quality standards and surveillance of the water supply system are essential for identification of potential health risks and disease prevention. These standards need to take into consideration current associated disease patterns in the community population, as well as exposure risks from not only drinking water, but also a wider range of water sources (WHO 2011a, p. 10). This enables a cost-effective allocation of resources to address public health risks (WHO 2011a, p. 36). Drinking water quality is assessed based on microbial, chemical and radiological hazards. The WHO Guidelines for Drinking-Water Quality provide minimum requirements for a list of chemicals and microbes, to ensure the safe practice and health of consumers (WHO 2011a, p. 2). Meanwhile, water consumers often assess the acceptability of water through their senses: the taste, odour and appearance of the water.

Microbial hazards (comprising of the majority of water-related health problems) include a wide variety of pathogenic bacteria, viruses and parasites that can put a community at risk of infectious disease outbreaks. Assessment of the microbial quality of the water is usually based on reference microorganism(s), such as *Escherichia coli* for faecal contamination. Reference pathogens are selected to facilitate an early detection of microbial pathogens in drinking water. This selection is dependent on local conditions such as source water characteristics, prevalence of the waterborne diseases or previous outbreaks, and health impact severity of the pathogens (WHO 2011a, p. 126). Knowledge Box (a) introduces several common waterborne diseases arising from such pathogens. Contamination can occur anywhere from the source water, rainfall run-off, faecal contamination, to bacterial growth in the piped distribution system. Thus, the arrangement of multiple barriers is necessary to reduce or prevent the water from being contaminated.

While microbial hazards are generally acute concerns, most chemical hazards in drinking water are of concern after long-term exposures (WHO 2011a, p. 26). Chemical guideline values specified by WHO generally reflect practically detectable limits that do not exceed the tolerable risks for the general population or vulnerable subpopulations over a lifetime of consumption (WHO 2011a, p. 27).

Knowledge Box (a): Common waterborne diseases

Waterborne diseases are caused by a wide range of bacteria, viruses and protozoa, of which the following are some common ones:

Bacterial diseases

Cholera (via *Vibrio cholerae)*: while many infected persons are asymptomatic, symptoms may include watery diarrhoea, vomiting, rapid loss of body fluids and severe dehydration.

E. coli: transmitted through exposure in recreational water as well as drinking water; symptoms include abdominal cramps, bloody diarrhoea, fever and vomiting.

Legionnaires' disease (via *Legionella*): grows in warm water and in piped distribution systems; symptoms include fever, dry cough, abdominal pain, diarrhoea, shortness of breath and muscle pain.

Leptospirosis (via *Leptospira*): common in tropical areas with heavy rainfall; symptoms include high fever and chills, headache, muscle pain, vomiting and jaundice.

Typhoid fever or the milder paratyphoid fever (via *Salmonella typhi*): while infected persons are occasionally asymptomatic, symptoms may include fever, headache, abdominal pain, constipation and diarrhoea.

Viral diseases
Hepatitis A or E virus: leading to infectious disease affecting the liver; symptoms include fever, nausea, loss of appetite, abdominal discomfort, jaundice, diarrhoea and in rare cases, acute liver failure.
Rotavirus: common cause of diarrhoea in children lasting up to a week; symptoms include self-limiting diarrhoea, fever and vomiting.
Norovirus: causes acute gastroenteritis among all ages lasting 1–3 days; symptoms include diarrhoea, vomiting, nausea, low-grade fever, abdominal pain and malaise.
Protozoan diseases
Cryptosporidium: microscopic parasitic infection; symptoms include watery diarrhoea, abdominal cramps, nausea and headaches.
Giardia: microscopic parasitic infection; symptoms include diarrhoea, malaise, weight loss, fatigue, excessive gas and vomiting.
(Adapted from Centre for Health Protection 2017)

In Hong Kong, water delivered from Dongjiang is assessed based on the national standards of China's Environmental Quality Standards for Surface Water (GB 3838-2002, Type II) and monitored for compliance. Raw water is then distributed to the city's 21 treatment plants and treated through a process of prechlorination, coagulation, sedimentation, filtration and completed with chlorine disinfection, pH correction and fluoridation (Advisory Committee on the Quality of Water Supplies 2002). The amount of chlorine is maintained at a 1.0 mg/l concentration when the water leaves the treatment plant. The treated water is then transported, via pumping stations if necessary, to service reservoirs before delivery to customers. Tested for the physical, chemical, microbiological and radiological dimensions, compliance of water quality with the 2011 WHO Guidelines for Drinking-Water Quality is monitored throughout the system, from the treatment plants to the consumer water taps (WSD 2017). Continuous maintenance of high standards has allowed governmental stakeholders to assert that Hong Kong's water is "one of the safest water supplies in the world" (WSD 2017) up unto the point of connection with private properties. However, the general public remain sceptical and have retained the practice of boiling water before drinking (ADM Capital Foundation 2017, p. 42).

15.2.3 Quantity and Reliability

The adequacy of water quantity for hydration, food preparation and hygiene has important impact on health. Although water requirements for consumption vary according to activity level, diet and climate, at least 7.5 L are required per capita daily for hydration and food according to WHO guidelines (WHO 2011a, p. 83). Additional water quantity is necessary for maintaining hygiene, doing laundry, generating income and other uses for maintaining well-being. In terms of reliability, interruptions to water supply should be minimized, as they may affect health as a result of degradation of water quality, or poorer hygiene due to reduced water quantity (WHO 2011a, p. 86).

When relying on local water yields, Hong Kong had a scarce and unreliable water supply. In 1963, an average of 36.4 litres per capita daily was considered relatively luxurious (Liu 2013). However, with the annual amount of Dongjiang River water supply established, the supply of water has risen dramatically, and Hong Kong people quickly adjusted to the other extreme of water usage. Consumption of freshwater alone reached a daily average of 130 litres per capita in 2011 (The Government of the Hong Kong Special Administrative Region 2011). Including seawater consumption used for flushing, which accounts for another 90 L, Hong Kong people's daily consumption of water reached a total of 220 litres daily per capita. Comparatively, the global daily per capita average is 110 L, while two comparable cities, Singapore and Shanghai, use an average of 150 and 106 L per capita, respectively (ADM Capital Foundation 2017, p. 62). This high rate of consumption per capita is neither wise nor sustainable. Compounded by the impacts of climate change and increasing industrialization and population growth of the Pearl River Delta region, the water consumption level of the city needs to be brought to attention.

15.2.4 Accessibility

Accessibility of water sources is crucial as it often determines the quantity of water obtained, which could compromise the hygiene practice and health of people. Accessibility is often measured by distance or time, particularly in developing countries, and whether the water source is improved (e.g. protected) or unimproved. In a vertically built dense city such as Hong Kong, the network of water pipes is crucial to maintaining an accessible healthy water supply to the residents. The total length of the water distribution network for freshwater in Hong Kong is 6,450 km, whereas the length for saltwater distribution is 1,550 km (Director of Bureau: Secretary for Development 2017, p. 240). These add up to a total network of over 8,000 km, which is approximately the distance of a 13.5 h flight from Hong Kong to Cairo, Egypt. According to the WHO Guidelines for Drinking-Water Quality (2011a, pp. 16–17), the plumbing system can also cause significant adverse health effects. Poor design, installation, alterations and maintenance of the plumbing system can lead to cross-connections with wastewater pipes and backflow or stagnation of water, which may result in cross-contamination of the drinking water or bacterial growth such as *Legionella*. Poorly maintained water storage systems, such as the rooftop tanks used in Hong Kong, could result in the intrusion of foreign contaminants. Improper material usage can provide suitable environments for bacterial growth or generate elevated amounts of heavy metal concentrations. Case Box demonstrates an incident of excess heavy metals found in Hong Kong's water supply system.

A main issue that has plagued Hong Kong's water supply system is the leakages and main bursts in its distribution network. In 2015, the amount of water lost was equivalent to one-third of the total water supply annually (ADM Capital Foundation 2017, p. 18), more than the amount of water produced by local yield. Around 15% of the water supplied was lost in the government-owned distribution network, a growing 14% from mains in private properties, 2% from inaccurate metres and 2% from illegal extraction (ADM Capital Foundation 2017, p. 54). The reasons for these water bursts, as cited by the government Water Supplies Department, include deteriorating pipes, and strain or stress caused by external weight or ground movement (Director of Bureau: Secretary for Development 2017, p. 240). About 20 and 16% of the freshwater and saltwater pipes, respectively, have been used for 30 years or more. Although the government implemented a 15-year Replacement and Rehabilitation Programme between 2000 and 2015 for approximately 37% of the government-owned water mains, the overall problem has not been significantly reduced.

Case Box: Hong Kong's water pipe lead pollution crisis in 2015

(a) *Discovery of the problem*

In July 2015, water samples in a newly constructed public housing estate were revealed to contain a lead level of more than 10 μg/L (micrograms per litre), exceeding WHO provisional guideline values. This led to fear of contaminated drinking water and a city-wide investigation of water quality. From July to September, the Water Supplies Department (WSD) and Housing Department tested water samples from 46 public housing estates and found a total of 106 samples in 11 estates with lead levels exceeding WHO standards. Drinking water samples were also taken from 772 kindergartens and 205 welfare units, revealing an excessive lead content in 8 kindergartens and in 366 water samples from wall-mounted water dispensers (Director of Bureau: Secretary for Development 2017, p. 286).

(b) *Health impacts of lead*

Lead is one of the most recognized environmental chemicals in terms of its adverse effects on human health. Exposure to lead can occur through various environmental sources, primarily ingestion of food, water, dirt, dust and air (WHO World Health Organization 2011b). Exposure to lead is associated with adverse neurodevelopmental effects such as impaired attention, activity, learning ability, memory and IQ, particularly among children (WHO 2011b). It is also associated with increased risk of cardiovascular mortality, impaired renal function, increased hypertension, reproductive dysfunction and increased risk of preterm delivery in pregnant women (WHO 2011b). Dose–response analyses conducted for lead exposure do not demonstrate any threshold indication for the chemical's key effects; thus, the WHO guideline limit of 10 μg/L of lead in drinking water is provisional only (WHO 2011a, p. 383). The groups most susceptible to lead poisoning include foetuses, infants, children (particularly those under 6) and pregnant women.

(c) *Investigation and departmental responses*
The presence of lead in tap water is largely attributable to improper materials used in household plumbing systems (WHO 2011a). These materials contain lead elements that become dissolved in the transported water. In Hong Kong's situation, water sampling by WSD and an independent team resulted in the conclusion that the source of lead came from the illegal use of leaded solders (metal alloy that is used to fuse a bond between two workpieces) in the freshwater supply plumbing system (Wai and Nin 2016, p. 24). In response, the government provided alternate sources of water to the affected residents in the form of bottled water and water tanks. It also advised the residents to take precautionary measures while the issue was being investigated and resolved. A task force was set up to investigate the causes of the incident, a review committee to provide quality control and monitor the water supply systems, and a commission of inquiry to review the governmental response.
Upon the unveiling of the lead-in-water scandal, Hong Kong's Department of Health and Hospital Authority arranged for blood testing for children under eight, breastfeeding mothers and pregnant women from the affected housing estates and kindergartens. Reference values for the blood lead levels were set by local experts after a review of local and international literature. For children, breastfeeding mothers and pregnant women, the normal level was set at <5 μg/dL, borderline raised level was around 5–44 μg/dL, and significantly raised level was set at >44 μg/dL (Centre for Health Protection 2015). By 30 September 2016, a total of 5,655 people were tested and borderline raised blood lead levels were found in 165 cases (Legislative Council Panel on Development 2017). Through continuous monitoring and follow-up blood reviews, many of the affected persons saw their blood lead levels return to normal, with 14 persons lost to follow-up. Children found with borderline raised blood lead levels also underwent a developmental assessment, in which 44 showed signs of developmental delay or mild developmental problems. The identified children were referred to appropriate rehabilitation services and their development monitored by the Department of Health.

15.2.5 *Affordability*

Affordability of water can be a hindrance to accessing a safe water supply. High costs can cause an unequitable access for low-income households. Too high of a

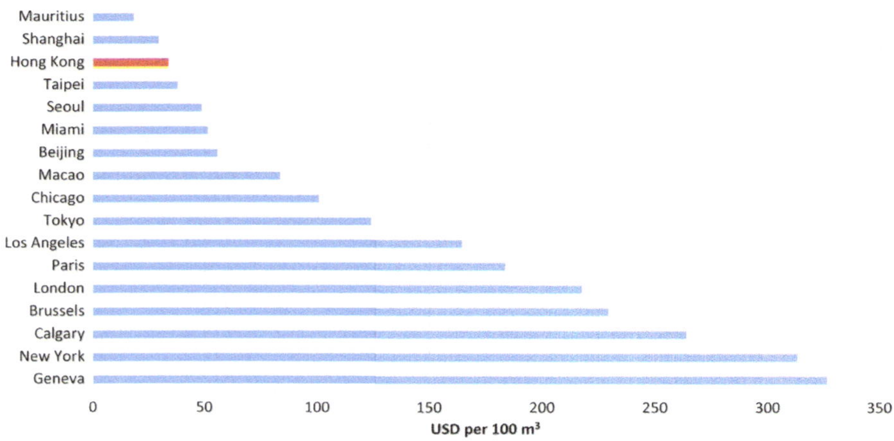

Fig. 15.1 Total drinking water charges for selected cities in 2015 (USD per 100 m^3) adapted from International Water Association Statistics and Economics (2015)

cost may even provoke households to choose alternative unimproved water sources or reduce water usage such that hygiene becomes poorly maintained (WHO 2011a, p. 85). However, a low cost may encourage an unsustainable consumption of water.

In Hong Kong, the total cost for the imported Dongjiang water has risen remarkably over the years from HK$2,494.80 million in 2006, to HK$4,222.79 million in 2015 (HK$7.8 = US$1) (Director of Bureau: Secretary for Development 2017, p. 224). The actual production cost of water at the 2015–16 price level was HK$8.7 per cubic metre (Director of Bureau: Secretary for Development 2017, p. 298). However, domestic water tariffs have not been revised since 1995 and kept artificially low through government subsidies (WSD 2016b). The freshwater charges for every 4-month cycle are levied in a four-tiered system, with the first 12 m^3 free of charge, and HK$4.16, HK$6.45, and HK$9.05 charged per cubic metre for the subsequent next 31, 19 m^3, and thereafter (WSD 2016b). As to the flushing water, the freshwater used for flushing is free for the first 30 m^3, and HK$4.58 per cubic metre thereafter, while the seawater is supplied free of charge (WSD 2016b). During 2015–2016, the average water tariff was HK$48 HKD per household per month, amounting to 0.3% of average monthly household expenditure (WSD 2016b). Comparatively, the threshold for affordable water set by the United Nations Development Programme is 3% of the median household income (Office of the United Nations High Commissioner for Human Rights, United Nations Human Settlements Programme, and World Health Organization 2010). According to the International Water Association Statistics and Economics (2015), the total drinking water charges for Hong Kong were the 13th lowest among 165 cities in 2015, costing US$33.71 per 100 m^3. The water tariffs in Tokyo and New York were 4 times and more than 10 times that of Hong Kong's. Drinking water charges for selected cities can be found in Fig. 15.1.

15.2.6 Summary of Water Supply

After looking at Hong Kong's water supply through various aspects of the Public Health framework, namely quality, quantity, reliability, accessibility and affordability, we can conclude that Hong Kong has set up a safe, reliable, easily accessible and affordable water supply. Complex infrastructure and water resources have been secured to adequately support the health of the densely inhabited urban population. Yet, this water supply is vulnerable as it is highly dependent on the provisions of a working system. Any major incidents can potentially have widespread effects, particularly if not managed properly. Additional issues such as ecological stability, industrial pollution and increasing drought due to climate change all pose risks to the water supply in the years to come. Figure 15.2 summarizes the five aspects of water supply, as well as the key stakeholders in urban water issues.

15.3 Drainage, Sewage and Water in the Environment

Drainage refers to the removal of unwanted water from the human environment. Unwanted water originates from various sources: used wastewater, rainwater, floodwater from overflowing rivers and other natural sources of surface water. In urban environments, proper drainage is necessary to maintain public health and reduce health risks, namely faecal–orally transmitted diseases and disease-bearing vectors. Faecal–oral diseases, as discussed in the earlier fact box on common waterborne diseases, are generally of minimal outbreak risk in areas with well-established water and sewage infrastructure. However, poor sewage treatment or a breakdown in the water system would quickly put the large urban population at risk. Furthermore, disease-bearing vectors, mainly mosquitos for Hong Kong, can easily find their way to breed and thrive in environments that are not well drained.

15.3.1 History of Hong Kong's Drainage and Sewage Systems

While working to secure a reliable water supply, the Hong Kong government also simultaneously developed the infrastructures for drainage and sewage. During its early colonial years in 1840s–50s, the need for drainage infrastructure was linked with the unsanitary and overcrowded living conditions, particularly in the local working-class residential districts. A combined drainage and sewerage system was constructed in parts of the city, which relied on rainfall to wash away sewage. The Public Health Ordinance was enacted in 1887 to set hygiene requirements for housing, drainage, public toilets and funerals (Drainage Services Department (DSD) 2015, p. 14). However, rapid population growth and intermittent heavy rains often exceeded the system's capacity and medical doctors warned the spread of infectious

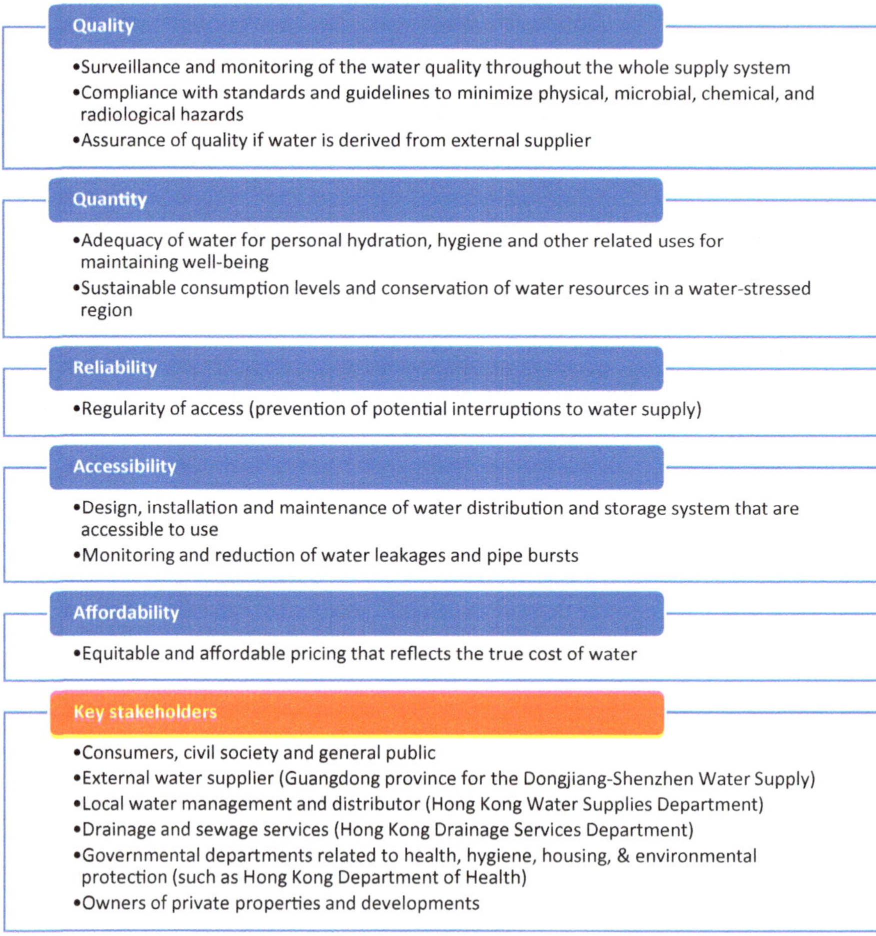

Fig. 15.2 Summary diagram on the five aspects of water supply and key stakeholders

diseases. In 1894, a bubonic plague outbreak led to the death of over 2,400 people and continued to affect the city annually for the next 30 years (DSD 2015, p. 14). This led to widespread governmental actions following a recommendation to build separate systems for drainage and sewage. With the separate systems, the overflow of water from heavy rainfall would not affect the capacity of the sewage systems, thus preventing untreated wastewater from entering the environment. The following sections address some issues in Hong Kong related to water found in the environment and look into the two separate systems in detail.

15.3.2 Water in the Environment

The quality of water in the environment is affected by sewage as well as other sources of pollution. Rapid industrial development starting from the mid-twentieth century increased the sewage and water pollution in Hong Kong. The increase in pollution rapidly deteriorated the marine and stream environments; as treatment of the wastewater and sewage was not a common practice (DSD 2008). Unregulated direct discharge added to the extent of water pollution as well. The government passed a legislation of the Water Pollution Control Ordinance in 1980, which controlled all discharges throughout Hong Kong using a licensing system and partitioned water control zones (Environmental Protection Department (EPD) 2017b). A water quality monitoring programme was later implemented by the government, with a total of 291 monitoring stations in 30 rivers, 10 water control zones, marine waters and beaches (Director of Bureau: Secretary for the Environment 2017, p. 301). Four key parameters of water quality, called Water Quality Objectives (WQO), are regularly measured: dissolved oxygen, total inorganic nitrogen, unionized ammonia nitrogen and *E. coli* bacteria (EPD 2015). Continuous monitoring has allowed the government to track its performance in improving environmental water quality. For example, since 2010, all government-managed beaches have continuously met the WQOs, particularly the target of preventing *E. coli* levels from exceeding 180 per 100 mL, to provide healthier environments for seaside leisure (EPD 2017a).

In addition to water quality, another concern is the quantity of water prevalent in the surrounding environment, particularly that of unwanted, yet undrained, water. Stagnant water can accumulate wherever water is not properly drained away. This creates potential breeding sites for vectors and increases the risk of vector-borne diseases, like dengue fever, malaria, Japanese encephalitis and Zika virus disease. Particularly, dengue fever is often at risk of becoming endemic in Hong Kong. However, the public's awareness of the stagnant water issue needs to be extended, as stagnant water develops not only from undrained residual water of floods and rainfall in the outdoor environment, but also through human activities in and around people's homes—see Knowledge Box (b) on stagnant water and mosquitos for more information.

Knowledge Box (b): Stagnant water and mosquitos

Retaining water in and around the homes can lead to unexpected breeding grounds for mosquitos and thus increase the risk of disease transmission. An enhanced surveillance programme as well as mosquito prevention and control work has been regularly implemented by the Hong Kong government (Food and Health Bureau 2017). Public announcements and health promotions remind residents in Hong Kong to help remove stagnant water sources around the home by:

- Tightly covering water containers;
- Keeping drains clear;
- Changing water for vases and clearing the water in the plates underneath flower pots;
- Properly disposing of empty containers;
- Preventing accumulation of water in air-conditioner drip trays;
- Placing mosquito larvae eating fish in fish tanks; and
- Smoothing out uneven ground surfaces to prevent puddles.

(Adapted from Food and Environmental Hygiene Department 2017a, 2017b).

15.3.3 Drainage System and Reduction of Flooding

Flooding generally occurs when there is an accumulation of surface water due to long-lasting rainfall or a rise of groundwater table above the surface (Chan 2017, p. 97). Other contributing factors to the accumulation of water include topographical (e.g. landscaping and vegetation), meteorological (e.g. saturated or solidified soil due to excessive precipitation or drought) and human factors (e.g. obstruction or lack of drainage basin and flood control measures) (Associated Programme on Flood Management 2013). In urban areas, vast impervious surfaces and lack of green infrastructure render flooding a common problem when there is a high volume of rainfall (Chan 2017, p. 98). Although high mortality is uncommon, health risks of flooding include waterborne diseases from contamination of drinking water sources and vector-borne diseases from stagnant water (Chan 2017, p. 98). Other common health risks include drowning, hypothermia, animal bites, injuries and electrical shocks. Knowledge Box (c) introduces additional water-related disaster risks threatening urban water infrastructure and several global platforms on the intersection between disasters and health.

Hong Kong experiences a high amount of annual rainfall, averaging 2,400 mm annually. Of even greater importance are the uneven temporal distribution of this rainfall and extreme heavy precipitation events, which can overwhelm the drainage systems beyond their design capacities in a short period of time. Impermeable surfaces such as roads and sidewalks, built-up districts near major waterways, old storm water drainage systems built with outdated protection standards, and the lack of green infrastructure to absorb precipitation create flooding risks throughout the city (DSD 2016b). Debris, either litter or natural materials, may build up around drainage inlets, further inhibiting efficient drainage. Additionally, Hong Kong is also a city built on and around steep slopes and hills, resulting in accumulation of run-off in downstream areas.

To tackle this problem, a flood prevention strategy has been implemented with long-term measures designed to expand the capacity of water drainage in three major ways. Drainage tunnels, designed to withstand a flood with a return period of 200 years (or a 1-in-200-year flood), have been constructed to intercept run-off at upstream locations and divert it directly to nearby water bodies(DSD 2016a). During extreme precipitation events, underground storm water storage tanks built in low-lying areas help to reduce the peak flow rate in the drainage system by temporarily retaining storm water (DSD 2016a). And finally, waterways have been reconstructed and revitalized with widened channels and ecological elements, which expand their flood control capacities (DSD 2016a). These drainage improvement works have enabled the city to be more resilient against heavy rainfall events, reducing the number of flooding "blackspots" from 90 in 1995 to 8 in 2016 (DSD 2016a).

Knowledge Box (c): Disasters and related global platforms

Flooding arising from heavy rainfall is only one of the potential disaster risks threatening the water infrastructure in Hong Kong. Other natural disaster risks include storm surges, tropical cyclones, landslides and even tsunamis or earthquakes. Other man-made emergencies may cause accidents or damage to the infrastructure. These disaster risks could have not only direct impacts on health, such as loss of life, injury, illness and disabilities (Chan and Murray 2017), but also indirect impacts as a result of water and drainage infrastructure breakdowns. Such breakdowns may lead to contamination of drinking water supply or discontinued water service. Bursting of water pipes could cause additional flooding in affected locales. These disruptions would reduce the amount of safe and reliable drinking water supplied to residents and such risks must be reduced to prevent a large impact on human health.

Globally, the Sendai Framework for Disaster Risk Reduction 2015–2030 urges substantial global reductions in disaster-related mortality, affected people, economic losses and critical infrastructure damage. It sets out strategies to help alleviate the impact of disasters and promote the need for health resilience. The fourth global target of the framework is applicable to water infrastructure, as it aims to "[s]ubstantially reduce disaster damage to critical infrastructure and disruption of basic services, among them health and educational facilities, including through developing their resilience by 2030" (United Nations Office for Disaster Risk Reduction 2015). However, despite health being mentioned throughout the framework, its core position in disaster risk reduction is not often adequately explored and recognized (Chan and Murray 2017). A growing interdisciplinary academic field, namely Health Emergency and Disaster Risk Management, stands at this health and disaster risk reduction intersection and will support implementation of health aspects in the Sendai Framework (Chan and Murray 2017).

15.3.4 Sewage Management

Whereas rainwater is comparatively cleaner, used wastewater with excreta disposal has a large load of infection-causing microorganisms such as bacteria, viruses, fungi and protozoa (Blom 2015). There is evidence that wastewater drains can be a source of microbial infection for households and healthcare services (Blom 2015). Poor sewage treatment results in deterioration of environmental water quality, as mentioned earlier, affecting people's recreational activities, drinking water, and subsequently, human health. The wastewater treatment process should include screenings to remove large objects and grit, and multiple treatments to separate sludge, smaller suspended solids, and chemicals from the water to improve the water quality. Sludge is a semi-solid by-product of sewage treatment, which is further treated through thickening, digestion and dewatering (Chiang 2013). The extracted sludge can then be properly disposed of in landfills, used as agricultural fertilizer, or incinerated.

In Hong Kong, a total of 300 sewage treatment facilities are in operation, treating over 1 billion cubic metres of sewage annually (DSD 2016a) or an average of 2.8 million cubic metres daily. The system services 93% of the city's population and consist of a 1,700 km-long network (DSD 2016a). To diminish the amount of sludge discharged into Hong Kong's main harbour and improve the water quality, a sewage system under the Harbour Area Treatment Scheme was constructed over a period of 20 years and began full operations in 2015 (DSD 2016a). Sewage from all the districts around the harbour is conveyed to a centralized sewage treatment plant for treatment and disinfection before being discharged at the western end of the harbour. The chemically enhanced primary treatment process (CEPT) is used, which utilizes ferric chloride and polymer to quicken the settlement of suspended solids. CEPT removes over 99% of *E. coli*, 70% of biochemical oxygen demand and 80% of suspended solids from the treated water, allowing up to 1,000 tons of sludge to be intercepted and prevented from entering the harbour (DSD 2016a). This improvement of sewage-discharged water quality has had, in return, a major impact on the environment, leading to the reopening of beaches in the western Tsuen Wan District, and the resumption of cross-harbour swimming races.

15.4 Continuing Challenges for Urban Water Issues

15.4.1 Water Stress, Ecological Limits and Sustainability

As mentioned in the beginning of this chapter, Asia is projected to experience severe water stress, particularly in the densely populated areas. As a densely populated city, Hong Kong will face difficulties in ensuring a sufficient water supply. Not only is Hong Kong naturally water scarce, but the Dongjiang River, which supports 80% of Hong Kong's water supply, is also the water source for seven other Pearl River Delta region cities, including Dongguan, Heyuan, Huizhou, Shaoguan,

Meizhou, Guangzhou and Shenzhen. Serving as a prominent example of inter-basin water distribution, the river sustains a total of 40 million people, with proportion of dependence varying among the cities (Liu 2012). However, the river has already experienced water stress, with water withdrawals since 2004 periodically exceeding the "ecological limit" established at 33% of the river's total water resources (Lee and Moss 2014; He et al. 2009). With continuing economic development in several less-developed cities, the demand for Dongjiang water is projected to increase, which will lead to increasing water competition among the region's cities.

Looking towards the future, Hong Kong should try to diversify its water sources to safeguard the health of its population. Since 2008, the Water Supplies Department has been discussing the adoption of a Total Water Management strategy. These include projects to develop seawater desalination and the use of reclaimed water in the upcoming years (Director of Bureau: Secretary for Development 2017, p. 244, 298). Such measures are necessary to secure a reliable water supply for the urban population.

15.4.2 Climate Change

The increasing impacts of climate change, which lead to changes in rainfall distribution, frequency and amount, will affect both the supply and demand of water (Shaw and Thaitakoo 2010). The increased frequency and severity of droughts, as well as higher intensity of rainfall and flooding events, will influence the quantity and quality of water available. In coastal areas, sea level rise has the potential to cause salination of water sources. Additionally, the altered rainfall and temperature patterns caused by climate change will influence the epidemiology of vector-borne diseases, as further discussed in Knowledge Box (d)

With a humid subtropical climate, Hong Kong is projected to be affected by climate change in the following ways by 2100 (Hong Kong Observatory 2015):

- Temperature: Mean annual temperatures are expected to rise by a total of 3–6 °C;
- Precipitation: Mean annual rainfall to increase by 180 mm and extremely wet years increase from 3 during 1885–2005 to 12 during 2006–2100;
- Sea level rise: Mean sea levels to rise by a total of 0.63–1.07 m; and
- Extreme weather events: Storm surges from tropical cyclones to increase in frequency and severity.

The challenge for Hong Kong and other urban areas globally is to plan and manage the water supply, and prepare the drainage and sewage systems, to mitigate any adverse impacts or disruptions. Adaptation measures should be taken to address potential variations in water quantity and quality. Since Hong Kong shares the same climatic setting with Dongjiang River and other Pearl River Delta region cities, reduced rainfall will diminish not only the quantity of Dongjiang water available for distribution, but also Hong Kong's local yield supplies at the same time (Director of

Bureau: Secretary for Development 2017, p. 225). Diversified water sources and a closed-loop system can be introduced to build up a climate resilient water supply. To maintain a high quality of drinking water, water treatment systems also need the flexibility to adapt to changes in the microbial and chemical content of incoming water.

Knowledge Box (d): The potential impact of climate change on rainfall, temperature and vector-borne diseases
The risk of vector-borne diseases largely depends on factors such as climate, altitude, population density and specific environmental requirements of the various mosquito species. With climate change, the transmission of vector-borne diseases is already shifting in geographical distribution and seasonality. Prolonged transmission periods can be expected, as warmer temperatures appear earlier in the year and extend for a longer period of time (Yang et al. 2012). In Hong Kong, as the summer weather becomes increasingly hot, the epidemic risk of dengue fever, transmitted by mosquito species *Aedes aegypti* and *Aedes albopictus*, is likely to increase (Lam et al. 2004). This is due to the increased survivability and biting frequency of the mosquitos, and shortened incubation period of the Dengue virus. However, malaria will experience a decrease in epidemic potential during the summer and an increase during the spring and fall seasons (Lamet al. 2004). The warmer temperatures during summer will be too hot for the survival of malaria-transmitting mosquito *Anopheline*, but the warming of the other seasons would enable it to thrive.

15.5 Research and Action Gaps for Urban Water and Health Issues

Although much research has been conducted on the topic of urban water and health, research gaps still exist. It remains an urgent need for good indicators in assessing and monitoring water to address the health dimension. A composite indicator, which comprehensively covers the various aspects of water supply through the public health framework, would be useful for monitoring the relationship between water and a population's health in urban settings. Additionally, such an indicator, if adopted globally, would be useful for multi-city comparisons of water supply systems. An improved comparison method among cities will allow these cities to learn from each other and thereby enhance the resilience of urban water systems. Furthermore, research is needed for the identification of appropriate microbial reference pathogens used to monitor water quality and reduce waterborne disease outbreaks. Water health aspects related to climate change, such as vector-borne diseases, constitute another area where further research is also needed.

As water is a multi-sectoral issue, tackling water and health issues will require a collaborative multi-sectoral approach and an appropriate platform. In urban areas, high population density, travel patterns, demographics and land use all contribute to a population's water-related health risk. As health risks are constantly changing with environmental and socioeconomic context, there needs to be an increase in monitoring and evaluation of the water supply system. Continuous monitoring, evaluation and update of reference pathogens for each urban community will support the human health dimension and allow for appropriate management and maintenance of the water supply system and its operations. Preventive and adaptive planning of the water supply system will also be needed to address increasingly stressed water sources, potential disaster incidents, and incoming climate change impacts.

15.6 Conclusion

This chapter has addressed the urban water and health issues using Hong Kong as a lens. Water impacts health through not only drinking water, but also wastewater in drainage and sewage systems, and water in the surrounding environment. The effects of water supply on health can be assessed through the public health framework covering quality, quantity, reliability, accessibility and affordability. In Hong Kong, the water consumption volume is high, attributable to low water tariffs and the illusion of stability created by the Dongjiang water supply agreements. Infrastructure quality and maintenance of water systems are crucial to ensure urban health, as demonstrated by the water pipe lead pollution crisis. Proper drainage systems are also required to prevent flooding of urban areas and the associated health risks. Undrained water in the environment can become a source of vector-borne diseases, while inadequate sewage treatment can pose a risk to waterborne (faecal–oral) disease. Continuous challenges to the provision of health-supporting water for the urban population include increasing water stress, disaster risks and impacts of climate change.

References

ADM Capital Foundation (2017) The illusion of plenty: Hong Kong's water security, working towards regional water harmony. Hong Kong

Advisory Committee on the Quality of Water Supplies (2002) ACQWS paper no. 11 publication of water quality data—2002, p 4. Retrieved 30 June 2017 from http://www.wsd.gov.hk/acqws/filemanager/common/p11.pdf

Associated Programme on Flood Management (2013) Flood forecasting and early warning. Integrated flood management tools series no. 19. World Meteorological Organization, Geneva, Switzerland. Retrieved 30 June 2017 from http://www.apfm.info/publications/tools/APFM_Tool_19.pdf

Blom K (2015) Drainage systems, an occluded source of sanitation related outbreaks. Arch Public Health 73(8):1–8

Bolton P (2006) Managing disasters and other public health crises. In: Pencheon D, Guest C, Melzer D, Gray JAM (eds) Oxford handbook of public health practice, 2nd edn. Oxford University Press, New York, pp 249–256
Census and Statistics Department (2015) Hong Kong population projections 2015–2064 (press release). Retrieved from https://www.censtatd.gov.hk/press_release/pressReleaseDetail.jsp?pressRID=3799&charsetID=1
Census and Statistics Department (2017) 2016 population by-census main tables: population density by district council district and year. Retrieved 30 June 2017 from http://www.bycensus2016.gov.hk/en/bc-mt.html
Centre for Health Protection (2015). Care plan for residents of public estates with elevated lead level in drinking water. Retrieved 30 June 2017 from http://chp.gov.hk/files/pdf/care_plan_for_raised_bll.pdf
Centre for Health Protection. (2017). Communicable diseases. Retrieved 30 June 2017 from http://www.chp.gov.hk/en/health_topics/9/24.html
Chan EYY (2017) Public health humanitarian responses to natural disasters. Routledge, London
Chan EYY, Murray V (2017) What are the health research needs for the Sendai Framework? The Lancet. https://doi.org/10.1016/S0140-6736(17)31670-7
Chiang MCK (2013) Innovative environmental friendly sludge treatment scheme with co-settling. Retrieved 30 June 2017 from http://www.dsd.gov.hk/EN/Files/Technical_Manual/technical_papers/ST21301.pdf
Director of Bureau: Secretary for Development (2017) Replies to initial written questions raised by Finance Committee members in examining the estimates of expenditure 2016–17: session no: 15. Retrieved 30 June 2017 from https://www.devb.gov.hk/filemanager/en/content_968/DEVB_W_2_e1.pdf
Director of Bureau: Secretary for the Environment (2017) Replies to initial written questions raised by Finance Committee members in examining the estimates of expenditure 2016–17: session no: 7. Retrieved 30 June 2017 from http://www.enb.gov.hk/sites/default/files/en/node4014/fcq_1617_e.pdf
Drainage Services Department (2008) Sewerage and flood protection: drainage services 1841–2008. Retrieved 30 June 2017 from http://www.dsd.gov.hk/EN/Files/publications_publicity/Publications/2008book.pdf
Drainage Services Department. (2015). Going the extra miles. Hong Kong
Drainage Services Department (2016a) Cleaner harbor, better life 2015–16 sustainability report. Hong Kong. Retrieved 30 June 2017 from http://www.dsd.gov.hk/Documents/SustainabilityReports/1516/common/dsd_full_report_2015_16.pdf
Drainage Services Department (2016b) Flood prevention: flooding problems. Retrieved 30 June 2017 from http://www.dsd.gov.hk/EN/Flood_Prevention/Our_Flooding_Situation/Flooding_Problems/index.html
Environmental Protection Department (2015) Marine water quality in Hong Kong in 2015. Retrieved 30 June 2017 from http://wqrc.epd.gov.hk/pdf/water-quality/annual-report/MarineReport2015eng.pdf
Environmental Protection Department (2017a) Beach water quality—water quality objective. Beachwq.gov.hk. Retrieved 30 June 2017 from http://www.beachwq.gov.hk/en/bwq_wqo.aspx
Environmental Protection Department. (2017b). Hong Kong Water Quality Resource Centre—legislation & guidelines. Wqrc.epd.gov.hk. Retrieved 30 June 2017 from http://wqrc.epd.gov.hk/en/legislation/wpco.aspx
Fant C, Schlosser CA, Gao X, Strzepek K, Reilly J (2016) Projections of water stress based on an ensemble of socioeconomic growth and climate change scenarios: a case study in Asia. PLoS ONE 11(3):e0150633. https://doi.org/10.1371/journal.pone.0150633
Food and Environmental Hygiene Department (2017a) Anti-mosquito poster: let's remove stagnant water eliminate mosquito for healthy living (home). Retrieved 30 June 2017 from http://www.fehd.gov.hk/english/safefood/pest_control_pamphlet.html

Food and Environmental Hygiene Department (2017b) Anti-mosquito poster: remove stagnant water to eliminate mosquito breeding (home). Retrieved 30 June 2017, from http://www.fehd.gov.hk/english/safefood/pest_control_pamphlet.html

Food and Health Bureau (2017) LegCo panel on food safety and environmental hygiene: enhanced mosquito prevention and control in face of threats from Zika virus infection (LC Paper No. CB(2)531/16-17(03)). Hong Kong. Retrieved 30 June 2017 from http://www.legco.gov.hk/yr16-17/english/panels/fseh/papers/fseh20170110cb2-531-3-e.pdf

He T, Ma X, Peng X, Li T (2009) Problems and contradiction in water resources environmental management of Dongjiang Basin (in Chinese). Water Resour Prot 25(6)

Hong Kong Observatory (2015) Climate projections for Hong Kong. Retrieved 30 June 2017 from http://www.hko.gov.hk/climate_change/future_climate_e.htm

International Water Association Statistics and Economics (2015) Total charge drinking water for 165 cities in 2015. Retrieved 30 June 2017 from http://178.62.228.131/graph/8

Lam KS, Hung WT, Fung WY (2004) Provision of service for characterizing the climate change impact in Hong Kong (Tender AM02—316). The Hong Kong Polytechnic University Department of Civil and Structural Engineering, pp 67–79. Retrieved 30 June 2017 from https://www.climateready.gov.hk/files/pdf/Climate _Change_Report_final.pdf

Lee F, Moss T (2014) Spatial fit and water policies: managing asymmetries in the Dongjiang River Basin. Int J River Basin Manag 12(4)

Legislative Council Panel on Development (2017) Administration's response to Hon CHAN Han-pan's paper on safety of drinking water and replacement of aged water mains dated 30 March 2017 as set out in LC Paper No. CB(1)761/16-17(01). Hong Kong. Retrieved 30 June 2017 from http://www.legco.gov.hk/yr16-17/english/panels/dev/papers/devcb1-896-1-e.pdf

Liu S (2012) A vulnerable Dongjiang is a vulnerable Hong Kong. Retrieved 30 June 2017 from http://chinawaterrisk.org/opinions/a-vulnerable-dongjiang-is-a-vulnerable-hong-kong/

Liu S (2013) Liquid assets IV: Hong Kong's water resources management under "one country, two systems". Civic Exchange, Hong Kong

Office of the United Nations High Commissioner for Human Right, United Nations Human Settlements Programme, and World Health Organization. (2010). The right to water fact sheet no. 35, p 11. Office of the United Nations High Commissioner for Human Rights, Geneva. Retrieved from http://www.ohchr.org/Documents/Publications/FactSheet35en.pdf

Shaw R, Thaitakoo D (2010) Chapter 1 Water communities: Introduction and overview. In: Shaw R, Thaitakoo D (eds) Water communities. Emerald Group Publishing Limited, pp. 1–13

The Government of the Hong Kong Special Administrative Region. (2011). LCQ9: Use of water resources (Press Release). Retrieved from http://www.info.gov.hk/gia/general/201111/09/P201111090226.htm

The Skyscraper Center (2017) Cities ranked by number of 150 m + completed buildings. Skyscrapercenter.com. Retrieved 30 June 2017 from http://www.skyscrapercenter.com/cities

United Nations, Department of Economic and Social Affairs, Population Division. (2014). World urbanization prospects: the 2014 revision, highlights (ST/ESA/SER.A/352). Retrieved 30 June 2017 from https://esa.un.org/unpd/wup/publications/files/wup2014 -highlights.pdf

United Nations Office for Disaster Risk Reduction. (2015). Sendai framework for disaster risk reduction 2015–2030. United Nations Office for Disaster Risk Reduction, Geneva. Retrieved 30 June 2017 from https://www.unisdr.org/files/43291_sendaiframeworkfordrren.pdf

UN-Water (n.d.) Scarcity. UN-Water. Retrieved 30 June 2017 from http://www.unwater.org/water-facts/scarcity/#

Wai ACH, Nin AL (2016) Report of the commission of inquiry into excess lead found in drinking water (the redacted version). Hong Kong. Retrieved 30 June 2017 from www.coi-drinkingwater.gov.hk

Water Supplies Department (2016a) Capacity of impounding reservoirs in Hong Kong. Wsd.gov.hk. Retrieved 30 June 2017 from http://www.wsd.gov.hk/en/publications_and_statistics/statistics/capacity_of_impounding_reservoirs_in_hong_kong/index.html

Water Supplies Department (2016b) Finance and water charges. Retrieved 30 June 2017 from http://www.wsd.gov.hk/filemanager/common/annual_report/2015_16/common/pdf/11_finance_and_water_charges.pdf
Water Supplies Department (2016c) Water from Dongjiang at Guangdong. Wsd.gov.hk. Retrieved 30 June 2017 from http://www.wsd.gov.hk/en/water_resources/raw_water_sources/water_sources_in_hong_kong/water_from_dongjiang_at_guangdong/
Water Supplies Department (2017) Water quality monitoring data: drinking water quality for the period of April 2016–March 2017. Retrieved 30 June 2017 from http://www.wsd.gov.hk/en/water_resources/water_quality/water_quality_monitoring_data/index.html
World Health Organization (2011a) Guidelines for drinking-water quality, 4th edn. Geneva
World Health Organization (2011b) Lead in drinking-water. Geneva
World Health Organization (n.d.) Water sanitation hygiene: diseases and risks. Retrieved 30 June 2017 from http://www.who.int/water_sanitation_health/diseases-risks/en/
Yang G, Tanner M, Utzinger J, Malone J, Bergquist R, Chan EYY et al (2012) Malaria surveillance-response strategies in different transmission zones of the People's Republic of China: preparing for climate change. Malaria J 11(1):426. Retrieved 30 June 2017 from http://dx.doi.org/10.1186/1475-2875-11-426

Chapter 16
Urban Water Issues in the Megacity of Tehran

Ali Ardalan, Mona Khaleghy Rad and Mahdi Hadi

Abstract Iran (the Islamic Republic of Iran) is a country in Western Asia considered as the second populous nation in the Middle East region. The main environmental issues in Iran are air pollution, water quality and quantity crisis, municipal and industrial wastes, and climate change. Tehran, the capital of Iran with a population of near nine million, as a megacity suffers from various environmental and social problems, such as severe air pollution, water crisis, and slums. Water management is particularly important for Tehran as it affects the daily function of the capital, its sustainable development, and health of the citizens. The per capita renewable water is now reached to a critical threshold level in Iran. Furthermore, the withdrawal to availability ratio, as a water scarcity index, is estimated to be more than eighty percent, which shows severe water stress. In such dry condition of Iran, Tehran's misallocation of water consumption through different categories of usage, i.e., agricultural, domestic, and industrial, has caused such important issue for the country that without resolution leads to an unfortunate chaos. Furthermore, the current per capita consumption of treated water in the city of Tehran is twice to three times of the international average. High-quality drinking water consumption for domestic and sanitary uses is another main challenge for Tehran. In addition, a significant fraction of treated water is leaked due to old water pipe systems. Additionally, sewage of the residential areas, surface runoff, and natural pollution due to geological structures can increase the chance of water resource pollution. Groundwater is one of the main sources of water supply in Tehran, but after it turns into wastewater, it is discharged away directly or with some treatments into surface waters and exits the city. This

A. Ardalan · M. Khaleghy Rad (✉)
Department of Climate Change and Health, Institute of Environmental Research, Tehran University of Medical Sciences, Tehran, Iran
e-mail: mona.khaleghy@gmail.com

A. Ardalan (✉)
Harvard Humanitarian Initiative, Harvard University, Cambridge, USA
e-mail: aardalan@tums.ac.ir

M. Hadi
Center for Water Quality Research, Institute for Environmental Research, Tehran University of Medical Sciences, Tehran, Iran

B. Ray and R. Shaw (eds.), *Urban Drought*, Disaster Risk Reduction,
https://doi.org/10.1007/978-981-10-8947-3_16

process decreases freshwater tables and reduces the water availability in the long term. Management of the nutrients is not advanced either. Nutrients of the soil are consumed in agriculture and discharged with the wastewater into the aquatic ecosystems, where they can trigger some environmental problems such as water pollution. Thus, sustainable water management (SWM) is crucial for Tehran. Applying a SWM in the megacity of Tehran may only become achievable by using an integrated water resource management, which is a process that manages the land, water, and related resources in a way that preserves the ecosystem while maximizes the social and economic comfort. Access to safe water, efficient economy, and sustainable ecosystem need to be revised in the water policies for Tehran. Engaging private sectors, creating revenue opportunities by water reuse activities, managing water-use allocation, and sustaining financial resources, above all, are the most important approaches toward a sustainable water management in it. In this chapter, we first look at the current water issues in Iran, especially Tehran. Then, by looking at the management strategies available for water resources and wastes, we analyze the shortcomings and the challenges. Finally, we summarize the opportunities and possible solutions to the water issues in the megacity of Tehran.

Keywords Megacity of Tehran · Water supply issues
Water and wastewater management

16.1 A Brief Summary for Tehran

16.1.1 Geography of Tehran

Tehran is the name of a province, a county, and a city in Iran; it is pronounced Tehrān in Persian. Tehran province is one of the 31 provinces of Iran. Its area is 18,909 km^2 (7,301 square mile) located in north of the center of Iran. Tehran province has borders with provinces of Mazandaran, Qom, Semnan, and Alborz in the north, south, east, and west, respectively. The county of Tehran is bordered by Shemiranat, Damavand, Eslamshahr–Pakdasht–Ray, and Karaj–Shahriar in the north, east, south, and west, respectively (Madanipour 1999).

Tehran province includes 13 townships, 43 municipalities, and 1,358 villages. Today, the province of Tehran has a population of 13.267 million people with more than 8.737 million living in the city of Tehran (SCI 2015). Currently, the megacity of Tehran is the capital of the province, and the economical, political, and industrial capital of the Islamic Republic of Iran. Tehran attracts a great deal of population, and it is increasing; currently, it is the 30th most populated city in the world. It is also the 27th largest city with a size as large as 730 km^2 with the population density of 12,896/km^2 (SCI 2015).

City of Tehran is located at the coordinates of 35° 41′N and 51° 25′E with an average altitude of 1,500 m above the sea level, on the southern slopes of the Alborz mountain chain, and at the foothill of a potentially active volcano, Mount Damavand with 5,678 m high, which is the highest peak in Iran and in the west-Asia. The city has been at the intersection of the historic east-to-west and north-to-south trading routes (known as Silk Road). The strategic location of this area explains the six thousand years antiquity of population, first as the ancient city of Ray, which was accommodating passengers crossed the desert, and then as the city of Tehran (Madanipour 1999).

16.1.2 Climate

Tehran has a semiarid climate due to its geographic location with Alborz mountains located in the north and the central desert in the south. According to the data of I.R. of Iran Meteorological Organization (IRIMO 2017), the mean annual temperature of Tehran at the Tehran station is around 17.4 °C. The hottest month of Tehran is July, with the mean maximum temperature of 40.8 °C and the mean minimum temperature of 18.7 °C, and the coldest month is January, with the mean maximum temperature of 13.8 °C and the mean minimum temperature of −6 °C (IRIMO 2017). Furthermore, the average annual precipitation for Tehran at the Tehran station is about 240 mm/year (Fig. 16.1). The precipitation peak is from late fall to mid-spring, and there is almost no rain from June to September (Fig. 16.1).

In a research on the city of Tehran by Roshan et al. (2010), parameters related to climatic change, such as daily mean temperature, humidity, rainfall, and wind speed, are studied for the period of 1953–2006. According to Roshan et al. (2010), the daily mean temperature has significantly increased, especially during the summer months. Furthermore, they showed that the annual relative humidity has significantly

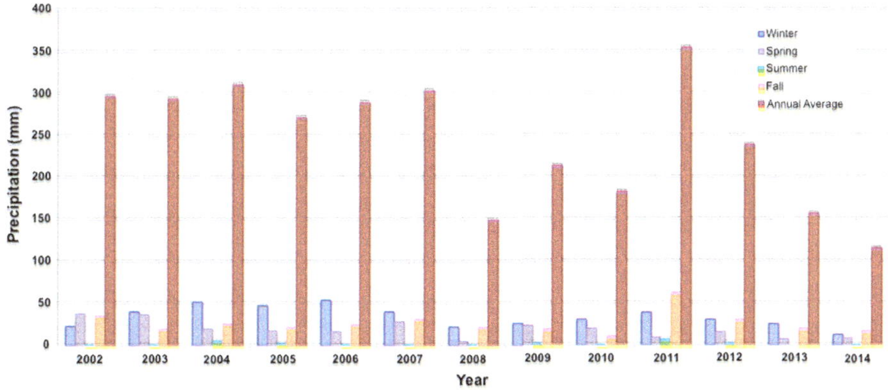

Fig. 16.1 Average seasonal and yearly precipitation in millimeter for 2002–2014. Data from IRIMO (2017)

increased during this period. Moreover, the trend of annual average rainfall is not statistically significant and the annual average wind speed has decreased since 1953.

In another research by Ranjbar et al. (2006), it was shown that the increase in the average annual minimum temperature has been more significant than the increase in maximum temperature during the 40 years of 1956–1995. Also, they showed that the rate of increase in the minimum temperature in Tehran is four times greater than the city of Varamin, which is small city near Tehran. They suggested that this difference is human sourced, as a result of urban heat islands and changes in the micro-climatic condition of Tehran (Ranjbar et al. 2006). From these studies, and others, it is evident that climate change is an important threat to the city of Tehran.

16.1.3 Environment

Tehran has been rated as one of the most polluted cities in the world—pollutions from water, air, and land to noise. An estimation of 1.1 million motor vehicles plus household fuel consumption and industries located in the south (70% of total) are the main causes of air pollution in Tehran (Madanipour 1999). Prevailing high atmospheric pressure and the encapsulating effect of the mountains aggravates these effects. Urbanization has brought issues of water and waste management to the city, as two main important measures of urban development (SOE 2012). Environmental pollution, population growth, uncontrolled industrialization, and ignoring sustainable development have caused extra stress on the water resources of the megacity of Tehran.

Urbanization grade of Iran in 2012 was measured as 60–80%, and Tehran's urbanization rate was more than 90% (Pahl-Weber et al. 2013). Urbanization means higher greenhouse gas emissions from transportations of the population in the cities. Also, urbanization results in higher energy and water demands, which consequently bring challenges of water and energy distribution to the cities. Furthermore, according to Gober (2010), the link of the energy and water usage, as known as the energy–water nexus, results in a hidden vulnerability factor, meaning that any shortages in one resource can cause issues in the other one. As he stated, "*The production of energy requires large volumes of water for cooling and turning turbines, while the treatment (including desalination) and distribution of water is highly dependent upon low-cost energy*" (Gober 2010).

Regarding energy consumption in Tehran, it suffices to mention the energy consumed for heating and cooling in this area. The data of Statistical Center of Iran (2011) shows that, on average, urban families in the province of Tehran, heat 91.7% and cool 90.4% of their total floor area (SCI 2011). These high percentages of heating and cooling are even far more than the Iranian urban family averages (77.1 and 67.7%, respectively). The major CO_2 emissions in Tehran are due to energy industries (30%), transportation (24%), and residential use (23%) (UNDP and DOE 2010; Pahl-Weber et al. 2013). The transportation sector's share is primarily due to the role of private cars plus an inadequate public transport (Pahl-Weber et al. 2013).

It is worth mentioning that among four popular means of transportation (bus, taxi, private car, and motorcycle), private cars have the highest (88%) contribution of CO_2 emission, in Tehran (Kakouei et al. 2012).

16.2 National Policies and Regulations for Water

According to the rule No. 45 of the Constitutional Law of Islamic Republic of Iran (Water Equitably Distribution Law), water of seas, rivers, natural streams, valleys, and any other natural route including surface and underground, flood, waste, seepage, lakes, lagoons, natural ponds, springs, mineral waters, and resources of underground waters are public properties, under the government surveillance and responsibility; the government uses them according to the public interests.

Ministry of Environment (MoE) sets policies related to the water and wastewater management in various sectors of urban and rural areas through the deputy ministry. The organization in charge of water resource management is the deputy ministry of water along with eleven regional water boards, and the one in charge of water pollution control is environmental protection organization. Drinking water quality standards are set, monitored, and enforced by Ministry of Health and Medical Education (MHME).

Furthermore, there are companies assigned for supervision and assistance to service providers, for example, national water and wastewater engineering company, which provides services regarding investment planning, human resources development, and establishment of standardized systems and procedures. According to the law of "establishing facilities for reconstruction of drinking water networks," water and wastewater companies can receive financial aid as part of their financial resources to reconstruct and develop water installations of cities and complete the incomplete wastewater plans and prioritize execution of wastewater plans. National economic council sets general tariff policies with some regional changes.

Due to the different approaches of water management in the various responsible sectors in the country, an integrated management system is essential for drinking water supply in both urban and rural regions. In a national document developed by the MHME, which is also ratified by the government for preparation of the related implementation packages by responsible organizations, this necessity is mentioned and a clarification of the responsibilities of any related governmental entities is included. The English version of this document is available at the Web site of Institute for Environmental Research (MHME 2011).

16.3 Water Resources and Water Infrastructures

The two main water resources of Tehran are subsurface and surface waters. The groundwater resources are wells (deep and semi-deep), springs, and qanats (aqueducts) (Habibi and Bernard 2005a). According to a recent report from Iran's statistical yearbook on the withdrawal from the groundwater resources of Tehran (ICS 2015), deep wells are the major (80%) suppliers of water for the city of Tehran (Fig. 16.2). The second major withdrawals (9%) are from Qanats, which have been part of water system for over 2000 years in Iran (English 1968).

On the other hand, Tehran province is surrounded by four watershed areas: Sefidrood River, Caspian Sea, Namak Lake, and the Central Kavir. However, the Namak Lake watershed area covers 69% of the total area of the province. The most important river resources of the province include Hableh, Abhar, Lar, Jajrood, Taleghan, Karaj, and Jajrood rivers. Drinking water of Tehran is currently supplied through modern infrastructures, such as dams, reservoirs, conventional water treatment plants, long-distance transmission pipelines, and water distribution systems (Ziller 2010). The map of Tehran and some of its surface resources are shown in Fig. 16.3.

There are nine under-operation dams in Tehran province that has a total storage capacity of 1699.3 MCM, and most of the dams are used for multiple purposes, such as hydropower, irrigation, flood control, and drinking water supply (MOE 2000). Mainly, these dams are built on the rivers. One of the most important dams of the province is Amir Kabir Dam (or Karaj Dam) built in 1961 in the northwest of Tehran with the watershed area of 764 km^2 and average annual water of 472 million m^3. The main purpose of the Karaj dam was to control spring waters, prevent floods, supply 340 million m^3/year of drinking water for Tehran, provide 130 million m^3/year of water for farming near Karaj, and generate 150,000 MW/h/year energy, especially as a help to the electricity network during high demand hours (Habibi and Bernard 2005b).

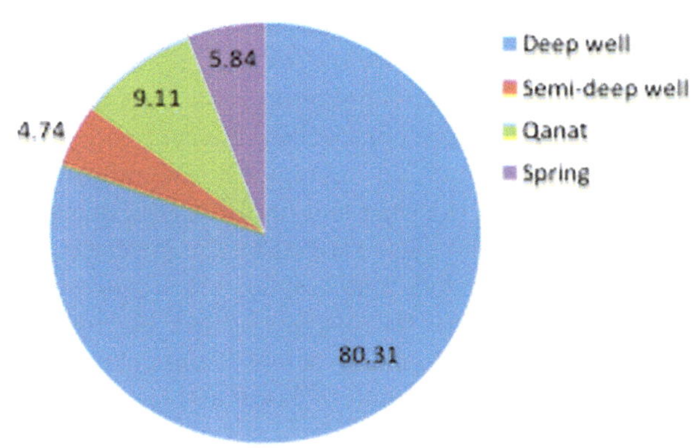

Fig. 16.2 Withdrawal of groundwater resources in comparison. Data from the Iran's statistical yearbook (ICS 2015)

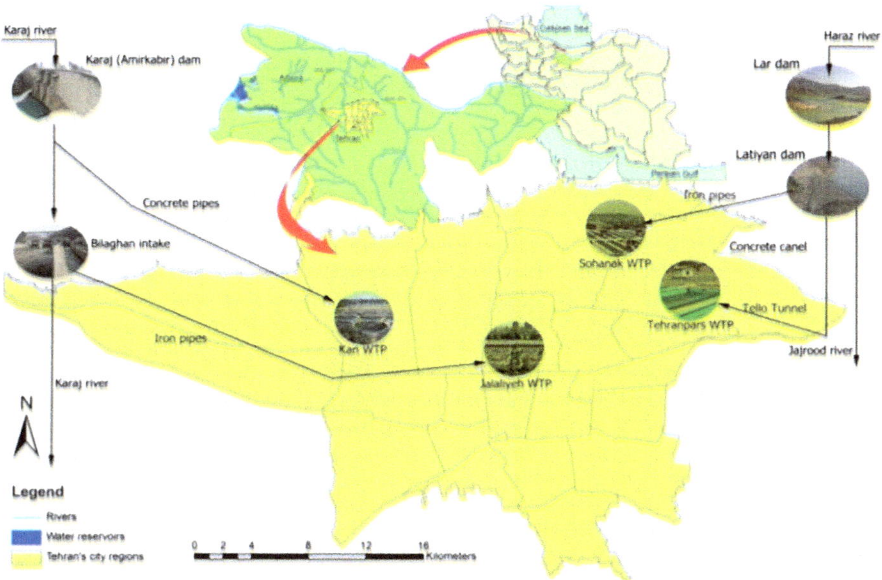

Fig. 16.3 Geographic location of surface water supplies and main drinking water treatment plants (WTPs) in Tehran (Hadi et al. 2016). Reprinted by permission from Springer Nature, Environmental Science and Pollution Research, Contribution of environmental media to cryptosporidiosis and giardiasis prevalence in Tehran: a focus on surface waters, Hadi et al. (2016)

16.3.1 Water Treatment Plants

There are five major functioning water treatment plants around Tehran (Habibi and Bernard 2005c). First, Jalaliyeh, established in 1955, is the oldest one in Iran and processes the Karaj river water. Second and the largest plant in Iran is in the west of Tehran, Shahre Ziba plant. The third and fourth plants are Tehranpars 1 and 2, both with the capacity of 4 m^3/s. These two plants process the Latyan dam water and supply for the east and some areas in the center, north, and south of Tehran. The fifth one is established in 2004, Sohanak, which is in the northeast of Tehran. This plant processes the Lar dam water by first chlorinating the water and then by controlling its color, smell, and taste. Tehran's another water treatment plant is called Mamlo processing the Jajrud river (Jahani and Reyhani 2007), which is located in the east of the city, within Pakdasht region. Mamlo provides the drinking water of the southern and southeastern resident populations of Tehran, and it is a backup resource for water management in Tehran.

More than 70% of Tehran's water is supplied from the surface water and the remaining 30% from groundwater resources (Aftabir.com 2016). To break down this statistics, around 2.5 billion m^3 of Tehran's water consumption is supplied from the groundwater resources, out of which 1.15 billion m^3 used in the agricultural sector, 800 million m^3 in drinking water sector, 170 million m^3 in industry sector (IWRMC

2017), 330 million m^3 in the service sector, and 50 million m^3 for irrigation of gardens and garden villas.

16.3.2 Sewerage System

The sewage system of Tehran was a traditional one, e.g., leaching pits, for a long time. By growing the population and the amount of sewage, the possibility of pollution in the water transfer to Tehran, through surplus, surface and subsurface sources, increased. Therefore, developing a sewage system has become a priority in the city plans. United Nations Developing Program and World Health Organization started the initial studies in 1971 (Habibi and Bernard 2005d). The first part of the plan was to collect the city's sewage and surface water. At the time of the revolution, this plan got suspended. Later, in 1985, the studies were updated and the plan was revised. Currently, many parts of Tehran are connected to the sewage system according to the updated plan. The capacity of the current wastewater treatment plants of Tehran is 787,000 m^3/d (Tsn.tpww.ir 2017).

16.4 Water and Sanitation Services

As mentioned in Sect. 16.2, the Ministry of Energy (MoE) sets the policies, and the provincial companies are in charge of services provision for water supply and sanitation in Tehran. Before the establishment of the branch of water and wastewater companies of the urban water and wastewater department, distribution and supply of water, as well as wastewater collection and treatment services were provided by various organizations, such as MoE, municipalities, private sectors, cooperatives, local councils, and municipal companies. In some cases, even there was no clear custodian. Rural water supplies also faced similar situation; they were also subject to many changes in management. In the year 1989, the urban water and wastewater department established in the MoE, and on January 01, 1991, the branch of water and wastewater companies ratified (Tashauoei et al. 2012). The approval of a law on establishment of water and wastewater companies opened a new chapter in the development and improvement of national water and sanitation services in Tehran.

Today, there are 12 affiliated companies and a specialized parent company, known as National Water and Wastewater Engineering Company (NWWEC), which provide comprehensive water and wastewater services in the urban and rural areas. The objectives in creating the NWWEC include defining MoE's duties for water and wastewater affairs, and efficiently using facilities available in the affiliated companies under approved policies. Moreover, the NWWEC has the main role in setting targets, steering, supporting and overseeing the executive affairs and the logistics, and supporting the water and wastewater companies (Tashauoei et al. 2012).

Some of the political and social conditions that affected the development of water resources management in Tehran include extensive citizens' participation, non-governmental organizations (NGOs), local participation in management, and other political changes in favor of water and agriculture sector (Ardakanian 2005). There have been also some developments in installations and equipment, such as in irrigation and drainage networks, water transmission pipelines, treatment plants, and urban water distribution networks. Furthermore, there has been improvement in institutional capacities, such as getting consultation from experienced national experts for construction advices, which reduced foreign exchange expenses, considerably (Ardakanian 2005).

16.5 Challenges of Water Supply in Tehran

Groundwater level decrease, land subsidence, water contamination, soil erosion, desertification, forced migration, and ecosystem damages are some of modern water-related issues of Tehran. Although we can blame the current crisis on climate change and frequent droughts, the main drought that Iran (and Tehran) is suffering from is socioeconomic drought, when it comes to water problems (Madani et al. 2016). Aggressive shortsighted regional development plans instead of integrated development plans have resulted in unintended water challenges whose long-term charges may be significantly higher than their short-term benefits.

Tehran is currently experiencing a rapid development and population growth, which make water challenge in Tehran different from other cities of Iran, as the depth and extent of their secondary impacts are greater. Along with changes due to the population increase and urban development, excessive water resource consumption, changes in quality and limit of water resources, and protecting water resources and aquatic environments against pollution are drivers of water crisis in Tehran. Some effective strategies toward sound management, in order to address this crisis, are discussed in the next sections.

16.5.1 Population Growth

As illustrated in Fig. 16.4, Tehran's population has increased in the last century, enormously. More precisely, it increased from 1.4 million in 1955 to 5.83 million in 1985, a 416% increase in only 30 years. This pattern of expansion from 1985 to 2015 continued with a lower rate of increase (144% of increase) from 5.83 to 8.43 million, respectively. The city's population has now exceeded 8.5 million. Despite the decreasing trend for the rate of population growth in Tehran, this megacity simply will be unable to manage growth rates because of the migration and new policies for population increase (Biswas et al. 2004). The population growth increased the needs

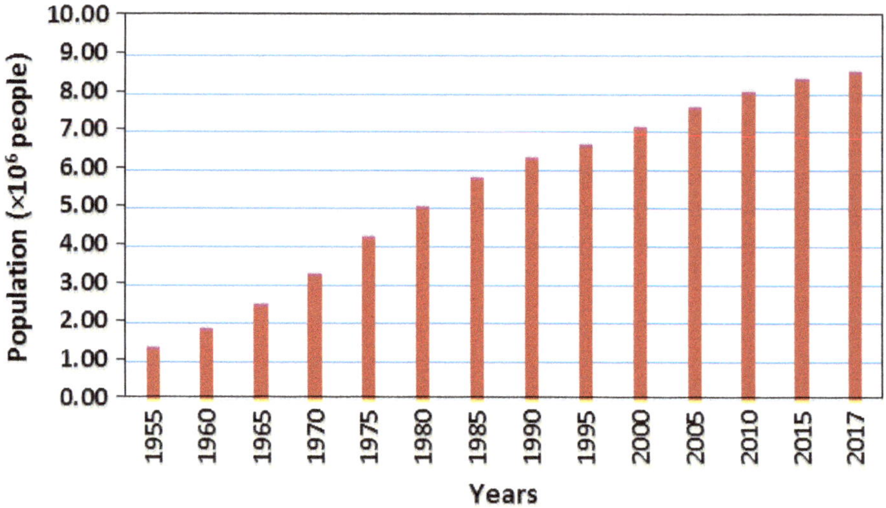

Fig. 16.4 Trend of population increase and population growth rate in the megacity Tehran. Data from the world population review (Worldpopulationreview.com 2017a)

for water supply and also more sophisticated managerial strategies (Worldpopulationreview.com 2017a).

According to 2015 census data, about 88% of 1,084,000 increased inhabitants in Tehran are immigrant population. Accordingly, the highest proportion (20.2%) of the total immigrant population of the country belongs to Tehran in comparison with other 30 provinces in Iran. Although Tehran in prior periods had also experienced challenges of immigrants, what already triggered the crisis of imposed population is the attractiveness of the capital and surrounding cities for immigrants compared to other metropolises of the country (Amar.org.ir 2016).

Tehran province accommodates 13.3 million people, 16.5% of the total population of the country. Out of this, about 12.2 (91%) million people are living in urban areas and 1.1 (9%) million people in rural areas. It is expected that this trend does not change in near future. The population living in Tehran is already around 5% more than that of half a decade ago (Amar.org.ir 2016; Donya-e-eqtesad.com 2017).

By definition, a megacity is a metropolitan area with more than ten million populations. The terms such as conurbation, metropolis, and motorplex are also equivalent to megacity (Wikipedia.org 2017b). The rapid growth of megacities has caused water planning and management challenges in developing countries, as out of 10 large cities of the world in 1994, only 3 were in developed countries. In 2017, the number of megacities in developed countries was expected to decline, one of which was Tokyo. While Tokyo's population was estimated to increase by less than 4% (Worldpopulationreview.com 2017b), cities like Jakarta, Karachi, Lagos, and Dhaka were expected to grow by greater than 60%. This increase for Tehran was estimated to be between 20 and 30% (Biswas et al. 2004).

As of 2017, there are 37 megacities in the world (Fig. 16.5a). Tokyo and Greater Jakarta are the largest of these megacities, while Shanghai has the world's largest city population (Biswas et al. 2004). The UN World Cities Report (UB-Habitat 2016) states that the number of "megacities" with more than 10 million populations has been more than doubled from 14 in 1995 to 29 in 2016. The UN also recognized that the management of these megacities is certainly difficult and if urban migration will not be controlled, it may cause serious issues. The "New Urban Agenda" proposed as a master plan by UN focuses on five guiding principles that include human rights, economy, democracy, education, and environment for cities to compensate the consequences of these migrations. Megacities can improve their potentials by considering this agenda and by recognizing what the UN calls transformative force (WCR.unhabitat.org 2016; Weforum.org 2016).

Water management is easier when the population density is not high. As presented in Fig. 16.5b, Tehran is the second most sensitized megacity. When population starts to increase and the rate of urbanization accelerates, clean water and safe sanitation

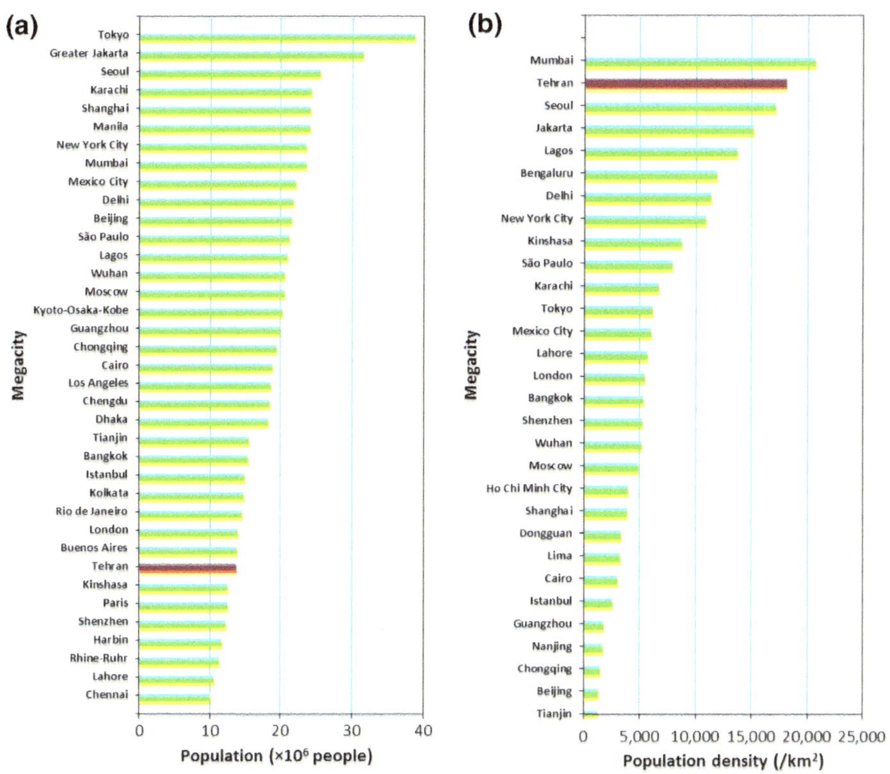

Fig. 16.5 Ranks of population (**a**) and population density (**b**) in megacities of world. Population data from the megacities (Wikipedia.org 2017b) and population density data from the list of cities proper by population (Wikipedia.org 2017a)

service provision, as well as control of storm waters, become increasingly complex and more serious for megacities similar to Tehran.

It need to be noted that when comparing the past and the present developments in developed and developing countries, the rate of growth and development in the megacities of the developed countries has been a slower process. Thus, the population growth in a megacity such as New York City has spread gradually over a century. This way, the city was able to develop its necessary infrastructures with a progressive manner and also manage all its water-related activities and services, effectively. In contrast, Tehran has witnessed explosive growths, e.g. 416% increase only in 30 years.

16.5.2 Rapid Urbanization Rate

As mentioned in Sect. 16.5.1, the rate of urbanization has increased significantly during recent decades in Tehran. The rates of urban development and urbanization in the megacity of Tehran have exceeded the capacities and resources of local government to plan and manage the urban needs in an efficient, equitable, and sustainable manner. Uncontrolled increase of population and urbanization requires urgent capacities of drinking water and sanitation services from government.

Figure 16.6 shows acquired false-color images from the Thematic Mapper sensor on NASA's Landsat 5 satellite of Tehran on August 02, 1985, and July 19, 2009. In Fig. 16.6, both images of a and b show vegetation in bright green, urban areas in color range from gray to black, and barren areas in brown. Non-urbanized areas in Fig. 16.6a are in fringe, and urbanization areas fill almost the entire frame of Fig. 16.6b. The main roadways and crisscrossing of the city in 1985 remained visible in 2009, while many additional roadways have been added, mostly in the northern parts of the city (Landsat.visibleearth.nasa.gov 2017).

Although urbanization has posed a major challenge in providing satisfactory water services to this megacity, its role and impact in development of a stronger and more robust national economy should not be ignored. The urban areas of the developing world, with around 30% of the total population in year 2000, contributed in nearly 60% of the total GDP and played an important role in social and cultural enhancements. Therefore, we need to recognize that urbanization process brings both opportunities and challenges at the same time (Lundqvist et al. 2005).

Tehran's population explosion and urban expansion may lead to various challenges, also with respect to storm water infrastructure systems. The storm water system may not keep pace with the rapid trend of the urbanization, and therefore, it may negatively affect the urban life and environment in Tehran today. Decrease of rivers' water quality threatens the life of the population exposed to flooding is an example of the damages may be resulted from inconveniences within the drainage system in Tehran. As illustrated in Fig. 16.6, generally the expansion of urban areas leads to significant changes from natural landforms and vegetative covers to unnatural and impermeable areas.

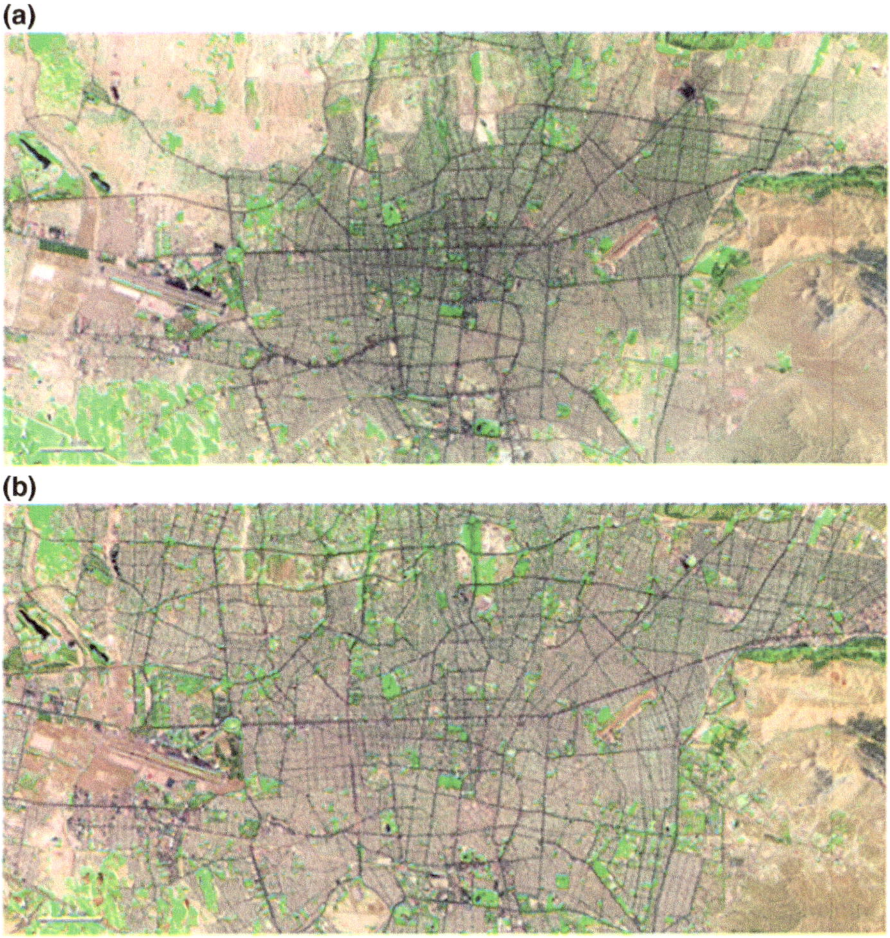

Fig. 16.6 Tehran's image from the "Thematic Mapper sensor on NASA's Landsat 5 satellite" captured on **a** August 02, 1985, **b** July 19, 2009. Reprinted from Landsat.visibleearth.nasa.gov (2017)

Even though Tehran has a semiarid climate, the existing flood control and drainage facilities are exposed to constriction and lack of sufficient freeboard. Therefore, Tehran is struggling with minor floods that occur mostly in the southern and central regions where flows are massive. Apart from increasing storm water runoff quantities, pollutants from non-point sources (e.g., debris, litter) are also accumulated on impervious areas, collected by storm water runoff and discharged into receiving waters. Therefore, different types of chemical and microbial pollutants concentrated in storm water runoff lead to the contamination of receiving water bodies. Pollution loads of point sources often contain very harmful substances; therefore, water quality of Tehran´s streams and rivers turns to be a hygienic issue for the city, especially in the case of flood events. In addition to increase of the likelihood for floods, another negative effect of the urbanization in Tehran is the fact that the impervious areas can prevent storm water to infiltrate into the ground and recharge the groundwater resources. Since Tehran's water supply system relies mainly on groundwater, the lack of recharge on the one hand and the increase in the water demand on the other lead to a considerable decline of the groundwater table (Ziller 2010). Considering all these challenges of urbanization and storm water management, adopting an infrastructure for Tehran's storm water to current development situations of this megacity seems essential.

16.5.3 Drought

Drought is a physical phenomenon that produces serious biological effects since it causes moisture imbalance and a complexity for plant–habitat relation (Subrahmanyam 1982). Thornthwaite (1947) has defined drought as a condition that when precipitation is not enough for the "established human and other biological activities" (Thornthwaite 1947). In a similar way, Mather (1975) believes that drought occurs when the moisture of soil or precipitation is not sufficient to satisfy the water need of plants.

In a study by Abounoori (2015), in which water-budget methods and Thornthwaite's aridity index are used, the intensity of drought effects during 1951–2013 in Tehran were assessed. The study shows that Tehran is suffering from moisture deficiencies in most part of the year and any type of water excess is not in the known climatic timetable. Rainfall occurs mainly in winter, declines to almost zero by May, and continues to be low until December. The rainfall water is evaporated very fast; thus, accumulation of moisture in the soil is not possible at any time of the year. The study showed that Tehran experienced 32 different types of droughts from 1951 to 2013, also revealed that the severity and intensity of droughts increased significantly from 2000 to 2012, at the time of disastrous droughts (Abounoori 2015).

Figure 16.7 shows meteorological drought zoning for recent seven years (up to the end of May 2017) of the megacity of Tehran based on Standardized Precipitation Evapotranspiration Index (SPEI) prepared by National Drought Warning and Monitoring Center (NDWMC) of Iran (NDWMC 2017). As reported by NDWMC, 16.6, 29.4, 36.7, 5.0, and 7.4% of Tehran province area are exposed to very severe drought, severe drought, moderate drought, mild drought, and normal drought, respectively. Also, respectively, estimates for Tehran city are 55.1, 44.4, 0.5, 0.0, and 0.0%. Furthermore, cities of Tehran, namely Qods, Shemiranat, Shahryar, Tehran, and Islamshahr, account for 100, 97.3, 72, 55.1, and 19.5% areas exposed to very severe drought, respectively.

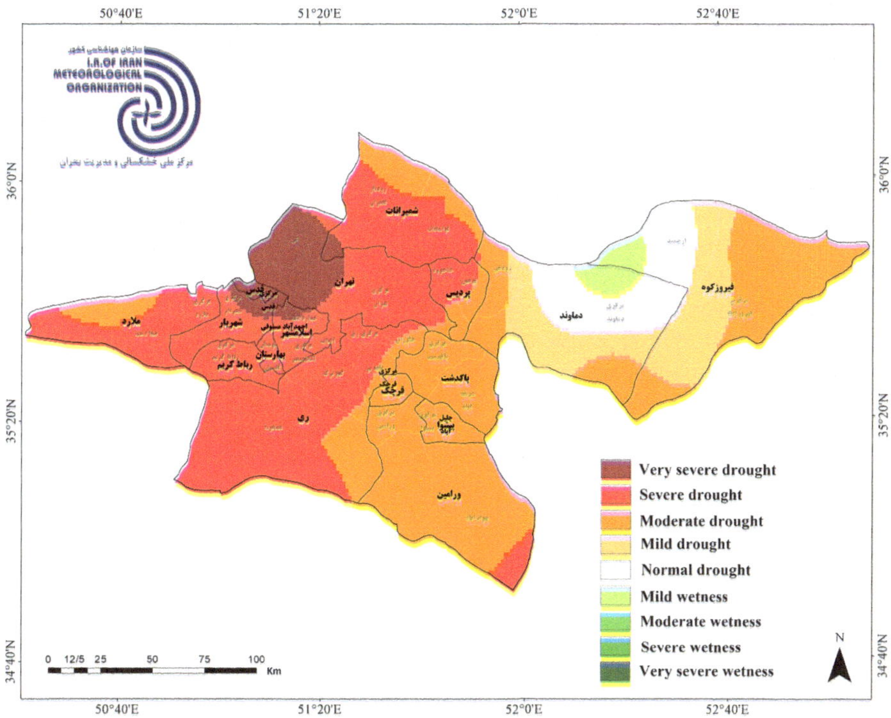

Fig. 16.7 Zoning of meteorological drought in Tehran megacity based on Standardized Precipitation Evapotranspiration Index (SPEI). Reprinted from NDWMC (2017)

16.5.4 Water Consumption Challenge

The per capita water consumption in Tehran is now more than ever before. The average individual water consumption in Iran is near 250 L/day (Jahani and Reyhani 2007), and as high as 400 L/day in highly populated areas, such as Tehran (Ravanshadnia et al. 2015). This puts Tehran's water usage at twice the world's average and may face the city to serious supply issues in future.

Perhaps the biggest problem of all in Iran and the megacity of Tehran is the massive water usage inefficiency inside the agricultural sector, which causes more than 92% of water consumption allocated to this sector in Iran. The annual rainfall in the province of Tehran is 4.5 billion m^3 from which 3.5 billion m^3 are consumed by the city (Ghatreh.com 2015). Out of these 3.5 billion m^3 of available water in Tehran, about 1.8 billion m^3 are consumed in the agricultural sector and 1.5 billion m^3 are consumed for domestic uses. About 51% of the water in the city of Tehran is consumed for agriculture, 5% for Tehran's public green spaces, about 23% for the domestic sector, and 15% for industrial and other uses. Therefore, although water use by agriculture sector in Tehran is lower than that in the country, it still has the highest percentage of water use in Tehran. At the same time, agricultural efficiency in Tehran is only 40% (Ghatreh.com 2015).

Furthermore, in Iran, most of the groundwater is used toward agricultural activities, where farmers use the highly subsidized energy for their digging expansion and use the water inefficiently. The current rate of irrigation efficiency in agriculture sector is just 35%. Also, only 5% of farmed areas are under pressurized irrigation (Azari 2017). Unlike the common pattern of water consumption in developed countries where most part of water use (around 50%) is devoted to the industry, in Iran only 2% is devoted to the industry and the agricultural efficiency is not even enough to be compensative. Moreover, the tap water quality is becoming unsatisfactory, in some cities of Iran, due to the nearby agricultural activities, inappropriate sewage collection and treatment systems, and discharge of domestic wastewater into surface or groundwater sources.

16.5.5 Water Stress Around the Country

One of the characteristic measures of water stress is the annual "withdrawals to availability ratio" (w.t.a). Based on this measure, increase in water withdrawal increases the water stress due to possible population and economic growth. Also, decrease in water availability, due to water pollution or climate change impact on water resources, increases the water stress (Kabat and Van Schaik 2003).

Based on Alcamo and Henrichs (2002), river basins with w.t.a ratio of greater than 0.4 are under severe water stress. World Water Commission (Cosgrove and Rijsberman 2014) and a Consortium of UN Organizations (Raskin et al. 1997) suggested to use this value as a threshold for "severe" water stress. Water stress is classified as "low" with w.t.a of lower than 0.2, "medium" with w.t.a of between 0.2 and 0.4, and "severe" with w.t.a of larger than 0.4.

Today about 40% of the world's population, 2.4 billion people, live in water-stressed areas and the UN projects that this will increase to 50% by 2030. Regionally, the Middle East, North Africa, and Asia are under the greatest stress (Hydrobioars.com 2016).

World Resource Institute (WRI) scored and ranked future water stress based on the combination of climate models with socioeconomic scenarios and measured the depletion of surface water in 167 countries by 2040 (Maddocks et al. 2015). A higher percentage means more water users are competing for limited supplies. WRI found that 33 countries face extremely high water stress in 2040. Iran with a score of 4.91/5.0 (w.t.a ratio of greater than 0.8) could face an extremely high stress by 2040.

Maddocks et al. (2015) showed that 14 out of 33 likely to be the most water-stressed countries in 2040 are in the Middle East, nine of which are considered highly stressed with a score of 5.0 out of 5.0: Bahrain, Kuwait, Palestine, Qatar, United Arab Emirates, Israel, Saudi Arabia, Oman, and Lebanon (Maddocks et al. 2015). The renewable water in Tehran province is around 4.5 billion m^3, out of which

3.5 billion m^3 are used by the city. Therefore, water stress index in Tehran province is 78% which is in severe stressed condition.

16.5.6 Financial and Educational Supports

International financial and educational supports for improving water supply and sanitation infrastructure in Tehran are inevitable, due to the challenges it faces and will be faced. During the first decade of twenty-first century, the main external partner of the Iranian water and sanitation sector in these regards was World Bank. Today, after imposing some international sanctions against Iran, the main external partners are the Islamic Development Bank (IDB), the United Nations, and NGOs. IDB has allocated more than 800 million euros of loans for Iran's water and wastewater projects, and 175 million euros to Tehran (Wikipedia.org 2017c).

The main part of the IDB water and wastewater funds goes to the Iranian wastewater. Iran is the third-largest shareholder of the IDB with 8.28% of share. There have been collaborations and trainings with international institutions, e.g., UNESCO-IHE in Delft, the Netherlands, together with the Power and Water University of Technology (Shahid Abbaspour University), are training Iranian professionals for water and wastewater technologies, planning and management. In addition, there have been 20 study tours to visit European water and wastewater companies organized for senior managerial, financial, and technical staff (Wikipedia.org 2017c).

There are also some private international non-governmental organizations in Tehran that pursue environmental or social goals. Foreign NGOs have also become active, despite the frictions inside the country. Organizations that were closed due to the frictions with ruling bodies before, through an intervention by the government encouraging participation of foreign NGOs, they began their activities, also some new startups in the areas of water and sanitation, have started projects in Tehran.

16.6 Discussions and Suggestions

16.6.1 Implementation of Sustainable Sanitation and Water Management (SSWM): A Short-Term Suggestion

The traditional urban water supply and wastewater management in Tehran is not sustainable when it comes to water and nutrients usage. It can cause environmental, economical, and cultural challenges that may also intensify through urbanization. Implementation of the SSWM principals for the water supply and sanitation system in Tehran is necessary. This tool should be emphasized by considering the reuse of water as definitive solution for water quantity challenges in this megacity. There

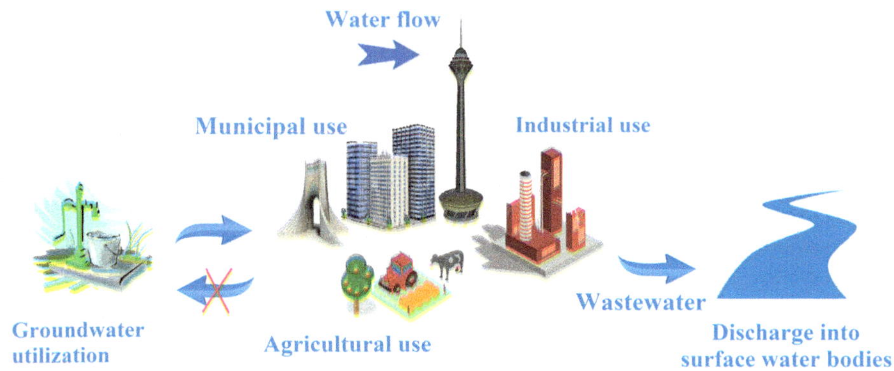

Fig. 16.8 Linear groundwater flow direction from aquifers to surface waters (adopted from Corcoran 2010) with some modification

are two reusing scenarios in SSWM, namely "water reuse" and "nutrient reuse" (Corcoran 2010).

As illustrated in Fig. 16.8 groundwater in Tehran taken from aquifers, after consumed in different sectors, it is discharged into surface waters to exit the city. This process leads to a significant decrease in the level of groundwater tables. It should be emphasized again that around 30% of Tehran water needs is supplied by groundwater. Thus in future, such process—transferring groundwater to outside of the city—will limit the water availability in a long term if no predictive plan for water reuse and groundwater recharge is designed.

Like water, nutrients are chemical elements that are required for plants' and animals' growth. In a sustainable ecosystem, there should be a nutrient balance in the natural cycle between the organism growth and their degradation (Jönsson et al. 2004).

In urban ecosystem, such as Tehran, nutrients are taken from the soil through agricultural products, and discharged with wastewater into aquatic ecosystems, where they can cause severe environmental problems (see Fig. 16.9). Such a process leads to contamination of downstream surface waters, and also deterioration of lands and soils at places that this water is used for agriculture activities.

This is the largest drawback of a centralized sewer system in Tehran where in most cases do not favor recycling of resources in a closing loop. Water (often groundwater—30%—and upstream surface waters—70%) enters the water distribution system then discharged into surface water, leading to groundwater depletion. Moreover nutrients, which come from the soil, are discharged into water bodies, leads to the city soil depletion, on the one hand, and probable eutrophication in downstream water bodies, on the other hand. This process is illustrated in Fig. 16.9. Moreover, most of the municipal wastewaters are used without sufficient treatment or with conventional treatment in agricultural fields, where they can, in addition to the environmental challenges, trigger some health concerns regarding exposure to microbial and chemical contaminations. Thus, conventional approaches to water management that concern wastewater as a waste have serious shortcomings (Corcoran 2010).

Implementation of a sustainable approach or tool in order to provide a sound management for water in this megacity is necessary. In a sustainable urban ecosystem, waste should be considered as a resource, and its management should be all-inclusive and form as part of an Integrated Water Resources Management (IWRM). IWRM is based on principles, such as considering freshwater a finite, vulnerable resource that is essential to sustain life, development, and the environment. Moreover, participatory approach for water development and management by recognizing water as an economic good is as an important principal of the IWRM (Tessendorff 1992). Hence, implementation of the IWRM systematic process, which is a coordinated framework for water, land, and other resource management, that ensures the economic and social well-being in an equitable manner, without compromising the sustainability of the ecosystems, is inevitable in Tehran megacity.

As presented in Fig. 16.10a, in a conventional urban water cycle, which applies to most cases in Tehran, reusing opportunities are not considered, which has lead to waste of water, nutrients, and energy. The linear urban water cycle (Fig. 16.10a) in Tehran may lead to missed reuse opportunities and unexpected impacts. Thus, only in a holistic way with a significant change in the water governance, water supply and sanitation systems can be managed. By using an integrated approach which includes various aspects of urban water management, such as water safety plans, sanitation safety plans, and other sectors such as health, housing, energy, and waste management, it will be easier to minimize the negative impacts throughout water and sanitation system in Tehran.

In a sustainable urban water and sanitation management (Fig. 16.10b), there are three reusing options, including reuse of energy by producing biogas from wastewater, reuse of water for irrigation and recharging of groundwater resources, and reuse of nutrients, such as phosphorus from human urine and excreta to recover for agriculture. Although, in a sustainable water reuse, gray water, rainwater, storm water,

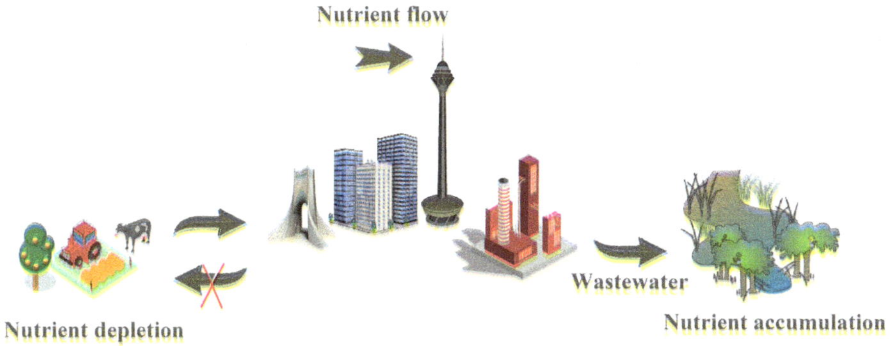

Fig. 16.9 Linear nutrients flow direction from soils to aquatic environments (adopted from Corcoran 2010 with some modification)

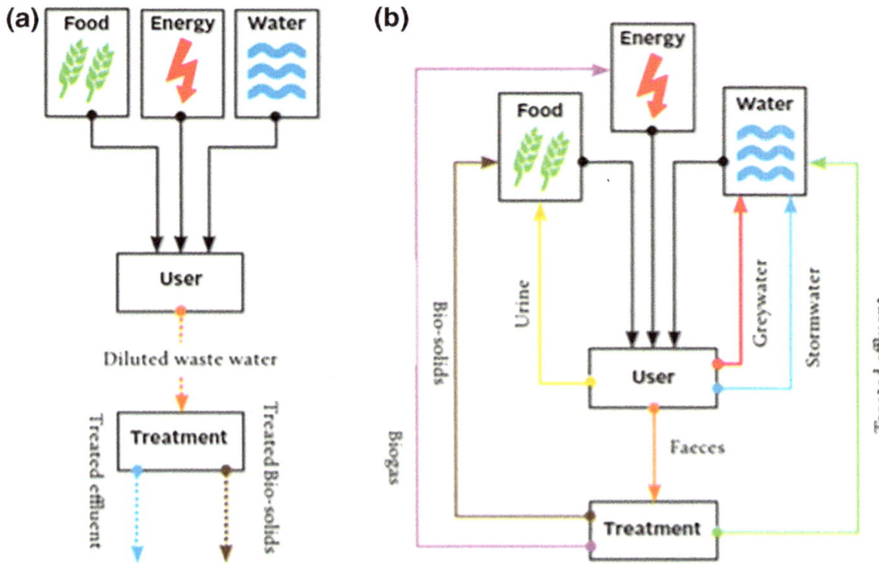

Fig. 16.10 Linear (**a**) in comparison with cyclical (**b**) wastewater management system. Reprinted from Howe et al. (2011)

and urine can be reused, logically (Fig. 16.10b), for Tehran, which has a different demographic, cultural, and urban complexities, also suffers from weak infrastructure, implementation of decentralized systems may not be feasible at the present.

Since the sewage collection system in Tehran is entirely centralized, perhaps the best option is to upgrade current centralized systems in such a way that the recycling of wastewater is done to a large extent with higher efficiency. In this regard, the use of advanced wastewater treatment systems after the conventional treatment system can produce a higher class of quality for water that can be used for various urban uses. This way, the discharge of treated effluent into the receiving water is prevented, the pollution of agriculture products with contaminated wastewater is reduced, water will return to its original location, and the balance between groundwater withdrawal and recharging will also be maintained.

In order to achieve an urban SSWM in the megacity of Tehran, implementing an integrated water resources management is necessary. As a management option, using additional treatment processes (coagulation and flocculation, filtration and disinfection) at the first step is required for producing a higher-class reclaimed water (class A). Class A can be used in public areas, such as parks and schoolyards. Thus to reuse the water for urban usages, the reclaimed water must receive extra treatments from the conventional secondary treatment in Tehran. The reclaimed water that is recharged into groundwater supplies requires extra treatments from Class A. Drinking water standards must be satisfied for this water to be able to reach the aquifer. Also, the degree of treatment depends on the way that the water is released: on the surface through the soil, or directly into the aquifer. If released to the surface, the percolation must also remove the nitrogen and meet other requirements, depending on the site condition, such as type of soil, and the distance from the aquifer, as well (EPA 2012).

As the second management option for water reuse in Tehran, application of decentralized treatment and reuse systems may be useful. In most megacities in the developed countries, application of centralized systems of water and sanitation services exists while they are confirmed to be ineffective and inefficient. Moreover, they are very expensive, which seems to be unreasonable for developing countries (Chocat 2002). Application of decentralized water supply and wastewater treatment systems potentially increases the opportunities for local stakeholder participations in planning and decision making.

Point-Of-Use (POU) systems, Point-Of-Entry (POE) systems, and Small-Scale Systems (SSS) are three main categories for decentralized treatment systems. However, the application of decentralized supply system mainly depends on the local conditions, and several other factors, such as easiness-to-use, maintenance, electricity and fuel supply dependence, and cost. They need to be well studied by considering all of their limitations and cost-benefit analyses. Moreover, engagement of the private sector for revenue opportunities from products of the sanitation systems, such as biogas, fertilizer, soil conditioner or irrigation water, and reuse opportunities by increasing number and quality of WWTPs, and also improvement of drainage and irrigation systems, can increase the economic efficiency of water and sanitation systems in return.

Allocation of more water to the industrial sector, improving irrigation systems efficiency in agriculture sector, and controlling non-revenue water in water supply systems are some other important aspects need to be considered for water supply management in Tehran. Moreover, supporting agricultural activities in rural areas bring job opportunities that might reduce the rate of migration from villages to cities and the pressure on the water supply systems in urban areas as well.

16.6.2 Implementation of Water and Sanitation Safety Plans: A Medium-Term Suggestion

Drinking water safety in a megacity such as Tehran can only be ensured when the performance accuracy parameters are well defined and considered while pursuing qualitative monitoring. The drinking Water Safety Plan (WSP) and Sanitation Safety Plan (SSP) are suggested by the World Health Organization (WHO) to address this matter. All parts of the drinking water supply system are included in the WSP in order to ensure the quality of drinking water. The SSP is also a well-defined risk-based approach that helps implementing the WHO guidelines for safe use of wastewater, excreta, and gray water. It can also be applied to all sanitary systems to ensure the system meets the health objectives.

According to the current laws and regulations, water and wastewater companies are mandated to supply safe and sanitary potable water. Ensuring water safety was the main concerns of these companies from their onset. This issue is addressed by expanding the network of laboratories with the purpose of identifying and measuring all agents that undermine the quality of water and ensuring the adequacy of sampling at most locations throughout the city, and by training specialists and interacting with monitoring bodies such as WHO and the Ministry of Health and Medical Education (MHME). Protection of water resources from chemical and microbial contaminants through sophisticated monitoring laboratories and approaches while assessing the unwanted effects of pollutants on the consumers have become the subject of WSP; some relevant issues to be addressed are as follows:

- Establishment of specialized safety committee units from ministries of health and MoE at national level to review issues and propose emergency action plans;
- Compiling guidelines on equipment and safety of water laboratories;
- Ensuring the adequacy of quality parameter measurement to guarantee the safety of water at consumption points with an emphasis on emerging pollutants;
- Increasing quality parameters to ensure the adequacy of water safety at consumption points.

16.6.3 Promoting Understanding of the Importance of Water at the Community and Government Levels: A Long-Term Suggestion

In terms of education, all stakeholders need to understand their roles regarding recycling and reclamation, the two new very necessary and important concepts in Tehran, in order to increase available water supplies, and break the population–water paradox. Dams only must be constructed to increase water storage capacity and prevent water shortages wherever there is no other option for controlling the water resources.

Furthermore, the irrigation systems in agriculture and water distribution networks management need improvements. It is also important to consider an effective pricing in order to help improving public understanding of water importance, and to control water consumption per capita per day. In Iran, other than government, religious organizations, educational entities, and mass media can play role in educating the public toward optimum water consumption, water recycling, and paying attention to water resources. The leading role in water management is for the industries, which can help both solving the water crisis and creating a sustainable development using the economic force.

Public education for environmental issues is also important for management of water at the point of use. Education, especially in the case of water consumption could provide an opportunity to increase local community knowledge on the value of treated water. By educating the community and children (in schools), their knowledge about the importance of water can be enhanced and they will learn to use the treated water for only high classes of use (drinking, cooking, and sanitary consumptions).

16.7 Summary and Conclusion

In this chapter, some of the important issues related to water supply system in Tehran as a complex megacity have been presented. This city is one of the most polluted cities on earth, suffering from different environmental issues. What makes the water challenge in Tehran different from other provinces in Iran is the depth and extent of the secondary impacts of its uncontrolled development. Population growth in the city, and more importantly, in slums increases the request for safe water supply and sanitation services. The rate of urbanization is also exceeded the capacities and resources of local government to plan and manage the urban needs in an efficient, equitable, and sustainable manner. As another important challenge, Tehran is now struggling with different types of droughts, which make everything harder for the sound management of water resources.

Management of water and nutrients through an unsustainable fashion in a conventional urban water supply and wastewater of Tehran can lead to critical environmental, economical, and social issues, which also intensifies by the process of urbanization. The centralized sewage systems in Tehran in most cases do not favor recycling of resources in a closing loop. The wastewaters in most cases with or without treatment are discharged into surface water, leading to groundwater depletion exiting the city. Moreover, nutrients, from soil, are discharged into water bodies, leading to soil depletion and possible eutrophication in downstream water bodies.

As a short-term suggestion, conventional approaches to water management that considers wastewater as a waste have serious shortcomings in Tehran. Thus, implementation of the Sustainable Sanitation and Water Management (SSWM) principles and an Integrated Water Resources Management (IWRM), and the reuse of water as definitive solution for water quantity challenges in this megacity should be considered. Improvement of wastewater treatment system by using additional treatment steps to produce higher level of classes for reclaimed waters could provide the possibility of water reuse consumption with higher classes of and also recharging of groundwater resources.

For a medium-term suggestion, the feasibility of the application of a decentralized water supply and wastewater treatment system needs to be assessed through local stakeholders who participate in planning and decision making. In the case of management of health issues related to water and sanitation, drinking Water Safety Plan (WSP) and Sanitation Safety Plan (SSP) need to be established and performed by water and sanitation authorities.

A long-term suggestion regarding the water challenges in Tehran is utilizing international financial and educational supports in order to improve water supply and sanitation infrastructure. Education can enhance public knowledge on the value of treated water, which is also important for controlling the high rate of water consumption in this megacity.

References

Abounoori A (2015) Analysis the economics of Water-Budget in Tehran using the Thornthwaite's method. Indian J Fundam Appl Life Sci 5:1000–1007
Aftabir.com (2016) Tehran's water bill was positive. http://www.aftabir.com/news/article/view/2017/02/13/1634952. Accessed 26 June 2017
Alcamo J, Henrichs T (2002) Critical regions: a model-based estimation of world water resources sensitive to global changes. Aquat Sci-Res Across Boundaries 64(4):352–362
Amar.org.ir (2016) Population and housing censuses. https://www.amar.org.ir/english/Population-and-Housing-Censuses. Accessed 26 June 2017
Ardakanian R (2005) Overview of water management in Iran. Water conservation, reuse, and recycling. In: Proceeding of an Iranian American workshop, The National Academies press Washington, DC
Azari N (2017) Helpless solutions on water issues: strategic lessons from the experience of developing irrigation systems under pressure. Network for Public Policy Studies (NPPS), N. 5385-2423, npps.ir/GetFile.aspx?id=223761
Biswas A, Tortajada C, Lundqvist J, Varis O (2004) The water challenges of megacities. Stockh Water Front. https://dlc.dlib.indiana.edu/dlc/bitstream/handle/10535/5254/WF2-04_Water_Challenges_of_Megacities.pdf?sequence=1. Accessed 26 June 2017
Chocat B (2002) Sustainable management of water in cities. Universidad Internacional Menendez Pelayo, Valencia. http://www.petus.eu.com/graphics/FRsusmanagecitiesBC.pdf. Accessed 26 June 2017
Corcoran E (2010) Sick water?: the central role of wastewater management in sustainable development: a rapid response assessment. UNEP/Earthprint
Cosgrove WJ, Rijsberman FR (2014) World water vision: making water everybody's business. Routledge
Donya-e-eqtesad.com (2017) Population inequality in Tehran. http://donya-e-eqtesad.com/news/1101783. Accessed 26 June 2017
English PW (1968) The origin and spread of qanats in the Old World. Proc Am Philos Soc 112(3):170–181
EPA (2012) U.S. Environmental protection agency. Guidelines for water reuse. Office of wastewater management, Office of water, Washington, D.C. EPA/600/R-12/618
Ghatreh.com (2015) Half of Tehran's water is spent on agriculture. http://www.ghatreh.com/news/nn27085105. Accessed 27 June 2017
Gober P (2010) Desert urbanization and the challenges of water sustainability. Curr Opin Environ Sustain 2(3):144–150
Habibi SM, Bernard H (2005a) Atlas of Tehran metropolis. Tehran's water reservoirs. http://atlas.tehran.ir/Default.aspx?tabid=247. Accessed 26 June 2017
Habibi SM, Bernard H (2005b) Atlas of Tehran metropolis. Important dams. http://atlas.tehran.ir/Default.aspx?tabid=246. Accessed 26 June 2017
Habibi SM, Bernard H (2005c) Atlas of Tehran metropolis. Water treatment plants. http://atlas.tehran.ir/Default.aspx?tabid=248. Accessed 26 June 2017
Habibi SM, Bernard H (2005d) Atlas of Tehran metropolis. Sewerage systems. http://atlas.tehran.ir/Default.aspx?tabid=249. Accessed 26 June 2017
Hadi M, Mesdaghinia A, Yunesian M, Nasseri S, Nodehi RN, Tashauoei H, Jalilzadeh E, Zarinnejad R (2016) Contribution of environmental media to cryptosporidiosis and giardiasis prevalence in Tehran: a focus on surface waters. Environ Sci Pollut Res 23(19):19317–19329

Howe C, Butterworth J, Smout I, Duffy A, Vairavamoorthy K (2011) Sustainable water management in the city of the future: findings from the SWITCH Project 2006–2011. UNESCO-IHE, Katwijk, The Netherlands

Hydrobioars.com (2016) Global water stress: not all regions are created equal. http://hydrobioars.com/index.php/2015/12/22/global-water-stress-not-all-regions-are-created-equal/. Accessed 26 June 2017

ICS (2015) Iran's statistical yearbook. Report No. 4897, 2015. Available from: Iranian Center for Statistics (ICS), Tehran (in Persian)

IRIMO (2017) "I.R. of Iran Meteorological Organization" Climatic data (in Persian). http://irimo.ir/far/wd/2703. Accessed 29 Sept 2017

IWRMC (2017) Iran water resources management company. Kholase Vaziate Masraf Ta 94–95 (in Persian). http://wrs.wrm.ir/tolidat-abzirzamini. Accessed 29 Sept 2017

Jahani HR, Reyhani M (2007) Role of groundwater in Tehran water crisis mitigation. In: Vrba J, Salamat AR (eds) Proceedings of the international workshop, Tehran, 9–31 October 2006: "Groundwater for emergency situations". IHP-VI Series on Groundwater No. 15, p 115

Jönsson H, Stintzing AR, Vinnerås B, Salomon E (2004) Guidelines on the use of urine and faeces in crop production. EcoSanRes Programme

Kabat P, Van Schaik H (2003) Climate changes the water rules: how water managers can cope with today's climate variability and tomorrow's climate change. WaterandClimate

Kakouei A, Vatani A, Idris AKB (2012) An estimation of traffic related CO_2 emissions from motor vehicles in the capital city of Iran. Iranian J Environ Health Sci Eng 9(1):13

Landsat.visibleearth.nasa.gov (2017) Tehran urbanization. https://landsat.visibleearth.nasa.gov/view.php?id=41308. Accessed 26 June 2017

Lundqvist J, Tortajada C, Varis O, Biswas A (2005) Water management in megacities. Ambio 34(3):267

Madani K, AghaKouchak A, Mirchi A (2016) Iran's socio-economic drought: challenges of a water-bankrupt nation. Iran Stud 49(6):997–1016

Madanipour A (1999) City profile: Tehran. Cities 16(1):57–65

Maddocks A, Samuel RY, Reig P (2015) Ranking the world's most water-stressed countries in 2040. http://www.wri.org/blog/2015/08/ranking-world%E2%80%99s-most-water-stressed-countries-2040. Accessed 26 June 2017

Mather JR (1975) Climatology: fundamentals and applications. MaGraw Hill

MHME (2011) National drinking water quality strategy. Ministry of Health and Medical Education 1. http://ier.tums.ac.ir/files/site1/pages/News_info/national_strategy_1.pdf. Accessed 25 Apr 2017

MOE (2000) Sedimentation in the reservoirs of large dams in Iran. Water Research Institute of Ministry of Energy (MOE). http://www.waser.cn/news%20and%20announcements/waser%20Technical%20Sessions/Sedimentation%20in%20the%20Reservoir%20of%20Large%20Dams%20in1final.ppt. Accessed 20 Apr 2017

NDWMC (2017) Percent of area exposed to drought. National Drought Warning and Monitoring Center reports. http://ndc.irimo.ir/far/wd/2957-%D8%AA%D9%87%D8%B1%D8%A7%D9%86.html. Accessed 26 June 2017

Pahl-Weber E, Ohlenburg H, Seelig S, Kuhla von Bergmann N, Schäfer R (2013) Urban challenges and urban design approaches for resource-efficient and climate-sensitive urban design in the MENA region. Universitätsverlag der TU Berlin

Ranjbar SA, Ali ABAA, Sadeghi HSA (2006) Effects of urban heat island and urbanization on weather and local climate of tehran greater area (based on Mehrabad and Varamin observations). J Environ Stud 32(39):59–68

Raskin P, Gleick P, Kirshen P, Pontius G, Strzepek K (1997) Water futures: assessment of long-range patterns and problems. Comprehensive Assessment of the Freshwater Resources of the World, SEI

Ravanshadnia M, Ghanbari M, Fardani MH (2015) Planning crisis management of water resources in Tehran in the next twenty five years. J Appl Environ Biol Sci 5(10):50–59

Roshan GR, Shahraki SZ, Sauri D, Borna R (2010) Urban sprawl and climatic changes in Tehran. Iran J Environ Health Sci Eng 7(1):43
SCI (2011) 1390 National census of population and housing. Statistical Center of Iran
SCI (2015) 1395 National census of population and housing. Statistical Center of Iran
SOE (2012) State of environment report of Tehran. http://en.tehran.ir/Portals/0/Document/The%20State%20of%20Environment%20%28SoE%29%20Report%20of%20Tehran%20.pdf. Accessed 26 June 2017
Subrahmanyam V (1982) Water balance and its applications: with special reference to India. Andhra University Press
Tashauoei HR, Vaghefi KA, Montazeri A, Mohebbi MR, Farhadpour Z, Alizadeh Z, Labafi S, Azhir T (2012) Integrated water quality management in urban and rural centres of Iran. IWA World Water Congress Exhibition 2012. http://www.iwawaterwiki.org/xwiki/bin/view/EventsExtra/IntegratedwaterqualitymanagementinurbanandruralcentresofIran. Accessed 26 June 2017
Tessendorff H (1992) The Dublin statement on water and sustainable development. Aqua AQUAAA 41(3)
Thornthwaite C (1947) Climate and moisture conservation. Ann Assoc Am Geogr 37(2):87–100
Tsn.tpww.ir (2017) Statistic of 1396. http://tsn.tpww.ir/. Accessed 26 June 2017
UB-Habitat (2016) World cities report 2016. Chapter 1: from habitat II to habitat III: twenty years of urban development. http://wcr.unhabitat.org/wp-content/uploads/sites/16/2016/05/Chapter-1-WCR-2016.pdf. Accessed 26 June 2017
UNDP and DOE (2010) Iran second national communication to UNFCCC. National Climate Change Office at the Department of Environment on behalf of the Government of the Islamic Republic of Iran. United Nations Development Programme (UNDP) and Department of Environment (DOE). http://unfccc.int/resource/docs/natc/snc_iran.pdf Accessed 26 June 2017
WCR.unhabitat.org (2016) WORLD CITIES REPORT 2016. http://wcr.unhabitat.org/main-report/. Accessed 26 June 2017
Weforum.org (2016) This map shows the incredible growth of megacities. https://www.weforum.org/agenda/2016/07/this-map-shows-the-incredible-growth-of-megacities/. Accessed 26 June 2017
Wikipedia.org (2017a) List of cities proper by population. https://en.wikipedia.org/wiki/List_of_cities_proper_by_population. Accessed 26 June 2017
Wikipedia.org (2017b) Megacity. https://en.wikipedia.org/wiki/Megacity. Accessed 26 June 2017
Wikipedia.org (2017) Tehran Province. https://en.wikipedia.org/wiki/Tehran_Province. Accessed 26 June 2017
Wikipedia.org (2017c) Water supply and sanitation in Iran. https://en.wikipedia.org/wiki/Water_supply_and_sanitation_in_Iran. Accessed 05 Oct 2017
Worldpopulationreview.com (2017a) Tehran population. http://worldpopulationreview.com/world-cities/tehran-population/. Accessed 26 June 2017
Worldpopulationreview.com (2017b) Tokyo population. http://worldpopulationreview.com/world-cities/tokyo-population/. Accessed 26 June 2017
Ziller A (2010) Assessment of structural storm water measures in Tehran through indicators of sustainable development

Chapter 17
Soil Sealing and Depletion of Groundwater in Rapidly Growing Peshawar City District, Pakistan

Atta-ur Rahman, Attaullah Khan, Noorul Haq, Samiullah and Rajib Shaw

Abstract The aim of this chapter is to explore the trend of soil sealing and its impact on water infiltration and loss of groundwater potential in Peshawar City District, Pakistan. The impacts of soil sealing on groundwater are a challenging research area, and this research will help in exploring the relationship between soil sealing by rapid urban growth and expansion and the resultant depletion of groundwater sources. In the study area, rapid population growth and urban expansion have engulfed agricultural land and the subsequent infrastructural development has caused soil sealing. Such human interventions have produced problems with rise in urban temperature, accelerated surface runoff, reduced fertile farmland, and largely halted percolation and infiltration. The direct and indirect impacts of soil sealing on human life and environs have also been highlighted in this chapter. This study is based on both primary and secondary data sources. Data pertaining to urban expansion has been acquired from the published sources as well as temporal Spot images of the past three decades. Similarly, soil sealing and its trend prediction for the year 2030 were calculated. Groundwater data for the entire district has been collected from Public Health Engineering Department, Peshawar. Field survey was conducted to explore the current status of water table and depleting groundwater sources. Population data has been compiled from district census reports and statistical profiles. The analysis reveals that rapid population growth has increased the abstraction of groundwater and the potential of these freshwater sources has been threatened.

Keywords Soil sealing · Groundwater · Water table · Infiltration · GIS
Remote sensing · Land use/land cover changes · Urbanization

A. Rahman (✉) · N. Haq · Samiullah
Department of Geography, University of Peshawar, Peshawar, Pakistan
e-mail: atta_urp@yahoo.com

A. Khan
Department of Geography, Government College Lund Khwar, Mardan, Pakistan

R. Shaw
Graduate School of Media and Governance, Keio University, Fujisawa, Japan

B. Ray and R. Shaw (eds.), *Urban Drought*, Disaster Risk Reduction,
https://doi.org/10.1007/978-981-10-8947-3_17

17.1 Introduction

Globally, soil offers various services and functions for sustaining civilization and human development (Cannon and Horton 2009; Munafò et al. 2010). However, rapid urban growth and expansion exert several irreversible impacts on the function, and structure of soil and their potential capacity are greatly threatened (Haygarth and Ritz 2009; Xiao et al. 2013). In urban areas, infrastructure and other physical developments are stimulating soil sealing. Among the common characteristic of cities is that they are covered by impermeable surfaces and one of its typical characteristics is the contiguous built-up area (Turok and Mykhnenko 2007; Wessolek et al. 2008; Scalenghe and Marsan 2009). Surface covering by impervious materials is the main cause of soil sealing (Arnold and Gibbons 1996; Prokop et al. 2011). Physical developments such as buildings, pavements and other infrastructural developments are the common surface covers in urban centers (Breuste 2011). It is an irreversible process of converting land under agriculture, forests, barren, and other ecological land into built-up environment (Sigbert and Freudenschuss 2002). As a result, it produces negative effects on human health and urban environment (Burghardt 2006; Salvati et al. 2011). It can rise surface temperature, accelerate surface runoff, and increase the risk of urban flood.

Worldwide, groundwater recharging rate is reduced by halting infiltration and percolation (Duley and Kelly 1939; Paul and Meyer 2001). Likewise, taking of agricultural land has turned into a potential threat to food security and urban ecology (Montanarella 2007). Rapid population growth, urban expansion, and infrastructural developments were identified as major causes of soil sealing from 1981 to 2014 in Peshawar City District, Khyber Pakhtunkhwa (KP) Province of Pakistan. Analyzing the impacts of urban growth and expansion on soil resources depends upon the availability of certain data (Imhoff et al. 1997). Geographical information system (GIS) and remote sensing (RS) have proved as the encouraging tools for detecting as well as monitoring of soil sealing (Munafò et al. 2010), which were applied in the study area.

In Peshawar, the urban expansion and the resultant industrial and infrastructural developments have increased soil sealing, which have reduced percolation and infiltration rate in response to the rate of discharge from groundwater more than its recharge. The analysis further reveals that in certain parts of Peshawar City District, the water table is lowering down at an alarming rate. This calls for the policy-makers to take special preventive measures and check the unprecedented effects of soil sealing on the depleting groundwater.

17.2 Overview of Soil Sealing

Globally, urbanization involves two aspects of rapidly growing population and the increasing trend in impervious surfaces. The growing population demand for the increasing freshwater supplies. Similarly, the expansion in built-up environment accelerates surface runoff from precipitation, reduces infiltration into the ground, and deteriorates water quality. Subsequently, these areas disturb the natural hydrological cycle by creating the problems of flooding, water pollution, and groundwater depletion (Heaney and Huber 1984; Singh and Singh 2002). The overexploitation of groundwater and reduction in its infiltration rate always creates pressure on freshwater resources.

Worldwide urban population growth and the trend of impermeable surfaces are rapidly increasing. The development of impervious surface covers mainly depends upon demographic and socioeconomic factors. The urban population of the world has increased from 746 million in 1950 to about 4 billion in 2016 (UN 2014 and 2016). It will further increase to 5 billion in 2030 and will cross the figure of 6.4 billion at the end of 2050 (Zhu et al. 2012). Not only the global urban population is growing with a rapid pace, but also urban land cover is increasing rapidly producing problems to the water resources. The total impermeable surfaces of urban areas are 0.579703 million square kilometers (Elvidge et al. 2007). In 2000, the global urban land cover was 0.6 million km^2, while the projected figure for 2030 is 1.25 million km^2 which will further increase to 2 million km^2 in 2050 (Angel et al. 2011). A catchment consisting of urban land cover is mostly impacted by the increasing trend of impermeable surfaces (Schueler 2003). Rapidly growing urban population requires an additional freshwater supply, and the increasing impervious surfaces impact surface runoff, evapotranspiration, and infiltration (USEPA 1993; Gleeson et al. 2010).

The impact of sealed surfaces in urban areas on surface and groundwater has been extensively investigated throughout the world. To evaluate the impact of land use change on runoff and recharging of groundwater as well as hydrological conditions of wetland, practical method has been devised in the USA (Harbor 1994). His study reveals that the phenomenon of changing land use pattern also affects the hydrological conditions of a specific region. With increasing impervious surfaces, not only surface runoff but also runoff volume increases to a considerable extent which produces flooding events locally and in downstream areas. Similarly, groundwater recharge is reduced creating problems to the aquifers and water table. Its impacts on wetland cannot be ignored. A systematic groundwater model approach of the land use changing trend and its possible impacts on the subsurface flow regime received in the Upper Roanoke Watershed (Southwest Virginia) was carried out (Cho et al. 2009). Other studies on Roanoke, Salem, and other adjacent cities were also carried out (Dietz 2000; Bosley et al. 2001; Crowder 2002), where they have pointed out that urbanization has increased in the region. These researches have assessed the economic, biological, and environmental impacts on the land development. Similarly, the increasing trend of urbanization has also impacted streams of the Upper Roanoke Watershed region.

In German, city of Leipzig applied the conceptual framework of driving forces, pressure, state, impact and response (DPSIR) concept for assessing the urban land use change on water balance (Haase and Nuissl 2007). They have also identified the consequences of the causes and policies of the impacts of urban growth on water balance.

It has been concluded that the surface sealing caused by the rapid urban growth and expansion has considerably impacted the water fluctuations. The impacts of urban water balance have been quantitatively determined which will become more imminent in the future in the city and on a regional scale. Wada et al. (2010) have analyzed the global groundwater depletion. They have pointed out that globally the growing population and increasing areas in irrigated farming and economic developments are continuously increasing the demand for freshwater supply.

Research on the impact of land use changing trend on the recharging rate of groundwater in the Chinese basin of River Guishui has analyzed that land use changing pattern has severely affected the groundwater recharging in the area, where dependence on groundwater is about 70% (Pan et al. 2011). It has been proved that urban expansion and the resultant reduction in farmland have reduced the groundwater recharge by $4 \times 10^{3*2}$ m^3 during the study period of 1980–2005. The urban land use changes are accompanied by the growing tendency in impervious surface cover (ISC) which has reduced the infiltration rate, resultantly depleted the groundwater recharging rate in the Chinese basin of River Guishui.

Döll et al. (2012) have carried out a worldwide analysis of effects about the water withdrawal on water storage variations. Their analysis was based on the global water resources for which they used WaterGAP model. They have analyzed that the global groundwater abstractions from 1951 to 2002 have grown alarmingly from the figure of 1615–4436 km^3/year. The intensive irrigated arid and semiarid regions of the world have greatly impacted the groundwater abstraction. However, extraction from groundwater is low in those regions where agriculture is higly dependent on surface water. The findings of their research can be concluded that the growing population and increasing demand for agriculture and industries have also increased the use of water and these anthropogenic factors have greatly impacted the global hydrological cycle.

A comprehensive review of researches has been carried out and to study the impact of land use changing trend on groundwater (Mishra et al. 2014). The analysis about the growing trend of urbanization has always resulted in reducing infiltration, which may affect the recharging rate and storage of groundwater sources. The growing population and the increasing trend of impervious surfaces have always negatively impacted the water resources. Although advances have been made in the identification of the impacts and aftermath of land use modifications on groundwater management, still the remaining gaps need to be filled.

Researches which have already been conducted related to the current study are the spatial analysis of the groundwater quality in the Peshawar district, Pakistan (Adnan and Iqbal 2014). They have attributed the degradation of groundwater quality due to urbanization and overexploitation of the groundwater sources. In Peshawar basin major factors behind the changes in status of surface and groundwater are rapid urbanization, industrialization and commercialization (Shahida 2006). Characteristics of industrial effluents and their possible impacts on quality of underground water (Tariq et al. 2006) are also related to the effects on the groundwater in Peshawar.

Urbanization and its effects on groundwater potential have been briefly described in the spatiotemporal variation on the groundwater potential resulted by urbanization in the Peshawar regime of Pakistan (Rahim et al. 2015a). Similarly, their work on the geostatistical approach using standardized precipitation index (SPI) and standardized water-level index (SWI) for assessing groundwater drought in Peshawar is also closely related to the current study (Rahim et al. 2015b). Navid et al. (2014) attributed decreased discharge capacity of Shahi Katta drain to rapid increase in population, urbanization and encroachment. A detailed study of the overexploitation of the groundwater resources and their effects in Peshawar valley has been carried out, in which it has been depicted that the groundwater sources are depleting (Khan et al. 2014).

Historically in Peshawar, the government has initiated certain policies for checking the unprecedented urban growth, its expansion, and to prevent the fertile farmland (Rahman et al. 2016). Master Plan of the City (1965), Town Planning Act of 1975, upgradation of the Municipal Committee to Municipal Corporation (1981), establishment of Peshawar Development Authority (PDA, 1987), designation of the Municipal Corporation to Peshawar City District (2001), and the establishment of Urban Policy Unit in 2013 are among the steps taken to stop the haphazard urban sprawl and to formulate certain strategies to develop the city in a systematic and planned manner. However, in Peshawar City District, multiple government agencies are working but there is no coordination between them. Besides, there is lack of effective urban land use policy and planning due to which the built up area is expanding haphazardly (Samiullah 2013).

The existing water supply system in Peshawar City District is completely based on groundwater with more than 1,400 tube wells providing freshwater for drinking, irrigation, and industrial purposes. The rapidly growing population and socioeconomic and infrastructural developments are continuously increasing the abstraction of water from the ground. These freshwater sources are under permanent stress which are causing their depletion. In order to reduce pressure on groundwater sources, the government has prepared a plan in 2008 to supply water from alternate sources. Warsak Dam was on the priority list (Rahman et al. 2016). However, this project still needs to be initiated.

17.3 Environmental Setup of Peshawar City District

Peshawar is the provincial capital and the largest urban center of KP Province. It has significant historic, socioeconomic, and geostrategic importance. Since 2000, the district of Peshawar has been designated as city district consisting of four towns and a cantonment. The city district is bounded by the Federally Administered Tribal Areas (FATA). The study area, Peshawar City District, is bounded on the northwest by Mohmand Agency, whereas Charsadda and Nowshera lie to the north, northeast, and east, Kohat and FR Peshawar lie to the south of Peshawar City District, and Khyber Agency lie to the west. Geometrically, the study area lies between longitudes 71° 22′ 00″ to 71° 42′ 00″ east and latitudes 33° 44′ to 34° 15′ north (Fig. 17.1). Total

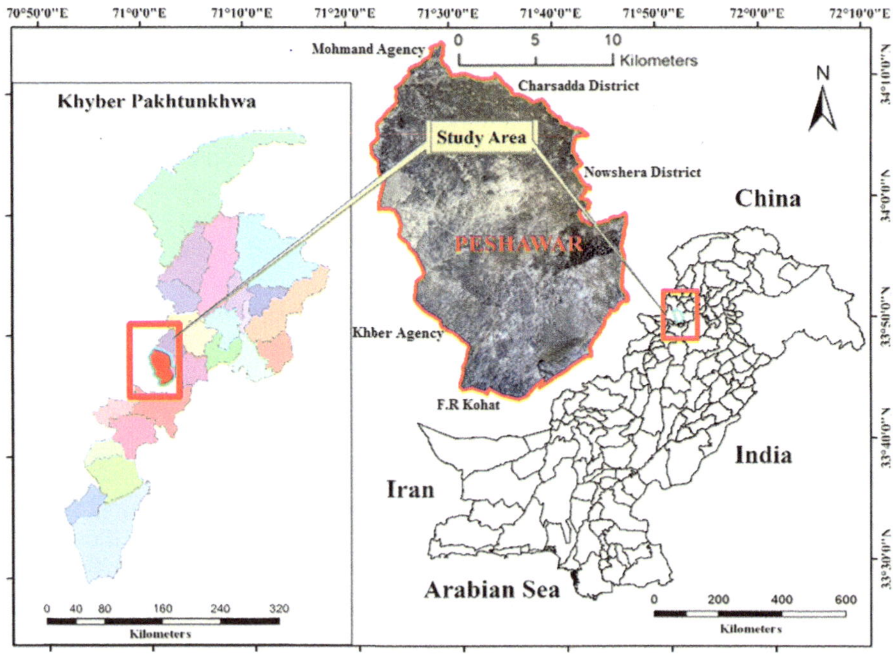

Fig. 17.1 Location of Peshawar City District, Pakistan

area of the city district is 1,257 square kilometers (km^2), which is about 1.69% of the total area of the KP Province (GoP 1999).

The study area is a featureless fertile plain, and its central part consists of fine alluvial deposits. The cultivated part of the city district has rich, light, and porous soil being a composition of sand and clay, which is favorable for the cultivation of a number of crops. Its average height above the sea level is about 358 m (1,173 ft; Fig. 17.2). Tarakai is the highest point, having an altitude of nearly 700 m (GoP 1999).

Climate of the study area is the subtropical continental type, where severe seasons of summer and winter are found. In Peshawar, during summer, the mean maximum temperature is above 40 °C, whereas during winter the mean minimum temperature is below 4 °C. In the study area, the average annual rainfall is around 400 mm (Fig. 17.3). It is estimated that in Peshawar, the average winter and spring rainfall exceed over the summer monsoonal rainfall. Climograph of Peshawar based on climate data for 36 years (1973–2009) has been shown in Fig. 17.3. Regarding other weather

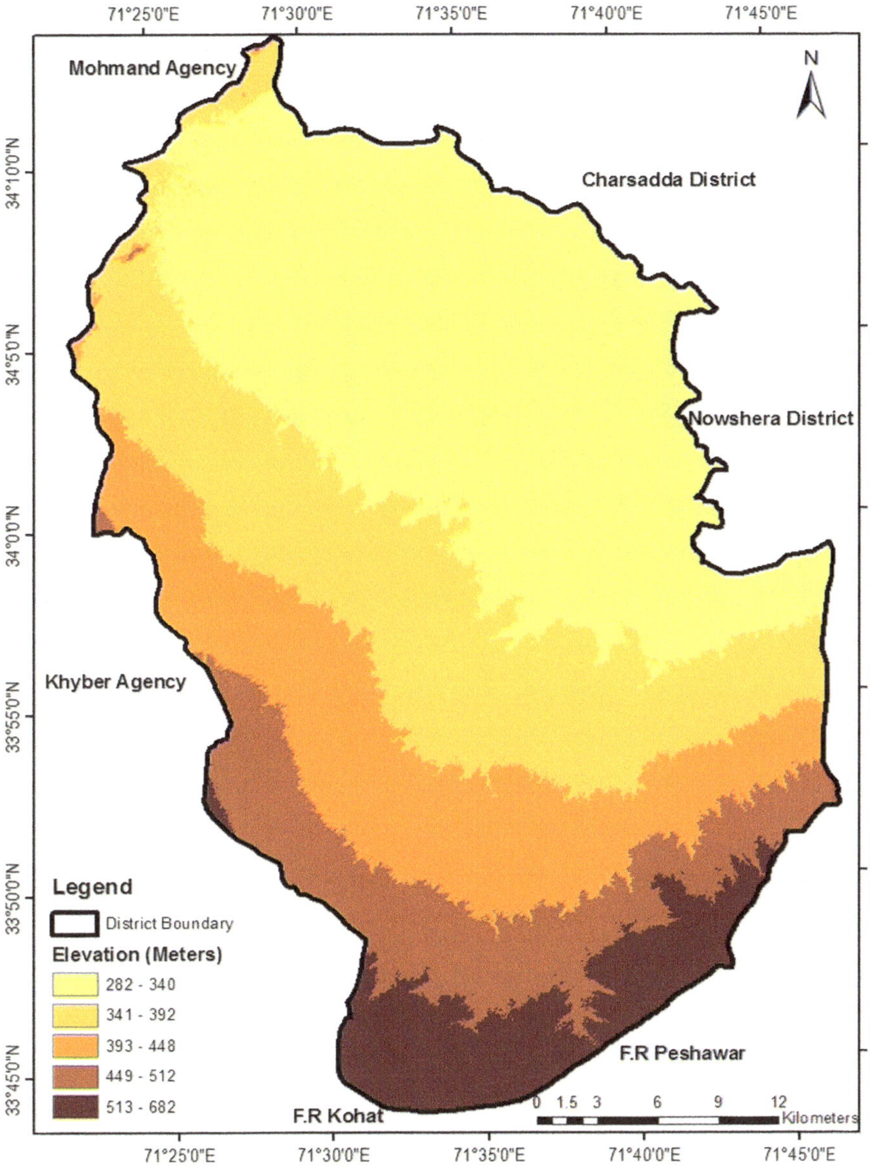

Fig. 17.2 Peshawar City District, digital terrain

parameters, the wind speed varies throughout the year that ranges from minimum 5 knots in winter to maximum 24 knots during summer. Similarly, the maximum relative humidity recorded in August remains higher than 70% (GoP 1999).

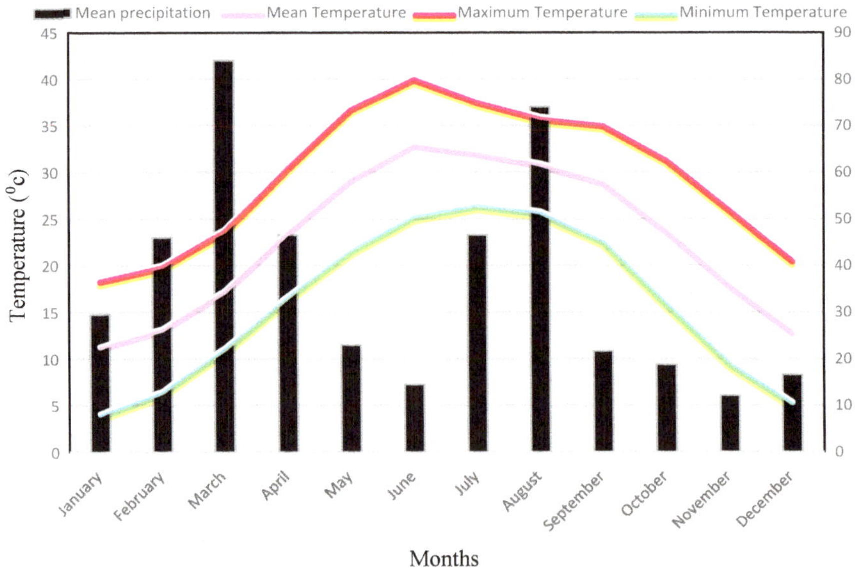

Fig. 17.3 Climograph of Peshawar City District

17.4 The Growing Peshawar City District

Peshawar is not only the provincial capital and largest city of the KP Province, but also the center of financial and administrative functions for the whole province and FATA due to which it has witnessed rapid population growth, urban expansion, socioeconomic, and infrastructural developments (Khan 2001). As a result, land take and soil sealing are also rapidly increasing. These human interventions are producing negative impacts on human health and environment by increasing surface temperature, accelerating surface runoff, and reducing percolation and infiltration to recharge the groundwater.

According to the 1998 census, total population of the city district was 2.019 million; however, the estimated figure for 2016 was 3.8 million and the projected population of 2030 will be 6.2 million. From 1951 to 1998, population has increased more than 5 times from 0.391 million to 2.019 million in 47 years (Table 17.1). During the intercensal period of 1951–1961, the increase in population was 35.29% with an annual growth rate of 3.08. While from 1961 to 1972, the increase was 52.55%, with a growth rate of 3.70. The population growth rate of the city district, during the intercensal period of 1972–81, was 3.89. Population of the city district has increased by 81.40% during the intercensal period of 1981–1998, with average growth rate of 3.56 per annum. The rapid growth in population needs greater demand for the supply of freshwater; as a result, the abstraction of groundwater has also been augmented. And pressure on groundwater sources is increasing continuously.

Table 17.1 Peshawar population, % intercensal increase, and growth rate (1951–1998)

Year	Population (million)	Intercensal increase (%)	Growth rate
1951	0.391	–	–
1961	0.529	35.29	3.08
1972	0.807	52.55	3.70
1981	1.113	37.92	3.89
1998	2.019	81.40	3.56
2016	3.8	–	3.56

Source GoP (1952), (1962), (1973), (1983) and (1999)

Table 17.2 Peshawar urban and rural population, % share, and density of urban population

S. No.	Year	Population (million)			Urban population (% share)	Urban population density (P/km^2)
		Total	Urban	Rural		
1	1972	0.784	0.273	0.512	34.8	217
2	1981	1.084	0.566	0.518	52.2	450
3	1998	2.019	0.983	1.036	48.69	782
4	2016	3.8	3.8	–	100	3050

Sources 1. GoP (1973), (1983) and (1999) 2. Estimated population for the year 2016

Peshawar City District has been experiencing rapid urbanization. Population of the city district has increased from 273,000 (34.79%) to 566,000 (52.21%), during the intercensal period from 1972 to 1981. According to 1998 census, urban population was 983,000 (48.70%), which was increased to 3,800,000 (100%: as Peshawar was designated as city district since 2000) in 2016. Density of urban population has also increased more than 14 times in 46 years, i.e., from 1972 to 2016, from 217 persons/km^2 (P/km^2) to 3050 persons/km^2, and thus made Peshawar as an overcrowded city district. Bulk of the urban population in Khyber Pakhtunkhwa Province is concentrated in ten cities, among which Peshawar is on top in terms of urban population share (Table 17.2; Fig. 17.4).

As a city district of a developing country, causes of urbanization for Peshawar can be attributed to people migration from rural areas to the largest urban center of the province. Rural–urban migration is rapid as people are thinking that urban areas provide better facilities of quality healthcare, education, and other basic services. Similarly, natural increase in urban population and the conversion of villages into towns indicate that urban living will be the way of life in the future (Arif and Hamid 2009; Kugelman 2014; Mehmood et al. 2016; Raziq et al. 2016). In KP during the last decade, 63% of all the internal migrants have moved to urban centers either from rural areas or from other urban centers, in which about 56% have settled in Islamabad and Peshawar (UN 2014).

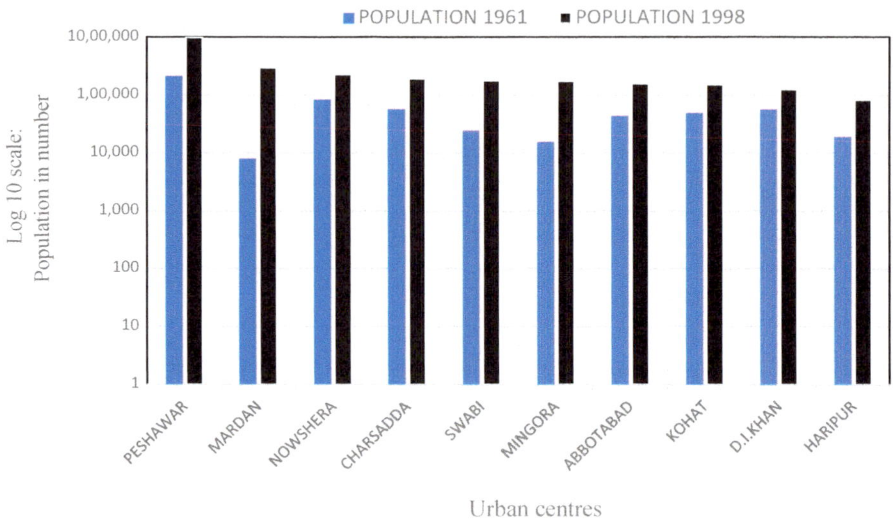

Fig. 17.4 Change in urban population of major urban centers in Khyber Pakhtunkhwa Province (*Source* GoP 1962, 1999)

Peshawar City District is more vulnerable to urbanization as well as soil sealing because of the increasing socioeconomic, physical, and infrastructural developments. Rapid urban growth and expansion of the city district have produced environmental and socioeconomic challenges. Similarly, infrastructural developments have replaced the natural ground cover by impervious material and have reduced the percolation and infiltration of rainwater to recharge the groundwater. Rapid urbanization has also produced water and sanitation problems, and the quality of water is rapidly deteriorating. Urbanization has reduced the agricultural land, which has been gradually used up by the built environment to 43,130 ha in 2013–14 from 44,468 hectares (ha) in 2001–02 (Ali 2014). According to Samiullah (2013), among the land use classes, residential area was the major sector that has consumed most of the farmland. It has been estimated that during 1991–2009 almost 8,700 ha (Samiullah 2013) of productive agricultural land has been consumed by the housing sector alone and it is expected that the same will further multiply if serious measures are not taken in time.

17.5 Spatial and Temporal Analysis of Built-up Areas

LU/LC classification analysis of satellite images of the study area for the period of 1981–2014 was carried out. Maps of 1981, 1991, and 2009 were taken from the published sources, while land use map of 2014 was prepared from Spot image 2014 by using the maximum likelihood classification algorithm, to know the spatiotemporal trend of soil sealing and its possible impacts on the groundwater depletion in the

Table 17.3 Built-up area changes in Peshawar City District (1981–2014)

S. No.	Year	Built-up area (ha)	Percentage	% increase in built-up area
1	1981	4,635	3.70	–
2	1991	7,182	5.70	54.95
3	2009	16,986	13.50	136.51
4	2014	20,451	16.27	20.40

Table 17.4 Land uses in Peshawar City District 2014

S. No.	LU/LC type	Area (ha)	Percentage (%)
1	Agriculture	84,596.1	67.30
2	Built-up	20,451.39	16.27
3	Barren land	15,234.84	12.12
4	Water bodies	4,588.05	3.65
5	Forests	829.62	0.66
	Total	**125,700**	**100**

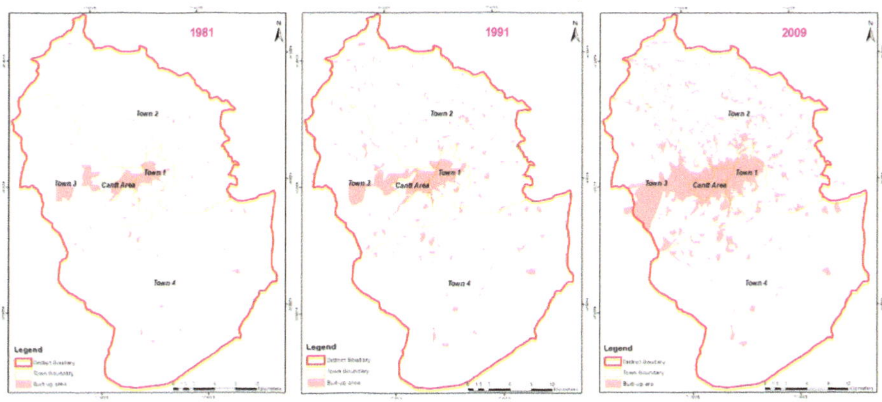

Fig. 17.5 Built-up area changes in Peshawar City District (1981–2009) after Samiullah (2013)

study area. Changes were detected in all land uses; however, the sealed surfaces have shown a rapid increase from 1981 to 2014. The built-up area has increased from 3.7% (4,635 ha) to 16.27% (20,451 ha) with an overall increase of 12.58% (Tables 17.3 and 17.4; Figs. 17.5 and 17.6).

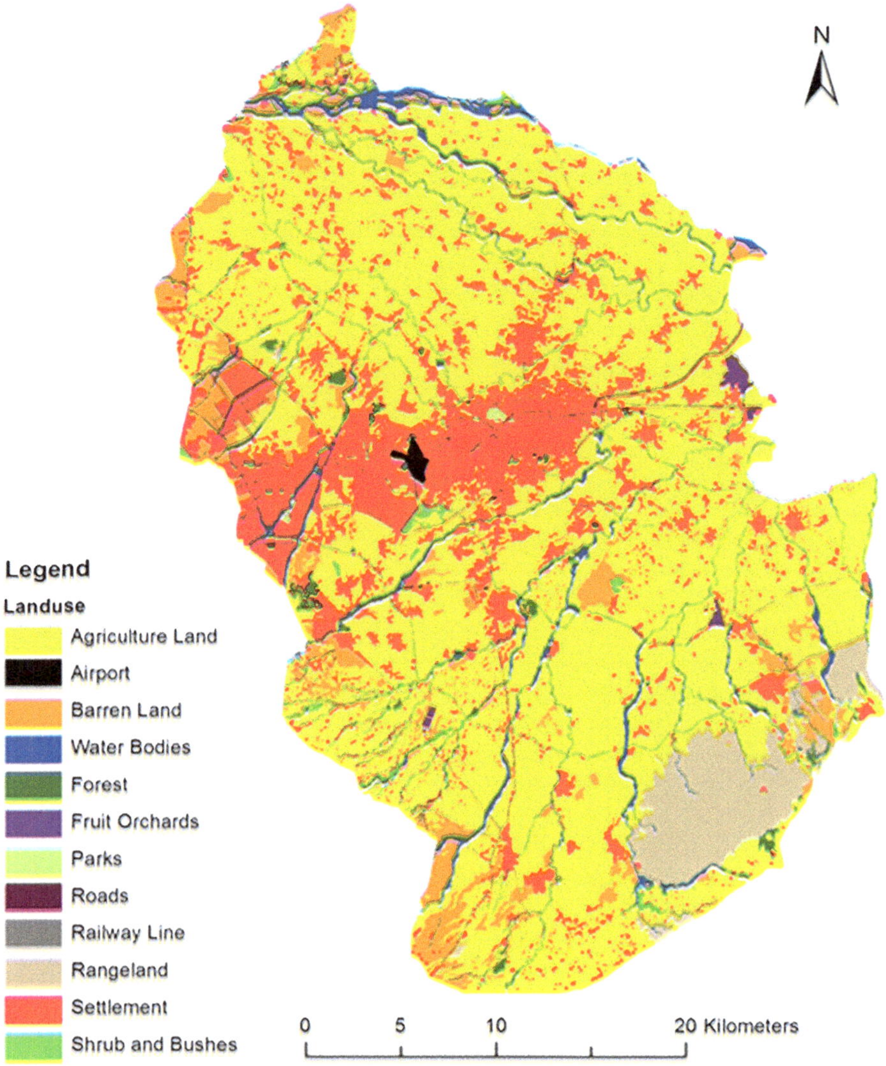

Fig. 17.6 Land use/land cover of Peshawar City District extracted from Spot image 2014

17.6 Soil Sealing Trend in Peshawar City District

The analysis reveals that increase in sealed surfaces during 1981–2014 period was mainly due to increase in urban population and rapid expansion in built environment. Similarly, the surrounding rural areas have been converted into an urban area. Migrants from other districts of KP and FATA to Peshawar City District have also affected the LU/LC. The rapidly growing Peshawar City District is experiencing socioeconomic and infrastructural developments at a faster pace. Vegetation and barren lands have been converted into built-up areas. These developments are favoring soil sealing, which has recorded a rapid increase during the last three decades. The overall change in built-up areas during the period of 1981–2014 was 15,816 ha (12.58%). If urban growth and expansion continued with the same pace, then in 2030 the sealed surfaces will be increased by more than 20% in Peshawar City District.

17.7 Groundwater Abstraction

Globally, groundwater extraction of all freshwater (1998–2002) was about one-third and was supplying water for industrial (27%), agricultural (42%), and domestic (36%) purposes (Döll et al. 2012). Since the mid of the twentieth-century changes in irrigation technologies, rapid population growth and industrial development have increased pressure on the groundwater sources (Kraft et al. 2012). During the last fifty years, a triple increase in the extraction of groundwater has been recorded in different countries with a growth rate of 1–2% per annum (Siebert et al. 2005; Margate et al. 2006; Van der Gun 2012). Climate variability in terms of precipitation volume and its distribution also affects groundwater (Healy 2010). Similarly, rapid urbanization and the resultant soil sealing have affected groundwater sources by halting percolation and infiltration to recharge the groundwater (Niemelä et al. 2010). Globally, these problems are not only confined to arid and semiarid areas, but humid areas equally suffer from such menaces (Macoun and El Naser 1999; Smedema and Shiati 2002).

In Pakistan for the most part of the year, arid and semiarid conditions prevail over the plain areas and dependency on groundwater for irrigation and other purposes is rapidly increasing. Canal irrigation system of the country provides about 40% of water needed for agriculture crops, and the remaining requirements are fulfilled from groundwater. Every year, surface water is supplemented by 59 billion cubic meters (BCM) of groundwater (Zardari 2008). Pakistan ranks fourth in the world in terms of groundwater abstraction (Margat 2008; Fig. 17.7). Pressure on groundwater is continuously increasing, resulting in the depletion of these unfrozen freshwater sources. The falling water table in the country will produce serious problems of freshwater supply.

17.8 Groundwater Abstraction in Peshawar City District

In Peshawar City District, groundwater is considered as an important natural resource of unfrozen freshwater, having a number of socioeconomic uses which is used for domestic, irrigation, industrial, and other purposes. River Kabul and Bara are not only the main sources of surface water, but also the recharging sources for the groundwater in the city district. Besides precipitation, a number of perennial and non-perennial streams flowing through the city district also contribute to the recharging process of

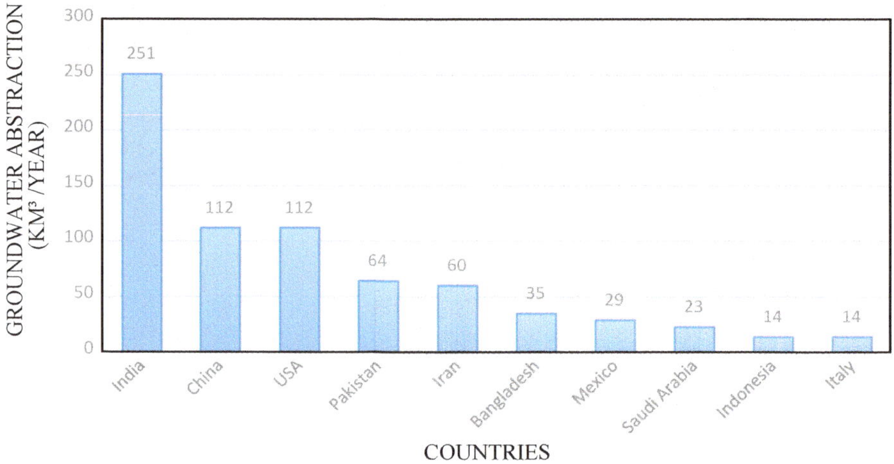

Fig. 17.7 Groundwater abstraction, trends in top ten groundwater abstracting countries modified after Margat (2008)

the groundwater (GoNWFP 2009). Water table in different parts of the city district varies from waterlogged conditions in the north near the major rivers, where it is less than 10 ft up to more than 250 ft in the south (Fig. 17.8). In the city district, pressure on groundwater is continuously increasing due to rapid population growth and supplementing groundwater for irrigation, thus increasing the abstraction of groundwater.

Similarly, urban growth and expansion, infrastructural and industrial developments, and the resultant soil sealing reduce the recharging rate of groundwater and the potential of these freshwater sources is greatly affected. In Peshawar City District, the problem of the falling water table is also a serious threat to the groundwater sources. Water table in some parts of the city district is decreasing at an alarming rate (Kruseman and Naqvi 1988). During the field survey in the months of October and November 2016 in different union councils of the city district, it was found that during the last three decades a number of tube wells have already dried up, which have been replaced by new and deeper tube wells indicating lowering down of water table.

In Peshawar City District, the existing water supply system is based on groundwater with more than 1,400 tube wells having a total discharge of about 100 million gallons per day (mg/d). The agencies responsible for water supply and sanitation are Town Municipal Administrations (TMAs), Public Health Engineering Department (PHED), Water and Sanitation Services Peshawar (WSSP), and irrigation department. Apart from these tube wells operated by the government departments, there are also commercial/community tube wells in operation. The average daily demand of the citizens in Peshawar City District is about 30 gallon per capita per day; however, their peak daily demand is 1.5 times more than the average daily demand.

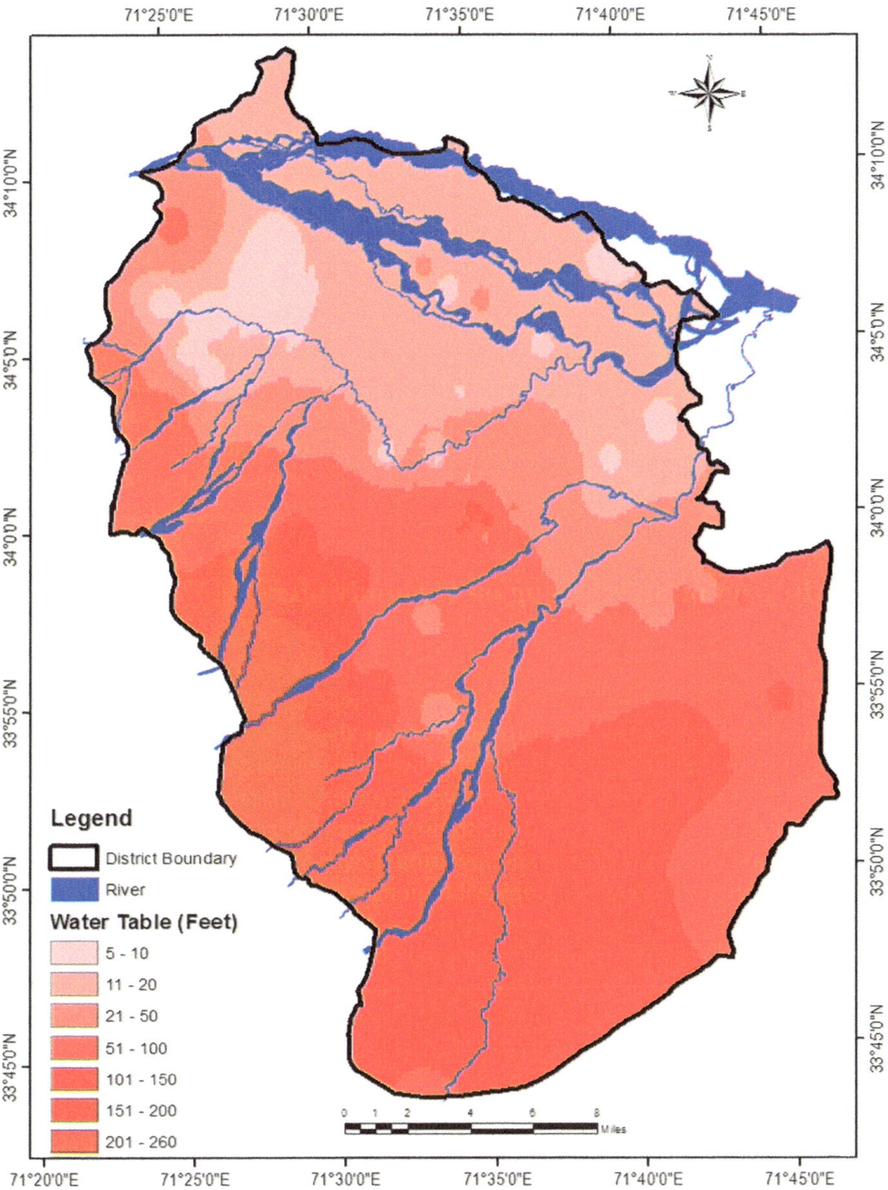

Fig. 17.8 Peshawar City District, water table

As population of the city district is rapidly increasing, abstraction of groundwater is also increasing to fulfill the growing needs of freshwater supply. In 1981, the population of the study area was 1.113 million, the daily demand for freshwater was 33 mg/d, while the peak daily demand was 50 mg/d. Compared to 1998 population of

Table 17.5 Population and daily water demand in Peshawar City District

S. No.	Year	Population (million)	Average daily demand for water (mg/d)	Peak daily demand (mg/d)
1	1981	1.113	33	50
2	1998	2.09	63	94
3	2016	3.8	114	171
4	2030	6.20	186	279

the study area has increased to 3.8 million in 2016, which was nearly twofold as and hence also increasing the daily average as well as peak demands to nearly twofold. However, projection for the year 2030 indicates that for the growing population the demand will be more than threefold as compared to 1998 (Table 17.5). These facts and figures indicate that pressure on groundwater is continuously increasing.

17.9 Impact of Soil Sealing on Groundwater

Human activities related to infrastructure development and urbanization are continuously replacing the natural landscape by impervious surfaces, which reduce infiltration and accelerate surface runoff. Similarly, it also increases runoff volume and decreases runoff time into streams and channels which result in peak discharge and increase the frequency of urban flood. However, increase in abstraction of groundwater and decrease in percolation and infiltration reduce their recharging rate resulting in the depletion of their potential and lowering down the water table.

Precipitation in different geographical regions varies with time and is affected by metrological factors and physical characteristics. Areas experiencing frequent precipitation, having soils with poor vegetation and abundance of sealed surfaces, reduce infiltration of rainwater into the soil and accelerate surface runoff (Booth 1991). Except precipitation, these characteristics are generally associated with urbanization, which increases soil sealing within urban watershed (Paul and Meyer 2001).

The study area has been experiencing rapid urban growth and expansion. During the last 33 years (1981–2014), the built-up area has increased from 4,635 to 20,451 ha with an overall increase of 15,816 ha accounting 12.58%. The same period has also witnessed a remarkable decrease in vegetation cover. When natural ground cover experiences 10–20% increase in impervious surface cover (ISC), runoff generation is doubled, while shallow and deep infiltrations are reduced by 4% each (US EPA 1993). Discharge from groundwater sources in the city district is 8 million gallons/hour (8.41 $m^3/s = 105$ mm/year). As the city district receives an average annual rainfall of 400 mm, infiltration from rain with 10–20% ISC is 21%, which is 91 mm/year, while discharge from tube wells is 105 mm/year indicating high abstraction and low recharging rate from precipitation. The recharging rate from precipitation was about

25% (100 mm/year) in 1981, when the sealed surfaces in the city district were only 3.70% (4,635 ha). Thus, it shows the fact that the recharging rate of groundwater sources has been reduced by soil sealing in the study area.

17.10 Conclusion

This chapter has analyzed the spatiotemporal analysis and the trend of soil sealing by urban growth and expansion in built-up environment in Peshawar City District, KP, Pakistan, and its impact on the potential of groundwater sources. Published maps of 1981, 1991, 2009 and Spot image of 2014 have been used to classify, identify, and monitor LU/LC classes and change detection of sealed surfaces in Peshawar City District. It has been detected the LU/LC changes from the last thirty-three years (1981–2014) observed a considerable increase in the built-up areas due to which most of the vegetation land has been converted into ISC indicating major soil sealing. The analysis has revealed that the sealed surfaces have increased from 3.70% (4,635 ha) in 1981 to 16.27% (20,451 ha) in 2014 indicating a total increase of 12.58% (15,816 ha). The projected ISC for 2030 is 20% (25,140 ha). When a particular watershed consisting of urban land cover undergoes modification with the increasing trend in sealed surface, groundwater is mostly affected. And not only the groundwater potential is threatened, but in some cases its quality has also been deteriorated. Similarly, the acceleration in surface runoff also increases the risk of urban flooding.

When natural ground is replaced by 10–20% increase in sealed surfaces, the runoff generation is doubled, evapotranspiration is reduced by 2%, and shallow as well as deep infiltration is reduced by 4% each. As in Peshawar City District, the increase in sealed surfaces is 12.58% from 1981 to 2014 confirming a decrease of 4% infiltration. Discharge from groundwater sources in the city district is 8 million gallons/hour ($8.41\ m^3/s = 105$ mm/year). As the city district receives an average annual rainfall of 400 mm, infiltration from rain with 10–20% ISC is 21%, which is 91 mm/year, while discharge from tube wells is 105 mm/year indicating high abstraction and low recharging rate from precipitation. The recharging rate from precipitation was about 25% (100 mm/year) in 1981, when the sealed surfaces in the city district were only 3.70% (4,635 ha). Thus, soil sealing has reduced water infiltration into the ground.

Rapid population growth of the city district has also increased the abstraction of groundwater as the existing water supply system is completely based on groundwater which has increased from 33 to 114 mg/d and the estimated demand will further increase to 186 mg/d in 2030; however, the daily peak demand is relatively more than the average daily demand. Rapid population growth combined with soil sealing has reduced the potential of the unfrozen freshwater sources, consequently causing their depletion and lowering down the water table.

Government has initiated certain urban policies to prevent the unprecedented urban expansion and to safeguard the fertile agricultural land. However, due to multiple agencies working in their jurisdiction within the city district and lack of coordination among the line agencies, these policies have not been fully implemented. Similarly, to reduce the pressure on groundwater source, the city government in 2008 has decided to supply water from Warsak Dam; however, this project has not yet been initiated and groundwater continues to be the major source for domestic, agricultural, and industrial requirements.

References

Adnan S, Iqbal J (2014) Spatial analysis of the groundwater quality in the Peshawar District, Pakistan. Proc Eng 70:14–22

Ali Z (2014) Peshawar loses vast agriculture land to urbanization. Dawn newspaper, Pakistan, 25 Oct 2014

Angel S, Parent J, Civco DL, Blei A, Potere D (2011) The dimensions of global urban expansion: estimates and projections for all countries, 2000–2050. Prog Plann 75(2):53–107

Arif GM, Hamid S (2009) Urbanization, city growth and quality of life in Pakistan. Eur J Soc Sci 10(2):196–215

Arnold CL Jr, Gibbons CJ (1996) Impervious surface coverage: the emergence of a key environmental indicator. J Am Plann Assoc 62(2):243–258

Booth DB (1991) Urbanization and the natural drainage system impacts, solutions, and prognoses. Northwest Environ J 7(1):93–118

Bosley G, Lohani V, Kibler DF, Lockard B (2001) Hydrologic evaluation of alternate land settlement patterns. In: Bridging the gap: meeting the world's water and environmental resources challenges, pp 1–10

Breuste JH (2011) Ecology in cities, manmade physical condition. In: Niemelä J (ed) Urban ecology: patterns, processes, and applications. Oxford University Press, New York, pp 71–72

Burghardt W (2006) Soil sealing and soil properties related to sealing. Geol Soc London Special Publ 266(1):117–124

Cannon WF, Horton JD (2009) Soil geochemical signature of urbanization and industrialization–Chicago, Illinois, USA. Appl Geochem 24(8):1590–1601

Cho J, Barone VA, Mostaghimi S (2009) Simulation of land use impacts on groundwater levels and streamflow in a Virginia watershed. Agric Water Manag 96(1):1–11

Crowder DW (2002) Reproducing and quantifying spatial flow patterns of ecological importance with two-dimensional hydraulic models, Doctoral dissertation

Dietz RW (2000) The use of GIS for integrated watershed analysis: integration of environmental models with GIS in the Upper Roanoke River Watershed. Virginia Polytechnic Institute and State University, USA

Döll P, Hoffmann-Dobrev H, Portmann FT, Siebert S, Eicker A, Rodell M, Scanlon BR (2012) Impact of water withdrawals from groundwater and surface water on continental water storage variations. J Geodyn 59:143–156

Duley FL, Kelly LL (1939) Effect of soil type, slope, and surface conditions on intake of water

Elvidge CD, Tuttle BT, Sutton PC, Baugh KE, Howard AT, Milesi C, Bhaduri B, Nemani R (2007) Global distribution and density of constructed impervious surfaces. Sensors 7(9):1962–1979

Gleeson T, Vander Steen J, Sophocleous MA, Taniguchi M, Alley WM, Allen DM, Zhou Y (2010) Groundwater sustainability strategies. Nat Geosci 3:378–379

Government of North West Frontier Province (2009) Environmental profile of NWFP, 2nd edn, p 4

Government of Pakistan (1952) Peshawar District census report of 1951, Population Census Organization, Statistics division, Islamabad, Pakistan

Government of Pakistan (1962) Peshawar District census report of 1961, Population Census Organization, Statistics division, Islamabad, Pakistan

Government of Pakistan (1973) Peshawar District census report of 1972, Population Census Organization, Statistics division, Islamabad, Pakistan
Government of Pakistan (1983) Peshawar District census report of 1981, Population Census Organization, Statistics division, Islamabad, Pakistan
Government of Pakistan (1999) Peshawar District census report of 1998, Population Census Organization, Statistics division, Islamabad, Pakistan
Haase D, Nuissl H (2007) Does urban sprawl drive changes in the water balance and policy? The case of Leipzig (Germany) 1870–2003. Landscape Urban Plann 80(1):1–13
Harbor JM (1994) A practical method for estimating the impact of land-use change on surface runoff, groundwater recharge and wetland hydrology. J Am Plann Assoc 60(1):95–108
Haygarth PM, Ritz K (2009) The future of soils and land use in the UK: soil systems for the provision of land-based ecosystem services. Land Use Policy 26:S187–S197
Healy RW (2010) Estimating groundwater recharge. Cambridge University Press
Heaney JP, Huber WC (1984) Nationwide assessment of urban runoff impact on receiving water quality. JAWRA J Am Water Resour Assoc 20(1):35–42
Imhoff ML, Lawrence WT, Stutzer DC, Elvidge CD (1997) A technique for using composite DMSP/OLS “city lights” satellite data to map urban area. Remote Sens Environ 61(3):361–370
Khan GD, Ali M, Akbar F (2014) Over exploitation of groundwater resources and their Influence on groundwater in Peshawar Valley. Civil Environ Res 4(6):85–90
Khan J (2001) Urbanization in Peshawar, making a case for healthy city project. Pakistan’s population issues in the 21st century. In: Conference proceedings 24th–26th Oct 2000 Karachi. Compiled by Population Association of Pakistan. Islamabad, Pakistan, pp 213–228
Kraft GJ, Clancy K, Mechenich DJ, Haucke J (2012) Irrigation effects in the northern lake states: Wisconsin central sands revisited. Groundwater 50(2):308–318
Kruseman GP, Naqvi SAH (1988) Hydrogeology and groundwater resources of N-W.F.P, Peshawar/DELFT
Kugelman M (2014) Understanding Pakistan’s unstoppable urbanization. In: Kugelman M (ed) Pakistan’s runaway urbanization: what can be done? Washington, DC, p 2
Macoun A, El Naser H (1999) Groundwater resources management in Jordan: policy and regulatory issues. World Bank Technical Paper, 105–116
Margat J (2008) Les eaux souterraines dans le monde. Orléans/Paris, BGRM/UNESCO
Margate J, Foster S, Droubi A (2006) Concept and importance of non-renewable resources. Non-renewable groundwater resources: a guidebook on socially-sustainable management for water-policy makers 10:13–24
Mehmood R, Mehmood SA, Butt MA, Younas I, Adrees M (2016) Spatiotemporal analysis of urban sprawl and its contributions to climate and environment of Peshawar using remote sensing and GIS techniques. J Geogr Inf Syst 8(02):137
Mishra N, Khare D, Gupta KK, Shukla R (2014) Impact of land-use change on groundwater—A review. Adv Water Resour Prot 2:28–41
Montanarella L (2007) Trends in land degradation in Europe. In: Sivakumar, Mannava VK, Ndiangŭi, Ndegwa (eds) Climate and land degradation. Springer, Berlin Heidelberg, pp 83–104. ISBN: 978-3-540-72437-7
Munafò M, Norero C, Sabbi A, Salvati L (2010) Soil sealing in the growing city: a survey in Rome, Italy. Scott Geogr J 126(3):153–161
Navid A, Shah SSA, Shahzad G (2014) Factors affecting the discharge capacity of Shahi Katta drain, Peshawar city Pakistan. Int J Comput Eng Res (IJCER) 12(4):17–21
Niemelä J, Saarela SR, Söderman T, Kopperoinen L, Yli-Pelkonen V, Väre S, Kotze DJ (2010) Using the ecosystem services approach for better planning and conservation of urban green spaces: a case study of Finland. Biodivers Conserv 19(11):3225–3243
Pan Y, Gong H, Zhou D, Li X, Nakagoshi N (2011) Impact of land use change on groundwater recharge in Guishui River Basin, China. Chin Geogr Sci 21(6):734–743
Paul MJ, Meyer JL (2001) Streams in the urban landscape. Annu Rev Ecol Syst 32(1):333–365

Prokop G, Jobstmann H, Schönbauer A (2011) Report on best practices for limiting soil sealing and mitigating its effects. European Commission, Luxembourg

Rahim A, Khan K, Jamal R, Tariq N, Akif A (2015a). The spatial and temporal variation in the groundwater potential due to urbanization in the Peshawar regime of Pakistan. Sci Int 27(3)

Rahim A, Khan K, Akif A, Jamal R (2015b). The geo statistical approach to assess the groundwater drought by using standardized water level index (SWI) and standardized precipitation index (SPI) in the Peshawar regime of Pakistan. Sci Int 27(5):4111–4117

Rahman A, Surjan A, Parvin G, Shaw R (2016) Impact of urban expansion on farmlands: a silent disaster 7. Urban Disasters and Resilience in Asia, 91

Raziq A, Xu A, Li Y, Zhao Q (2016) Monitoring of land use/land cover changes and urban sprawl in Peshawar City in Khyber Pakhtunkhwa: an application of geo-information techniques using of multi-temporal satellite data. J Remote Sensing GIS 5(174):2

Salvati L, Bajocco S, Ceccarelli T, Zitti M, Perini L (2011) Towards a process-based evaluation of land vulnerability to soil degradation in Italy. Ecol Ind 11(5):1216–1227

Samiullah (2013) Expansion of built up area and its impact on urban agriculture: a case study of Peshawar, Pakistan. Ph.D. thesis submitted to the Department of Geography, University of Peshawar, Pakistan

Scalenghe R, Marsan FA (2009) The anthropogenic sealing of soils in urban areas. Landscape Urban Plann 90(1):1–10

Schueler TR (2003) Impacts of impervious cover on aquatic systems. Watershed Prot Res Monogr 1:142

Shahida N (2006) Monitoring of surface water groundwater, air and soil in Peshawar Basin against time the 3rd Dimension. Doctoral dissertation, Ph.D. thesis submitted to University of Peshawar, 238p

Siebert S, Döll P, Hoogeveen J, Faures JM, Frenken K, Feick S (2005) Development and validation of the global map of irrigation areas. Hydrol Earth Syst Sci 9(5):535–547

Sigbert H, Freudenschuss A (2002). Development of indicators for soil quality at the European level. In: 17. World congress of soil science, Bangkok (Thailand), 14–21 Aug 2002

Singh DK, Singh AK (2002) Groundwater situation in India: problems and perspective. Int J Water Resour Dev 18(4):563–580

Smedema LK, Shiati K (2002) Irrigation and salinity: a perspective review of the salinity hazards of irrigation development in the arid zone. Irrigat Drain Syst 16(2):161–174

Tariq M, Ali M, Shah Z (2006) Characteristics of industrial effluents and their possible impacts on quality of underground water. Soil Environ 25(1):64–69

Turok I, Mykhnenko V (2007) The trajectories of European cities, 1960–2005. Cities 24(3):165–182

United Nations (2014) World urbanization prospects: population division of the Department of Economic and Social Affairs of the United Nations Secretariat, New York

United Nations (2016) The world's cities in 2016: Data Booklet (ST/ESA/SER.A/392). Population Division of the Department of Economic and Social Affairs of the United Nations Secretariat, New York

United States Environmental Protection Agency (1993) Guidance specifying management measures for sources of nonpoint source pollution in coastal waters United States Environmental Protection Agency 840-B-92-002. USEPA Office of Water, Washington, DC

Van der Gun J (2012) Groundwater and global change: trends, opportunities and challenges. United Nations Educational, Scientific, and Cultural Organization, Paris, France, p 44. ISBN 978-92-3-001049-2

Wada Y, van Beek LP, van Kempen CM, Reckman JW, Vasak S, Bierkens MF (2010) Global depletion of groundwater resources. Geophys Res Lett 37(20)

Wessolek G, Schwärzel K, Greiffenhagen A, Stoffregen H (2008) Percolation characteristics of a water-repellent sandy forest soil. Eur J Soil Sci 59(1):14–23

Xiao R, Su S, Zhang Z, Qi J, Jiang D, Wu J (2013) Dynamics of soil sealing and soil landscape patterns under rapid urbanization. CATENA 109:1–12

Zardari NUH (2008) An improved multicriterion analysis approach to avoid subjectivity in irrigation water allocation decisions, Doctoral dissertation, University of New South Wales, Sydney, Australia

Zhu T, Melamed M, Parrish D, Gauss M, Klenner LG, Lawrence M, Liousse C (2012) Impacts of megacities on air pollution and climate. World Meteorological Organization, Geneva/International Global Atmospheric Chemistry, University of Colorado, Global Atmosphere Watch Report No. 205, p 309

Chapter 18
Adaptive Policy Responses in Managing Urban Water Crisis in Sri Lanka

A. J. M. K. K. Aviruppola and K. W. G. Rekha Nianthi

Abstract Sri Lanka is one of the few countries of the world that had been flourished with a remarkable civilization which dates back to 6 BC. Astonishingly, the civilization was developed in the drier part of the country where the irrigation system was highly sophisticated. Since the system was structured in a way to distribute water for agriculture, drinking, and sanitation covering all the day-to-day needs of man through a complex distribution and management mechanism, the concept of water management appeared to have been deep rooted in Sri Lankan ancient society. However, industrialization paved the way to the establishment of enormous industries in the country which highlighted the distinction of urban and rural areas. Since industries targeted the urban areas, they emerged as urban cities subjecting to cause many positive as well as negative outcomes to the country. Due to the emergence of climatic changes through global warming, improper water management practices in the urban cities became a pressing issue that needed to be addressed comprehensively. Understanding the gravity of the issue, this paper highlights the policy responses of Sri Lanka toward managing the issue of urban water crisis as it needs to be addressed in a comprehensive manner. Then, the paper examines the overall position of Sri Lanka in urban water crisis management, and finally, it provides a profound understanding as to the prospective policy formulations on urban water crisis management of Sri Lanka.

A. J. M. K. K. Aviruppola (✉) · K. W. G. Rekha Nianthi
University of Peradeniya, Peradeniya, Sri Lanka
e-mail: kithi_v@hotmail.com

K. W. G. Rekha Nianthi
e-mail: rekhanianthi@yahoo.com

B. Ray and R. Shaw (eds.), *Urban Drought*, Disaster Risk Reduction,
https://doi.org/10.1007/978-981-10-8947-3_18

Keywords Crisis · Water · Urban · Policy · Climate change

18.1 Introduction

Sri Lanka is an island located in the Indian Ocean, surrounded by the tropical beaches near the southern tip of India between latitude 5° 55′–9° 50′ and longitude 79° 31′–81° 53′ (CBSL 2016a, b). The country is allocated to 25 administrative districts, and the physical features of the land evince the diverse topography which has resulted in many adverse effects of climate change within the island's small-scale landmass of approximately 65,610 km^2 (CBSL 2016a, b). Comprising approximately 20 million people (Department of Census and Statistics 2012) and an annual population growth rate of 0.73% (ADB, South Asia Working Paper Series: June 2015), Sri Lanka is the 57th most populous nation in the world. Hence, the total population is expected to increase up to 21.9 million in 2020 (ADB, South Asia Working Paper Series: June 2015). Among them, 74% of the population is concentrated in the southwest and the highest density is recorded in Colombo. Most importantly, the demand for safe water increases by 8–10% each year in Sri Lanka with the increasing level of population. Comprising 75% of the population, Sinhalese constitute the largest ethnic group, while comprising a little over 11% of the population, Sri Lankan Tamils stands as the second major ethnic group in Sri Lanka (ADB, South Asia Working Paper Series: June 2015).

During the period of 1953–1981, municipal councils (MCs), urban councils (UCs), and town councils (TCs) were treated as urban areas (Sri Lanka National Water Development Report 2006). Nevertheless, since 1992, municipal councils, urban councils, and Pradeshiya Sabhas were identified as the local government institutions, but municipal council and urban council areas were the only areas that were considered as urban areas (DCS 2003); thus, Monaragala, Mullaitivu, Mannar, Polonnaruwa, and Kilinochchi districts that were under town councils are now being recognized as rural areas whereas; Colombo, Kandy, and Galle are the major urban cities of Sri Lanka, while Kurunegala, Negombo, Jaffna, Trincomalee, and Matara are the other urban cities of Sri Lanka. Therefore, in this stance, urban water supply can be identified as, water supply line, within Greater Colombo, within all urban and municipal council administrations, and urban areas especially in Colombo, specified by the Urban Development Authority (UDA), and within towns and Pradeshiya Sabhas (lowest level of elected local authority) where the population is more than 6,000 (Sri Lanka National Water Development Report 2006). Since the population growth in Sri Lanka is rising, it is inevitable that Sri Lanka faces immense water supply challenges in the urban sector raising the importance of addressing the policy responses in managing urban water crisis in Sri Lanka.

With the enormous support being given by multiple external development agencies, the water supply and sanitation sectors in Sri Lanka have achieved constructive results over the years (ADB, South Asia Working Paper Series: June 2015). The development of water and sanitation facilities for the people located in the north and east parts of the country was triggered with the ending of 30 years of civil conflict in 2009. Furthermore, in the aftermath of Tsunami in 2004, multiple grants were also received in order to rehabilitate and improve water and sanitation facilities in coastal regions (ADB, South Asia Working Paper Series: June 2015).

However, national figures highlight some significant variations across the country. The Household Income and Expenditure Survey (HIES), 2011, presents sector data according to sub-sectors. There are three sub-sectors: urban sub-sector (primary and secondary cities/towns and communities with populations above 6,000); rural sub-sector (low-density village areas); and estate sub-sector (plantations with small yet high-density cluster communities).

On a national level, currently almost 85% of the population has access to safe water since they directly reach water supplies from piped water systems, protected wells, or rainwater systems (ADB: South Asia Working Paper Series, June 2015). Therefore, over 9 million people (about 44% of the population) have access to piped water and more than 0.6 million people (about 3%) have access to hand pump tube wells. Moreover, when it comes to the rural population, while 36% of them have access to safe drinking water through protected dug wells, only 1% of the population uses rainwater harvesting systems (ADB: South Asia Working Paper Series: June 2015).

However, as per the statistics, 15% of the population still remains with the inability to access a safe water source within 200 m of their residence (ADB: South Asia Working Paper Series: June 2015) which leads to understand the challenges Sri Lanka face despite the significant positive outcomes that have been achieved. Hence, understanding the gravity of the urban water crisis in Sri Lanka, this chapter intends to address Sri Lanka's adaptive policy responses in managing urban water crisis under three subsections.

(i) Firstly, the challenges that Sri Lanka face in addressing the urban water crisis
(ii) Secondly, the adaptive policy responses in managing urban water crisis in Sri Lanka
(iii) Finally, the conclusion making recommendations to overcome the issue with a comprehensive policy response.

18.2 Challenges in Addressing Urban Water Crisis in Sri Lanka

In spite of the achievements that have been made in managing the urban water sector of Sri Lanka, a significant level of challenges still remain to be addressed due to the increasing levels of (Fig. 18.1).

(i) Urbanization
(ii) Water scarcity and degradation
(iii) Water rights issues
(iv) Inadequate investment in water resources development
(v) Poor urban planning

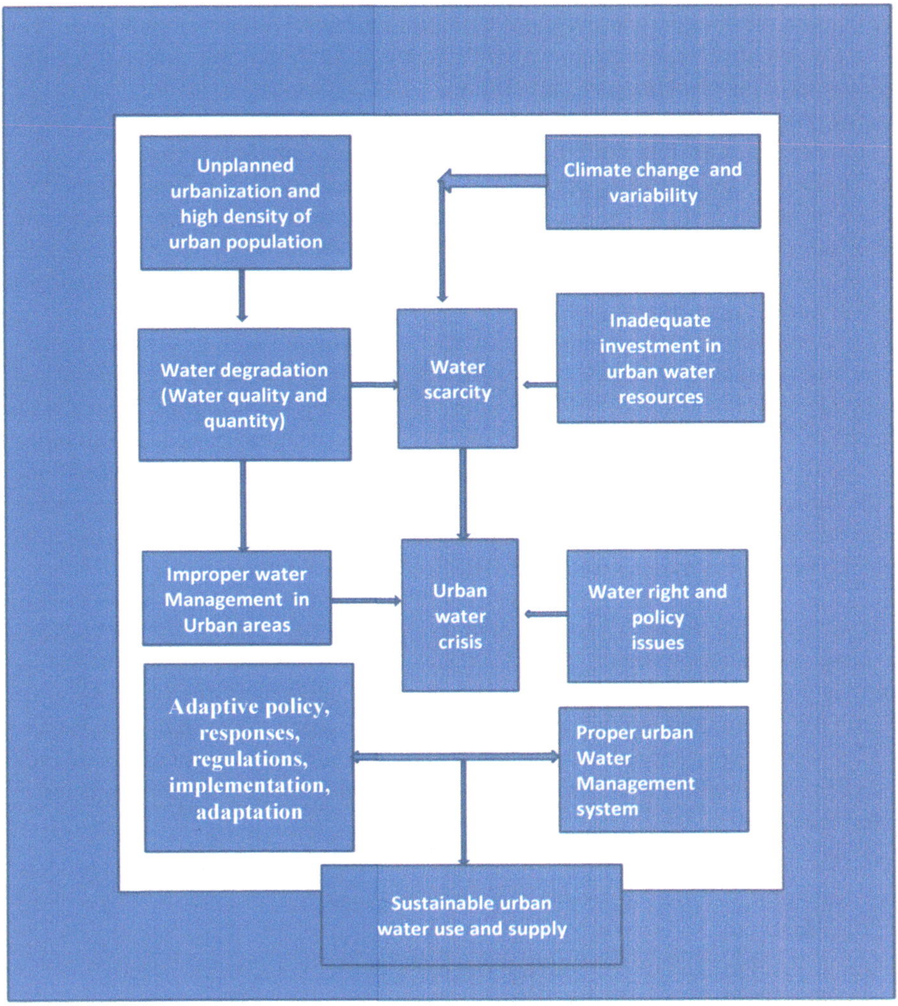

Fig. 18.1 Conceptual framework of urban water crisis in Sri Lanka. *Source* Author (2017)

18.2.1 Increasing Level of Urbanization

With the rapid growth of population, urbanization has become one of the world's most challenging issues (ADB: South Asia Working Paper Series, June 2015). Globally, urban population is expected to be increased by over a million people a week; thus, the possibility of creating massive challenges for the delivery of infrastructure and basic services such as housing, electricity, water, and sanitation stands at a higher position (ADB: South Asia Working Paper Series: June 2015). Mostly, the increasing levels of population growth and industrial activities being based in the urban sector have resulted in urbanization of Sri Lanka. People tend to move from rural areas to urban areas in order to fulfill employment requirements and that has resulted in urbanization which has posed a challenge to the government, as the government should take up measures to strike a balance between the growing needs of its people and the scarcity of resources. These issues may continue to intensify since the effects of economic development and the rural migration of people continue to exist. Hence, it is inevitable that the country is in dire need to provide more facilities based on the increasing levels of population growth in the urban sector, especially water supply for drinking, household activities, and sanitation. Further, the issue has acquired global attention as under the requirements of fulfilling the sustainable development goals, the quality of life in the developing world's rapidly urbanizing cities will not constitute sustainability without consistent and safe access to clean drinking water and satisfactory sanitation facilities. As we all are aware, the number of people without access to safe drinking water sources is rising while drainage systems and flood control systems also require rapid improvement. While most of the provinces in Sri Lanka have exceeded 80% in access to drinking water in safe conditions, unfortunately, some provinces still have not come to a satisfactory level (Drinking Water Status in Sri Lanka, Department of Census and Statistics 2012/13).

A viable and operative sewerage system can be only seen in the city of Colombo, and it is hardly adequate to face the rapid population growth (ADB: South Asia Working Paper Series: June 2015). The rapid economic growth has attracted more people to the urban areas, further increasing the necessity of taking measurements to provide clean water and implement proper wastewater management systems. Thus, the urban water crisis due to the increasing levels of population growth in the urban sector has been an immense challenge to the government in addressing the urban water crisis in Sri Lanka.

18.2.2 Increasing Level of Water Scarcity and Degradation

In Sri Lanka, the average island wide rainfall is close to 1,900 mm, and the total annual precipitation is approximately 132 BCM (billion cubic meters), while the total surface runoff is approximately 50 BCM (ADB: South Asia Working Paper Series: June 2015). The internally renewable water supply capacity per year has been estimated to be 43.2 BCM, and withdrawals amount annually to be 8.7 BCM (ADB: South Asia Working Paper Series: June 2015). Thus, in principle, there is no shortage of water to be seen, but the competition between the water users, incompetent levels of compliance with regard to pollution control and inadequate land use policies have become a menace to watersheds.

Furthermore, groundwater is widely used in Sri Lanka for domestic, industrial, commercial, agricultural, and other purposes, but the quality of the groundwater differs all over the country depending on various aspects. Usual rainfall, landscape, geology, and hydrogeological conditions (ADB: South Asia Working Paper Series: June 2015) are few of them to be precise. Although groundwater resources are to be seen throughout the country, highly productive aquifers are only to be found in the sedimentary limestone zone extending from Puttalam to the Jaffna Peninsula and Mullaitivu most of which have not been recognized as urban areas in Sri Lanka. The rest of the parts of the country have local and sporadic weathered and fissured hard rock aquifers of slightly low productivity (ADB: South Asia Working Paper Series, June 2015). Therefore, despite the drilled wells of around 35,000 throughout the country in managing water distribution, systematic mapping of groundwater resources related to their quality or quantity has only been started; thus, it still remains a challenge in overcoming the issue at hand.

Moreover, due to mining of sand, clay, and also due to dumping of waste near rivers, lakes, canals, and other water reservoir systems, it has resulted in degradation of river environments, damaging river health and water resource systems (ADB: South Asia Working Paper Series: June 2015). Mitigating this issue requires a strong political intervention and political will which has so far been limited, particularly in the Western Province where most of the industrial development of the nation is taking place.

Water quality has been deteriorating in some rivers especially when rivers cross the urban areas. Dumping of domestic waste, industrial waste and untreated wastewater into water ways are the main causes of water pollution in Sri Lanka. Rural and agriculture-based areas mainly agrochemicals are main pollutant sources. Kelani River is one of the main water sources for Colombo, but unfortunately, its water quality is predicted to be endangered by untreated or inadequately treated wastewater and solid waste (Ratnayake 2010).

Therefore, due to the inadequate levels of administration and planning in the water sector, the distinction between urban–rural areas, environmental degradation, and the unequal distribution of resources has exacerbated the water scarcity and degradation which has become a challenge in managing urban water crisis in Sri Lanka.

18.2.3 Increasing Levels of Water Rights-Related Issues

When it comes to the right to extract water, it is generally connected to land rights; thus, appropriation and mining are based on land ownership (Sri Lanka National Water Development Report 2006). However, the issue of water rights has become a challenge due to the increasing amount of water-related conflicts that have been occurred. Since most developed water resources are utilized by national agencies, administrative water allocation is the prominently practiced system in Sri Lanka (ADB: South Asia Working Paper Series, June 2015). Parliamentary acts have been formulated for the effective governance of water, and it is the duty of agencies and relevant ministries to comply with government policies (Sri Lanka National Water Development Report 2006).

As the owners of water rights and water allocations, Irrigation Department and the Mahaweli Authority of Sri Lanka play a major role in water allocation purposes. These agencies, in fact, act according to a programmed and duly planned strategy centered on the needs of project recipients and are also assigned with the task of distributing appropriate water from any source (ADB: South Asia Working Paper Series, June 2015). But, the issue lies within the system as they lack substantial provisions to allocate water rights to new users. Therefore, it is of paramount importance that the government amends the existing practice of water distribution and promotes a more transparent system which encourages equity in terms of water rights and thereby ultimately addresses the urban water crisis of Sri Lanka.

18.2.4 Increasing Levels of Inadequate Investments in Water Resources Development

As a countervailing measure in addressing the rapid urbanization and the population growth, promoting and paving the way for investment opportunities is of paramount importance. With the increasing demand for water, the need to handle the existing water scarcity and degradation issues arises. The Government of Sri Lanka has estimated that[1] SLRs 700 billion ($5.38 billion) worth investment will be required for projects on water and sanitation facilities in 2020 (ADB: South Asia Working Paper Series, June 2015) from which around 80% (SLRs 564 billion) is estimated as being necessary to reach the sectoral targets of 60% piped water supply coverage and 7% piped sewerage coverage by the year 2020 (ADB: South Asia Working Paper Series, June 2015). Nonetheless, marking an unfortunate reality of the funding process by local banks, both the government and foreign donors are expected to provide between SLRs 35 billion and SLRs 50 billion per annum with an additional sum of SLRs 33 billion leaving a significant financial gap, if all the proposed programs are to be implemented completely (ADB: South Asia Working Paper Series, June 2015).

Hence, it shows a major lacuna in the investment procedures of Sri Lanka which ultimately has resulted in poor administration and distribution of water to urban sector. There, the government has been unable to strike a balance between the rapid population growth and the limited water resources availability. Since the government plays a vibrant role in this regard, it is important that the water sector management is to be developed through the enhancement of investment opportunities in Sri Lanka in order to overcome the urban water crisis.

18.2.5 Poor Urban Planning

In major urban cities of Sri Lanka, namely Colombo, Kandy and Galle, rivers are the sources for the main water supply. Water is pumped to Colombo through surface water from Kelani and Kalu rivers, and water is pumped to Kandy through Mahaweli River, whereas water is pumped to Galle through Gin River. Usually, water is pumped

[1]SLRs: Sri Lanka Rupees.

from rivers to storage reservoirs, and replenished water availability is the availability of river flow where in the dry season salt water intrusion takes place (Integrated Urban Water Resources Management in Sri Lanka Background, Issues and Training Needs, University of Moratuwa, Sri Lanka). These storage reservoirs have limited capacity, and no expansions have been done in the recent past (Integrated Urban Water Resources Management in Sri Lanka Background, Issues and Training Needs, University of Moratuwa, Sri Lanka).

Furthermore, due to the unlimited industrial activities, constructions and condominiums are being built mostly in the Western Province; the need for a sophisticated urban plan including water distribution is needed in order to comply with the changing industrial and economic activities taking place. Therefore, since poor urban planning remains a challenge in addressing urban water crisis, it is important that the government addresses these issues to overcome the threats and challenges in the sector.

Despite the challenges in managing urban water crisis in Sri Lanka, in the long run Sri Lanka has accorded high priority to water supply and sanitation sectoral development by setting up targets to be achieved in providing safe drinking water and improving sanitation facilities to all the residents by the year 2020. In terms of access to safe water, it is targeted to reach 100% in 2020, whereas, in 2005, 2009 and 2015 the targets were to reach 80, 85 and 94% (National Water Supply and Drainage Board's projected Development Policy Framework 2010, Colombo). In terms of access to pipe-borne water, it is targeted to reach 44% in 2020, whereas, in 2005, the target was to reach 29%, and in 2009, the target was to reach 37% (National Water Supply and Drainage Board's projected Development Policy Framework 2010, Colombo). Moreover, it is also targeted to reach 7.0% in terms of providing coverage to pipe sewerage systems in 2020, which is drastically different with the targeted figures in 2005, 2009 and 2015 (National Water Supply and Drainage Board's projected Development Policy Framework, 2010, Colombo). Hence, these data in fact stand as proof of Government's high priority toward water supply and sanitation development sector in Sri Lanka.

18.3 Adaptive Policy Responses in Managing Urban Water Crisis in Sri Lanka

It can be said that water-related policies of Sri Lanka have been evolved with time as Sri Lanka has a history of been subjected to water management. The proclamation by King Parakramabahu the Great (1,153–1,186 AD) "In my kingdom are many paddy fields cultivated by means of rainwater, but a few indeed are those which are cultivated by means of perennial streams and great tanks (Arumugam 1969 quoted from Mahavamsa) and "In such a country let not even a small quantity of water obtained by rain, go to the sea without benefiting man" (Arumugam 1969 quoted from Mahavamsa) can be recognized as one of the ancient statements of water policies of the state. Hence, it is evident that the ancient policies on water management had given weight on making water available for livelihood needs and other requirements. However, with the growth of industrial and commercial activities, growing urbanization and pollution of water resources have altogether resulted in placing developed water resources at a risk (Sri Lanka National Water Development Report 2006).

Establishing policies in Sri Lanka stems from various resources: starting from the Constitution, extending to the Governing Party's manifesto, Policies adopted by the Cabinet, the Public Investment Programme and the Annual Budget (Mosley 1994). The provisions of the Constitution of the Democratic Republic of Sri Lanka 1978 have accorded the foremost place in the protection of environment as the duty of the state.

18.3.1 Constitution of the Democratic Socialist Republic of Sri Lanka 1978

Constitution of the Democratic Socialist Republic of Sri Lanka 1978 has following articles:

(i) Article 27(14) of the Constitution states that "the State shall protect, preserve and improve the environment for the benefit of the community"
(ii) Article 28 (f) states that the fundamental duties of the State are "to protect nature and conserve its riches."

Furthermore, under the directive principles of state policy and fundamental duties,

(i) Article 27(2) (c) recognizes that "The realization by all citizens of an adequate standard of living for themselves and their families including adequate food, clothing housing and continuous improvement of living conditions."
(ii) Article 27 (9) of the Constitution further states that, "the State shall ensure social security and welfare of the people."

All the above-mentioned provisions cover water and sanitation as a basic human need; hence, it asserts the importance accorded to water and sanitation. This shows that, uninterrupted water services should be available to people living in both urban and rural areas and in case if they are deprived of it that amounts to a violation of the basic law of the country.

In addition, in order to address the urban water crisis of Sri Lanka, The list 1 of the ninth schedule of the 13th Amendment to the 1978 Constitution has spelled out the functions of the provincial councils.

(i) Function: 9:2 Rehabilitation and maintenance of minor irrigation work
(ii) Function: 1.9 Irrigation planning, designing, implementation, supervision, and maintenance of all irrigation schemes relating to rivers running through more than one province or inter-provincial irrigation and land development schemes.

Additionally, the concurrent list (list iii) also refers to the water management which should be handled by both provincial and the central government:

(i) 17: Irrigation
(ii) 17:1: Water storage and management, drainage and embankments, food protection, and planning of water resources

These can be recognized as clear indications of the obligations that have been accorded to either the central government or the provincial councils in managing water resources understanding the gravity of water management, especially in the urban sector. The fact that water resources management has been granted constitutional protection is a vivid explanation to one of the many adaptive policy responses in managing urban water crisis in Sri Lanka.

18.3.2 National Drinking Water Policy of Sri Lanka

The National Drinking Water Policy of Sri Lanka provides an outline for addressing the fundamental issues and challenges which the country is facing in supplying safe water to the residents of Sri Lanka (National Drinking Water Policy: Sri Lanka). Competition over water uses as a result of growing urban population has caused environmental degradation, increased levels of inadequate policies and institutional constraints and increasing cost of development which have negatively impacted on the water sector to meet its targets. Hence, the policy is directed at addressing these issues through a comprehensive framework of strategies and a vision of a broader scope (National Drinking Water Policy: Sri Lanka).

The policy highlights that despite Sri Lanka's better record in safe drinking water supply to urban and rural areas, the challenges remain to be apparent on maintaining the services provided to current consumers and extension of such services to the unserved population of Sri Lanka which is approximately over 3 million at present (National Drinking Water Policy: Sri Lanka). Therefore, the policy provides a comprehensive framework and a policy guideline in order to improve the national service coverage in providing safe water. Moreover, the policy highlights the Government's recognition of access to safe drinking water as a basic human right. Following are few of the adaptive policy principles laid in the National Drinking Water policy in dealing with the management of urban water crisis:

(i) Access to safe drinking water is a basic right where beneficiaries have corresponding responsibilities.
(ii) As a result of a planned and developed water supply, a people centered, actively involved and demand responsive approaches will take place.
(iii) Irrespective of the social position and geographical location, investments to develop drinking water supply will be centered on priority needs.
(iv) Complying with the social, economic, and environmental necessities of present and future generations, the Government of Sri Lanka will act as the guardian

of water resources and in handling such resources on behalf of everyone under operative, efficient, and equitable means.

(v) By way of adopting suitable equipment, it is mandatory that each and every water services institution and agency ensure supplying of safe drinking water to the people.

These policy principles provide a comprehensive framework in addressing urban water crisis as it has directly stated the responsibility of the government in providing water services to people. Moreover, the policy is divided into two sub-sectors in order to cover both the rural and urban areas in delivering water supplies. There, under the "urban water supply sub-sector," urban water supply facilities and services are primarily centered on provision of pipe-borne water supply to urban areas as defined by municipalities, capitals, urban centers, industries, and some residential areas (National Drinking Water Policy: Sri Lanka). Urban water supplies will fall under the pursuit of the ministries which are accountable for such services and for their implementation by the appropriate local authorities (National Drinking Water Policy: Sri Lanka). Hence, it shows how the policy has addressed to minimize the effects of urban water crisis through a comprehensive policy approach.

Additionally, the policy also identifies the principle agency responsible for development, operation, and maintenance of drinking water supplies as the National Water Supply and Drainage Board (NWSDB) which has been established under an Act of parliament in 1975. As per the policy, NWSDB is tasked with:

(i) Providing technical support to other stakeholders regarding every technical component.
(ii) Being accountable for designing, planning, building, and operating small, medium and main pipe-borne water supply schemes.
(iii) Promoting devolution of responsibilities to intensify the management of water supply at the lowest suitable level for transparent and effective water management.
(iv) Maintaining and formulating a database regarding the access to safe water supplies around the country, water quality, water equity, etc.

Apart from the NWSDB, non-governmental organizations, provincial councils, and local authorities are also vested with the responsibilities in managing water supply to the urban and rural sector to minimize water crises.

18.3.3 Urban Water Supply Policy of Sri Lanka (National Water Supply and Drainage Board Website)

In order to act as an umbrella policy to cover the rising urban population and fulfill the growing demands of economic sectors of the country such as shipping, tourism, commerce, industry, the National Policy for Urban Water recognizes the imperativeness of developing facilities which provide a satisfactory level of service to all consumers including tourists and customers (Urban Water Supply Policy of Sri Lanka). According to the Policy, urban water supply has been defined as water supply:

(i) Within Greater Colombo,
(ii) Within urban and municipal council administrative areas,
(iii) Within the areas newly selected by the Ministry of National Planning,
(iv) Within the areas where complex technology is adopted.

Further, the following procures are regarded as important in developing and practicing a smooth operation of the schemes as mentioned in the Policy.

18.3.4 Preventive Maintenance Procedure (Urban Water Supply Policy of Sri Lanka)

This facilitates trouble free operation of civil mechanical equipment as preventive maintenance programs can eradicate unexpected breakdowns and massive costs incurred during sudden break downs (Urban Water Supply Policy of Sri Lanka).

18.3.5 Quality Control Procedure (Urban Water Supply Policy of Sri Lanka)

Proper quality control mechanisms can eradicate water-borne health hazards in the relevant area of service which would ultimately strengthen the treatment process by detecting unit operational systems (Urban Water Supply Policy of Sri Lanka).

Additionally, the Policy provides for large-scale investment opportunities to augment the water supply in the urban areas. According to the Policy, a selection criterion shall be established under the investment selection criteria, to rank various projects by their priority (Urban Water Supply Policy of Sri Lanka). Based on vital parameters like per capita investment, population density, subsidy ratio pertaining to the area, these criteria shall determine the priority of districts and then the division of urban centers (Urban Water Supply Policy of Sri Lanka).

Hence, from the above-mentioned information, it is apparent that the Urban Water Supply Policy of Sri Lanka has specifically focused on the matters affecting the water supply to the urban sector and the comprehensive document has been able to distinguish the urban sector with the rural sector in where a standadized definition has been introduced to contrast the difference between the two.

18.3.6 National Policy on Water Supply and Sanitation 2002

As stated in the vision of the Policy, the Sri Lankan government is concentrating on improving various aspects of standard of living while promoting economic prosperity as well as preserving the environment. Providing the Sri Lankans with clean drinking water and necessary sanitation facilities is seen as key necessities to achieve this target (National Policy on Water Supply and Sanitation 2002). As per the stated goals of the Policy, it is aimed at

(i) By 2025, providing 100% of the population with safe drinking water.
(ii) Providing piped water supplies to 100% of the population living in urban areas.
(iii) Achieving national standards for the quality of water and service levels in both rural and urban areas (National Policy on Water Supply and Sanitation 2002)

The government aims to achieve the aforementioned targets mainly by providing drinking water to the consumers through piped networks and also by other sources, including but not limited to tube wells, dug wells, and tankers (National Policy on Water Supply and Sanitation 2002). This shows the systematic approach toward addressing water management to overcome the urban water crisis of Sri Lanka which has become a challenge in the twenty-first century due to the rapid population growth, urbanization, and industrial activities being based on the urban sector.

Population density in the country, in general, has increased in keeping with the increase in population. In 1981, the population density was 230 persons per square kilometer, but in 2001, it had increased to 300 persons per square kilometer. As of 2012, the figure has increased to 323 (population density (persons per square kilometer) in Sri Lanka, Department of Census and Statistics 2012).

There is a high variation of population densities among districts. Colombo, Gampaha, and Kalutara districts in the Western Province, coastal districts of Galle and Matara in the Southern Province, Kandy District in the Central Province, and Jaffna in the Northern Province have high population densities. According to the current Census, the highest level of population density is prevalent in the Colombo District. The population density in the Colombo District, which stood at 2,605 persons per square kilometer in 1981 and increased to 3,330 persons per square kilometer in 2001, has increased further to 3,438 (population density (persons per square kilometer) in Sri Lanka, Department of Census and Statistics).

Therefore, by going through all the above-mentioned information, it is apparent that the government of Sri Lanka has taken appropriate measures to address the urban water crisis which has occurred due to enormous industrial activities taking place in the urban sector, urbanization, and population growth. Notwithstanding the policy responses, Sri Lanka still lacks a profound legal framework in the management of urban water crisis addressing the changes that have been occurred in the twenty-first century. Hence, the next sub-section will be describing the recommendations to improve the current position in managing urban water crisis in Sri Lanka.

18.4 Urban Water Supply Versus Rural Water Supply in Sri Lanka

18.4.1 Scope of the Rural Water Supply and Sanitation (RWSS)

Grama Niladhari (GN) Divisions within a Pradeshiya Sabha (PS) area, except the ones situated in former Town Council areas that also have a population over 6,000, are classified as "rural areas." The central government, local authorities, non-government organizations, and community-based organizations immensely contribute to enhance water supply and sanitation sector activities in rural areas as well as urban areas; since in both urban and rural areas, scarcity of water is a tremendous challenge that people face. As water and sanitation can be regarded as essential aspects of community life that requires the participation of the communities at all stages of development and management. Therefore, it is of paramount importance that a strong institutional framework which regulates the respective roles, mandates and responsibilities, authority, legal provisions, systems, and procedures of all the stakeholders of the sector is established (National Policy For Rural Water Supply and Sanitation Sector 2001).

18.4.1.1 Local Authority

Local authorities play an important role in performing the relevant functions regarding Rural Water Supply and Sanitation Sector, using their powers and authority that stems from relevant ordinances, acts, and regulations. Apart from that, they are expected to perform the following functions which are within their scope of authority (National Policy for Rural Water Supply and Sanitation Sector 2001).

(i) Strengthening the capacity for provision of services and ensuring the quality and standards of the services.
(ii) Facilitate planning and implementation of sector activities.
(iii) Promote investment, development, and sustainable management of the sector activities.
(iv) Determining relevant tariff according to the tariff structure formulated by the government.
(v) Facilitating external agencies to participate in sector development and promoting them for maintenance and sustainable operation of the facilities.
(vi) Encouraging the private sector by providing necessary protection and encouragement for investment.
(vii) Facilitate the community-based organizations in managing and implementing the sector development activities.
(viii) Make sure that the principles of the policy in implementation are adhered.

(ix) Protecting the environment in all sector development activities priority to the conservation of water resources.
(x) Provide services if necessary.

18.4.1.2 Community-Based Organizations

Community-based organizations (CBOs) are groups formed for providing and managing water supply and sanitation facilities to the members of such groups.

The CBOs are sometimes registered as a Development Society, a Trust, an NGO or as a Company under the provisions of the Companies Act No. 7 of 2007. They are recognized as authorized institutions that take part in the development of the sector. Raising funds, receiving grants, obtaining loans, levying tariffs, developing services, and management of facilities usually fall within their scope of authority. CBOs are subject to various regulations and standards imposed by the central Government as well as other legislative authorities such as provincial councils and the local authorities (National Policy for Rural Water Supply and Sanitation Sector 2001).

The functions of CBOs include (National Policy for Rural Water Supply and Sanitation Sector 2001):

(i) Assessing the needs, aspirations, and the demand of the communities for water supply, sanitation facilities, and services.
(ii) Assessing the economic viability and technical feasibility of different methods for providing sanitation facilities and water supply.
(iii) Arrange internal funding.
(iv) Planning, designing, preparing proposals, properly implementing, and sustainably managing the facilities and assets.
(v) Ensuring the participation of user community and other partners at all stages of the process.
(vi) Managing the facilities in a sustainable manner.
(vii) Preserving the environment and protecting the water sources and watershed areas

18.4.2 Analysis of Urban Water Supply and Rural Water Supply

Provincial councils, municipal, urban councils, Pradeshiya Sabhas, and CBOs have been delegated responsibility of delivering water and sanitation services in certain areas at the subnational level. While some of the functions of these agencies are devolved, others remain centrally managed (ADB, South Asia Working Paper Series: June 2015). In some instances, there is an overlapping mandate of agencies and target clientele and only limited coordinating mechanisms exist between agencies and decentralized units. There are three important features which ought to be noted when analyzing the overall water supply of Sri Lanka (urban water supply and rural water supply) (ADB, South Asia Working Paper Series: June 2015).

(i) Firstly, even with the rise of community-based organizations, the quality of services provided to consumers in rural areas is not in line with the services provided to consumers living in urban areas (ADB, South Asia Working Paper Series: June 2015).
(ii) Secondly, despite the fact that substantial progress has been shown in improving access to safe water supplies and sanitation facilities in post-conflict areas in the Northern Province and the Eastern Province, more work has to be done (ADB, South Asia Working Paper Series: June 2015).
(iii) Finally, providing safe water to estate communities living in areas such as Nuwara Eliya District has to be improved (ADB, South Asia Working Paper Series: June 2015).

Therefore, it is abundantly clear that maintaining and improving the water supply service levels for current consumers are equally challenging as providing services to people who are not served currently, which amounts to approximately 3 million (ADB, South Asia Working Paper Series: June 2015).

18.5 Recommendations

Although Sri Lanka has been successfully able to address urban water crisis through adaptive policy measures as discussed above, in order to develop a comprehensive legal document on addressing the urban water crisis in the coming future following recommendations are of paramount importance:

(i) Formulating a policy incorporating objectives for setting up an internal monitoring and regulatory unit in the National Water Supply and Drainage Board (NWSDB) including its mandate, functions, and authority. The internal monitoring and regulatory unit's initial activity would be (ADB, South Asia Working Paper Series: June 2015):
    (a) Setting up norms, aims, and standards for performance of services.
    (b) Evaluating the performance of regional support centers monthly based on vital performance indicators set in monthly monitoring sheets or templates.
(ii) Introducing institutional and governance reforms to meet the government's objectives and targets for the water and sanitation which will require support for water and wastewater utilities in two main aspects, i.e., (ADB, South Asia Working Paper Series: June 2015):
    (a) Sector-wide governance support
    (b) Organization-specific support. This will include both capacity development and asset development initiatives.

(iii) Introducing a system to build cisterns or storage tanks in our houses to collect rainwater falling on roofs and flowing in gutters to be used for watering our plants, washing cars, clothes, pots and pans and even for bathing an ablutions (Dr. J. B. Kelegama, Managing Water Resources to Ensure Equitable Economic and Social Development).

18.6 Conclusion

Water management is an essential requisite of social and economic development. As it was discussed earlier, it is apparent that, in order to develop a comprehensive policy approach, the[2] NWP and other water policies have to be updated to make sure that it complies with the economic, social, and political changes that are occurring in the country. Therefore, by considering all the above-mentioned information, it is evident that Sri Lanka needs to address the issues and challenges in the water sector in order to overcome the urban water crisis as it has become one of the most striking challenges coupled with the effects of climatic changes. Hence, Sri Lanka should address the following key issues to develop its current position in urban water crisis management:

(i) Lack of consensus on key policy issues
(ii) Lack of water sharing methodologies
(iii) Inadequate databases and early warning systems
(iv) Inadequate responses in addressing poverty alleviation and controlling water-related diseases
(v) Increasing levels of climatic changes which has resulted in water-related risks
(vi) Competition among water users
(vii) Absence of and fragmentation of responsibility for water resources
(viii) Unclear lines of demarcation of responsibility in respective authorities
(ix) Lack of adequate supervision by the respective authorities
(x) Interference by politicians and failure to involve the community in managing water resources.

References

ADB (Asian Development Bank) (2015) South Asia working paper series information, Sri Lanka's water supply and sanitation sector: achievements and a way forward

Arumugam S (1969) Water resources of Ceylon-its utilization and development. Water Resource Board Publication, Colombo

CBSL 2016a, b (Central Bank of Sri Lanka) (2016) Economic and social statistics of Sri Lanka, Sri Lanka

CBSL 2016a, b (Central Bank of Sri Lanka) (2016) Annual report, Sri Lanka

[2]NWP: National Water Policy.

DCS (Department of Census and Statistics) (2003) Statistical abstract 2002, Sri Lanka
DCS (Department of Census and Statistics) (2012) Poverty statistics/indicators for Sri Lanka, Sri Lanka
DCS (Department of Census and Statistics) (2012/13) Drinking water status in Sri Lanka, Sri Lanka
DCS (Department of Census and Statistics). Population density (persons per square kilometer) in Sri Lanka, Sri Lanka
Mosley P (1994) Institutional issues, Working Paper 1 in comprehensive water resources management in Sri Lanka. volume II: annexes. ADB and USAID
National Water Policy for Rural Water Supply and Sanitation Sector (2001) Ministry of urban development, Construction & Public Utilities
National Water Supply and Drainage Board's Projected Development Policy Framework (2010) National water supply and drainage board. Ministry of City Planning and Water Supply, Colombo
National Policy on Water Supply and Sanitation (2002) National Water Supply and Drainage Board, Sri Lanka
Ratnayake RMSK (2010) Urbanization and water quality control for the source of water in Colombo City, Sri Lanka. Paper submitted to the 1st WEPA international workshop, Hanoi, Viet Nam, 8–9 Mar. http://www.wepa-db.net/policies/state/srilanka/overview.htm
Sri Lanka National Water Development Report (2006) 2nd UN world water development report

Chapter 19
Preparing for the Future: Challenges in Water Management in Colombo, Sri Lanka

Deepthi Wickramasinghe

Abstract Water has become one of the key natural resources which are highly vulnerable to degradation in both quality and quantity under the circumstances of urban expansion and population explosion observed in major cities. Thus, adequate water management measures are considered as essential components in attaining sustainable development. The city of Colombo, the commercial capital of Sri Lanka, is increasingly experiencing a water crisis as a result of a rapidly growing demand that has outstripped existing resources. Two major sources of water are used to meet the needs of water supply to the city: rivers and reservoirs. Numerous factors could be held responsible for water shortages which appear from time to time, and the results are obvious: absence of or mostly, insufficient water in need. This paper highlights the issues in continuous water supply to Colombo with a special emphasis on current threat factors including pollution by domestic sewage, municipal wastewater and industrial effluents as well as salinity intrusion in Kelani River and other water sources. Potential effects of climate change could pose on water supply and are also discussed. Finally, it examines the available options in enhancing water security such as water conservation and proper planning for a sustainable water future.

Keywords Water supply · Shortage · Drought · Urban environment

19.1 Water: A Scares Resource?

Water is a spatially and temporally variable resource on earth. It is increasingly becoming a limited resource (Fabre et al. 2016), with the accelerated growth and expansion of world population, changing consumption patterns and variation in global environmental systems, especially under the effects of climate change. Lack of sufficient available water resources to meet water needs within a region is defined

---

D. Wickramasinghe (✉)
Department of Zoology and Environment Sciences, University of Colombo, Colombo 03, Sri Lanka
e-mail: deepthi@zoology.cmb.ac.lk

B. Ray and R. Shaw (eds.), *Urban Drought*, Disaster Risk Reduction,
https://doi.org/10.1007/978-981-10-8947-3_19

as "water scarcity" (UNEP 2012). Last century has seen a growth of water use which is more than twice the rate of population expansion and even though there is no shortage of water as such on a global scale, many regions in the world are increasingly becoming persistently short of water (UN 2012a). It is reported that two-thirds of the global population (4.0 billion people) face severe water shortage for at least 1 month of the year (Mekonnen and Hoekstra 2016). In this context, water has been given adequate attention by several key international fora. Water was a major point of discussions at 21 June Rio + 20 roundtables. The United Nations General Assembly recognizes water and sanitation as one of their principal concerns and endorses the fact that availability of adequate supply of drinking water and sanitation is central for a healthy living thereby asserting it as a human right (UN 2012b).

Water scarcity is not merely what is evident on surface. It goes beyond immediate physical availability. A complex cycle exists where fractions of water are found in abiotic and biotic components of earth. This cycle is not always visible (Mekonnen and Hoekstra 2016). Insufficient natural water resources to supply a region's demand are considered as physical water scarcity, while economic water scarcity results from poor management practices of adequately available water resources (Mancosu et al. 2015). Yet, again many human interactions play a significant role in availability, quality and quality of water resources (Cook 2012).

19.2 Drought

Drought is a common, recurring feature of climate which could occur in any climatic region. It is difficult to define a drought (Henley et al. 2013). In simple terms, it could be described as paucity of precipitation over an extended period, causing a shortage of water resources for any activity, group or environmental sector (Changnon 2002). Climatologists worldwide always struggle to identify the beginning of a drought and the basis (i.e. criteria) for declaring a time period for a particular drought remain to be an ongoing argument among scientists and policy-makers (Wilhite 2000). Perhaps, a drought could be one of the longest disasters that humankind faces, since they could be of long durations starting from several weeks to even a couple of years (Parry et al. 2016). Yet again, the areas affected by a drought condition could be hard to define, making the demarcation of such boundaries vague. These features make droughts complex to understand and manage (Wilhite et al. 2007). Driven both by climatic changes and inefficient management of water resources, drought disasters are becoming more frequent and water scarcity is increasing in severity across the globe (IWA 2016).

Drought should not be treated exclusively as a physical phenomenon since water scarcity could be a result of reduced or poor water supply which is unable to meet existing demands. Significant water shortages could occur when a complicated array of social and political factors influence limited water resources (Keshavarz et al. 2013). Therefore, the socioeconomic definitions of drought are commonly linked with "the demand and supply of essential economic goods with elements of meteorological, hydrological and agricultural drought" (Wilhite and Glant 1985). This explanation differs from other popular definitions of drought, as it identifies or classifies droughts, depending on the time and space processes of supply and demand.

Such socioeconomic droughts have been increasingly evident across the globe in the recent past. Rapid urbanization is threatening the water security as sprawling cities cause "urban drought" due to the disparity between supply and demand of water. For instance, the growth of megacities has placed unbearable stress on water resources in Asia where most states are already experiencing water shortages (Bastiaanssen and Harshadeep 2005; Srinivasan et al. 2013). This situation will only exacerbate as the water demand for many activities in cities is on the rise. Moreover, the impacts of climate change influencing water cycle with frequent floods and droughts are increasingly reported (McDonald et al. 2014; Koop and van Leeuwen 2016).

There could be many contributory factors that could lead to an urban drought condition. As indicated, all of these factors are unfortunately of more or less anthropogenic origin, and in many instances, human activities could "trigger" a drought.

This chapter discusses "urban drought" and challenges in water management in Colombo District, Sri Lanka. More specifically, this chapter will articulate the different reasons why water has become a scarce resource, community reactions to solve their water issues and responses of the government. It also discusses the additional burden of climate change on water resources. Finally, the chapter presents an outline on how to manage drought risk to prepare better for future challenges.

19.3 City of Colombo

Even though the smallest district in terms of size, the Colombo District remains the most populous and most urbanized in Sri Lanka. City of Colombo is the most significant urban centre in the district which is divided into 13 administrative divisions (Table 19.1). City of Colombo is the economic capital and the administrative hub of Sri Lanka. This district is a home for 2,375,000 of population which represents around 14% of the total population of the island. Colombo City has attracted a large proportion of internal migration as well.

Table 19.1 Administrative and demographic information of the district of Colombo

Information on Colombo District	
No. of administrative divisions	13
Total urban population	2,219,782
Total population (including rural and estate sectors)	2,375,000
Total area (ha)	69,900
Population density (individuals/ha)	34.0

Source Department of Census and Statistics (DCS) (2015)

19.4 Water Supply in Colombo

The singular authority responsible for water supply across Sri Lanka is the National Water Supply and Drainage Board (NWSDB). Since its establishment in 1974, several urban water supply and sanitation schemes operated by local authorities were transferred to NWSDB. Apart from few rural and town supply services, NWSDB manages the complete process from production to trade of water to the user. Its expenditure every year on production and distribution of safe drinking water to Colombo and suburbs is huge. All water connections are metered and the NWSDB charges a fee from each water consumer, based on water volume they consume. The treated water from these sources is used for consumption as well as for the other uses including cleaning and even for flushing toilets.

In this chapter, Colombo refers to the district of Colombo, unless otherwise specified. In the last few decades, there is a significant improvement in the water supply to Colombo. For instance, the promotion of piped water into the premises has risen from 37% in 1990 to 70% in 2010 (DCS 2012). Nevertheless, nearly 25% of the households depend on wells to meet their water demands. Table 19.2 presents a summary of access to drinking water in Colombo.

As in many other expanding cities, Colombo faces an ever-rising problem of slum populations. The biggest slum population in the country is recorded in Colombo District. Sevanatha (2003) provides information on the water supply to these urban dwelling which is underprivileged. It states that only 44% of these families have individual house connections and availability of 24 h of piped water covers only 56%. Thirty per cent of families have difficult access to drinking water.

The district gets water from four different sources (Table 19.3). The major source is the Kelani River (Fig. 19.1) which reaches the Colombo City and flows to the sea at Mattakkuliya, Colombo 15.

The maximum available capacities from Labugama and Kalatuwawa reservoirs are already extracted by the NWSDB. Further expansion of capacities of these reservoirs is improbable. Hence, the only option NWSDB that is left with is to withdraw more water from the Kelani River mainly from the Ambathale intake and treatment plant which is situated 14 km from its point of discharge to the Indian Ocean.

19.5 Issues in Water Supply

With rapid urbanization, provision of safe drinking water to urban communities has become an important and challenging task of governments in developing countries such as Sri Lanka. Among the key issues in the water supply in Colombo is the intermittent supply of pipe-borne water mainly due to low and fluctuating system pressure. Distribution network of pipes in many areas is very old and sometimes more than a century. This has caused many unacceptable disturbances to daily life of the city dwellers. In addition, there are limitations to continuous water supply:

Table 19.2 Principal forms of access to drinking water in Colombo, 2011

	Well				Pipe born Water				Other			
Administrative unit	Number of households	Protected well within premises	Protected well outside premises	Unprotected well	Tap within unit	Tap within premises but outside unit	Tap outside premises	Rural water supply project	Tube well	Bowser	Bottled water	River/tank/stream/spring/other
Colombo district	558,755	121,297	11,862	7,026	350,327	26,942	18,202	17,600	1,754	58	682	3,005
%	100	21.7	2.1	1.3	62.7	4.8	3.3	3.1	0.3	0	0.1	0.5

Adopted from Department of Censes and Statistics (DCS) (2012)

Table 19.3 Sources of water supply to Colombo City

Source	Location
Kelani River	Fourth longest river in the island (145 kilometres long). It originates from the Sri Pada Mountain Range and flows to Colombo
Kalu River	A main river in the island (129 kilometres long). It originates from the Sri Pada Mountain Range and flows to Kalutara
Kalatuwawa Reservoir	Artificial reservoir in Avissawella
Labugama Reservoir	Artificial reservoir in Avissawella

Fig. 19.1 Kelani River flowing by Colombo

some areas of the city get water only during off-peak hours of the day. Furthermore in some other locations, water supply is limited to an average of 6–10 h per day.

ADB report (2016) highlights that due to many reasons, the NWSDB is facing continuous challenges in providing water supply and wastewater services in Greater Colombo.

19.5.1 Non-Revenue Water (NRW)

Non-revenue water (NRW) is one of the major challenges that found in water supply sector. The gap between the volume of water supplied through a system and the volume of water which is billed as authorized consumption is identified as the NRW. According to Malathi (2016), this revenue loss is due to three factors: physical water loss due to leakages and wastages, etc., commercial loss due to unauthorized consumption as well as unbilled authorized consumption such as water supply to the poor urban communities for free. She also reports that currently the island-wide NRW stands at 30% while it is 49% in Colombo City alone. Supply of water for free of charge, especially for urban poor, is essential for a developing country such

Fig. 19.2 Public toilets in Weli Park (a community leisure park), Nawala

as SriLanka (Peiris et al. 2008). For an example, in 2005 the NWSDB had supplied treated water free of charge for consumption of 64,700 families out of which 50% consisted of low-income settlements (LIS) (Dharmapala and Ranasinghe 2006). This included approximately 1600 stand posts, 1450 common bath taps and 276 toilet taps where a considerable amount of water gets wasted. Many public toilets in Colombo are supplied with water free of charge (Fig. 19.2) with manual flushing techniques where wastages are high.

19.5.2 Pollution

The major water supplier to Colombo, the Kelani River, is one of the most polluted rivers in the country. The river has been subjected to extensive pollution continuously in the recent past because of numerous causes.

Land-based pollutants including treated and untreated industrial effluents, agricultural run-off, domestic and municipal effluents have been identified as the principal sources of water pollution in Kelani River. Water quality measurements of Kelani River, which meets a major part of the demand for pipe-borne drinking water supplies of Colombo, are testimony to the magnitude of the prevailing water pollution (CEA 2014). Yet, sewage generated from low-income dwelling units and industrial effluents from many industries, especially from tanning, metal finishing and processing industries are released to the Kelani River very often.

Moreover, excessive sand mining and lowering of river beds at the lower reaches in the Kelani River, have caused saline water intrusion from the ocean. This results in making the water saline and non-potable, where sand mining is particularly severe, and the salinity intrusion has extended inwards to Ambatale on several occasions. Since a large part of the potable water supply for Colombo is produced from the Kelani River at Ambatale, this is has become a serious issue.

It is also increasingly evident that irrational dumping of municipal, industrial and domestic solid wastes in the river banks as well as in adjacent low lying areas adjacent. Additionally, untreated or undertreated sewage from different sources mainly from the industry along the waterways is discharged into the river. Among many, one key reason behind this is the poorly maintained sewage system established by the British during colonial days has been overwhelmed and doesn't function in full capacity. Many toxic substances including textile dyes, heavy metals and pesticides have been recorded from the river water. High amounts of organic and microbiological pollution have been reported too (Athukorala et al. 2013). Yet, the pollution levels vary depending on the seasons, rainy and non-rainy. In rainy seasons, storm waters in various quantities reach the river with much sediment loads. In the dry season, saltwater intrusion and enhanced concentrations of other pollutants become a burden to achieve water quality standards for the NWSDB.

19.5.3 Saltwater Intrusion

Saltwater intrusion is commonly observed in many river basins in the island, and it is a major problem. The induced flow of seawater into freshwater rivers and aquifers primarily due to the lowering of the water level or river bed is "saltwater intrusion" (Atwater 2008). Especially, the drop in water levels in the river and in aquifers, which are in hydraulic connection with the sea, induces gradients that result in migration of salt water from the sea towards the river. This causes increased salinity of water which makes it unsuitable for consumption (Ratnayake et al. 2012). During the dry season, it is a common problem in most rivers including both Kelani and Kalu (NWSDB 2015). In the lower reaches of the Kelani River, water contaminated with high salinity has become a continually growing problem.

19.5.4 Problems with Alternative Water Resources

Groundwater could be an alternative to meet the demand for water in Colombo. However, extracting groundwater is not common in Colombo City but in some other parts of the district dependence on groundwater for domestic purposes could be seen. Groundwater could be regarded as a resource which is inefficiently utilized (ME &

NR 2004). Multiple issues including the ownership (current groundwater ownership is with the individuals), unavailability of reliable information on groundwater quality, poor monitoring and increased pollution loads especially microbial pollution could be recognized as reasons for this (Imbulana et al. 2006). The absence of a specific government authority in charge of groundwater management adds on to these problems (Karunaratne and Pathmarajah 2002). Currently, this resource is considered as a free good and some industries continue to extract groundwater immensely which obviously contribute towards depletion of the resource (NWSDB 2015).

So far, minimum attention has been given to recycling and reuse of water in the country even though in some Asian countries like Singapore successfully practises these measures. Given the high cost of technology and human resources, at least the alternative of using partially treated water in some industrial and commercial activities should be highly considered.

19.6 How Do Communities Manage Water Problems?

Depending on the use and the user, the type and intensity of water demand vary. The link between water insufficiency and economic status of users too is significant and influence on the solutions. Most of the public and private offices and institutions often possess storage tanks where adequate water is stored and hence, no drops in supply are experienced. Nevertheless, many mid-income level and poor communities suffer from inadequacy of water for their daily needs. In particularly, slum dwellers and residents in low-income housing schemes often face severe threats due to unavailability of water in acceptable quality and in quantity. In most cases, these communities lack capacity, expertise and adequate infrastructure to develop their own supply of water such as wells. Often though, such communities are provided with common public taps where water supply is free of charge to meet their basic demands. It is not uncommon to evident people stay in queues till the turn comes for them to collect water even during early mornings and night time. In many instances, women and female children bear the responsibility of waiting in queues and collecting water for the household.

Alternative sources to exist in different contexts. As mentioned in Table 19.2, some dwellers extract groundwater through dug wells which is not common in the city but in suburbs. One key challenge is to find a suitable place to construct the well since land is crucial resources especially in highly populated areas of the city. Moreover, in the absence of adequate rain many well dry up dragging the users into trouble. There are instances where deep groundwater aquifers are exploited using tube wells which are only practised in people with moderate- and high-income levels. Some private parties construct their own tube well for industrial and commercial purposes. While the variation of water table in an area could pose threats to the continuous supply, lack of regulations on over-extraction of groundwater is another issue that needs attention of the authorities.

Water has become a commercial product and many institutes, offices especially restaurants and hotels purchase bottled water for drinking purposes. Due to a common belief that bottled water which is extracted from springs and deep tube wells is safe and pure, there is a huge demand for this commodity. Nevertheless, it does not justify the harm that the plastic bottles generate as discarded bottles are becoming a growing problem in solid waste management in the city.

19.7 Response from the Government

Government of Sri Lanka is taking steps to provide an improved water supply in many ways (DNP 2014). The development plan and framework “Vision for a New Sri Lanka 2010–2020” focus on many sustainable development agendas and highlight priorities. Firstly, it envisages to increase the population’s access to improved water supply and achieves national standards for service and quality. The second objective is to improve sanitation facilities, increase the population’s access to those facilities and set high national standards for service and quality of water. Additionally, the framework wishes to enhance the service by increasing the piped water supply in the country from the current 37 to 60% by 2020 and sewerage coverage from 2.5 to 7%. While some of these measures have already implemented by the government, many issues have continued to evolve. High-rising buildings and the expansion of industries, including tourism, and many new developments have posed threats to water supply.

The DNP reports and NWS & DB further elaborate on certain policy and action which is presented below under several categories.

19.7.1 Quality of Surface Water Supply

To increase cost-effectiveness of Ambathale water intake in the Kelani River water purification plant, a new project is proposed targeting to enhance the productivity in purification process. Another remedial measure is to construct an effective salinity barrier at Kelani River for uninterrupted water supply with less salinity burden.

19.7.2 New Governance for Groundwater

Since the groundwater resources have gained much less attention when compared with the surface water, it is proposed to launch more investigations to cover aquifer basin geometry and characteristics, aquifer properties, groundwater potential, aquifer recharge assessment and quantification, water chemistry and its variation in the coastal sandy aquifer extending from Colombo to Negombo area.

19.7.3 Water Quality and Wastewater Treatment

In the framework, several new projects have been proposed to enhance the quality of water by using proper wastewater treatment. This included projects with new technological approaches. "Establishing and Operating a Pilot—Scale Electron Beam Facility to Treat Wastewater from Textile Dyeing Factories" and "Sustainable Land Use and Management Strategies for Controlling Soil Erosion and Improving Soil and Water Quality" are few projects worth mentioning here.

19.8 Adaptations to Climate Change Impacts

Climate change is expected to influence on the water cycle causing unreliable and unpredictable weather, which would obviously affect the availability of water in time and space.

In managing design, construction and maintenance of water sector infrastructure, meeting the objectives to minimize the impact of climate change is an important factor. In the recent past, unusually high temperatures have been recorded in many districts including Colombo and the Climate Change Secretariat (CCS 2000) reports an increase in temperature from 1.4 to 2.7 °C in the past few decades. As a coastal district in the wet zone of the country, Colombo is experiencing a change in climate. Increase of natural disasters related to the water cycle is evident. For instance, devastating floods occurred in both in 2016 and in 2017 and made significant influence on the water supply. On the other extreme, a prolonged period of drought which persisted for many months in 2017, affected water sources including Kelani river and dug wells.

In this context, the water sector warrants special attention. Diversifying sources of water supply will become necessary to conserve available water resources and to secure the water supply. Few of the potential solutions include making available of new storage facilities, sustainable extraction of groundwater, different measures taken towards water conservation and improved recycling and desalinating water. Table 19.4 depicts some suggestions on adaptive measures in combating the impacts of climate change targeting Colombo.

19.9 Managing Urban Drought Risk—Preparing for the Future

19.9.1 Proactive Versus Reactive Approach

The combination of the probability of an event and its consequences is a risk according to disaster management science. Managing or reducing risk is essential in social,

Table 19.4 Climate change adaptations suggested for drought risk reduction in Colombo

Climate change impact	Results	Adaptive measures
Increase in ambient temperature	Increased evaporation, water loss in rivers and reservoirs	Enhance green cover in the watershed Integrated watershed management
Reduced water flows due to changes in water cycle	Water stress Inadequate water supply Intermittent water supply	Soil conservation measures Ecosystem-based approaches
Increased water demand to cope up with changes in environment	Water stress—more demand for water Inadequate water supply Intermittent water supply	Community-based water conservation programmes Public awareness programmes Alternative water supply including rainwater harvesting Conservation of wetlands to store more water
Deteriorated water quality due to pollution and saltwater intrusion	Pressure on water treatment Loss of value of water	Enforcement of pollution control laws Pollution monitoring Enhanced water treatment infrastructure

economic and environmental security. Throughout the world risk management is often practised for many disasters, yet urban drought risk management is still in preliminary stages. In the global context, many cities follow a reactive approach to manage droughts rather than a proactive attitude (Rossi 2000). Hence, many unplanned measures are carried out only when a drought hits to cope with water shortages using temporary solutions. Many short-term and ad hoc solutions could be available to minimize the immediate impacts of water shortages. For an example, providing additional water sources such as water bowsers could help with water needs of communities. But in the long run, these options are less practical as well as costly. These solutions are mostly driven by social and political pressure. Growing demand for water in expanding urban areas is pushing water managers to direct their attention on drought risk analysis and for an attitude change towards proactive risk management.

Therefore, an overview of the current practices is important.

19.9.2 Community Participation

Community participation is a tool that can help align the economic, social and environmental needs in conservation. It bridges the gap between the governed and those who govern and offer promising insights into conserving any valuable resource.

Even though very often water crisis management is largely viewed as the responsibility of the government, the role of the communities too is significantly high.

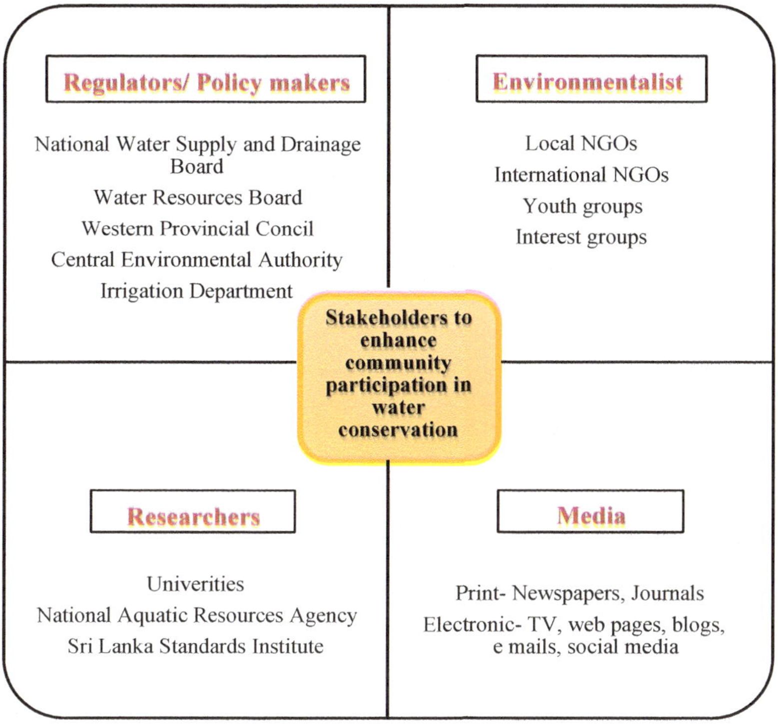

Fig. 19.3 Stakeholders to enhance community participation in water conservation

Historically, water has been regarded as an unlimited resource and was never a scarce commodity that needs special measures to conserve. On the other hand, in many instances, it is common that when the water is readily available, the community forgets the value of water conservation and the impacts of drought. In many instances, the community claims their right for water yet fails to report a leaking tap. Sensitive behaviour towards water conservation, contribution towards maintaining infrastructure, demand reductions and several other drought measures are essentially linked to awareness and increased participation of the community (Fig. 19.3).

Communities should actively take part in exploring new avenues including popularization of rainwater harvesting especially to reduce the cost of non-revenue water supply to public places such as schools, religious places, hospitals. Rainwater harvesting is somewhat popular in many other parts of the country but not very common in Colombo. Yet, there is a potential of constructing rainwater harvesting tanks especially by institutes with many employees where a large amount of water is needed. Some government institutes as well as schools have already constructed their own tanks as a supplementary water source. Figure 19.4 shows a rainwater harvesting tank in hospital premises in Kandy District where water collected and is used in

Fig. 19.4 Rainwater harvesting in Kandy

toilets. Thus, dependence on billed water is lessened. The same solution could be adopted in needy places in Colombo as well.

19.9.3 Policy and Governance Gaps

Even though technical challenges of addressing water scarcity are often highlighted, most of those are not that difficult to manage. What is much more difficult to deal with is poor governance. In many parts of the world, governments are often slow to act, unaccountable, and deny or underestimate manifest harm. This is not different in Sri Lanka too. As described early, the major source of Colombo water supply is highly polluted. Nevertheless, many policies and laws exist in books.

One good example is discharge of industrial wastewater. Such activities are regulated by Environmental Protection Licences (EPL) scheme under the National Environmental Act of 1980 (CEA 2014). It is thus mandatory to obtain an EPL to discharge wastewater into any terrestrial or aquatic environment. Therefore, treating wastewater to meet the prescribed standards in the Act is mandatory to obtain the EPL which describes the stipulated criteria. Nevertheless, practical reality of adhering to the stipulations is questionable. In many instances, pollution loads discharged to rivers are not controlled according to laws and water deterioration continues. In some areas of the lower reaches of the Kelani River, illegal discharge of untreated wastewater and dumping of solid wastes are pervasive.

19.10 Conclusion

Water is becoming a scarce resource worldwide and emergence of drought conditions is not unusual. Due to rapid impoverishment of water resources, Sri Lanka's most populated and urbanized district Colombo is facing an "urban drought" condition. It hampers the continuous water supply for diversified needs of an expanding population.

In Colombo, nearly 70% of households are equipped with water connections while 25% depend on wells. National Water Supply and Drainage Board (NWSDB) holds the authority of water supply to Colombo. It charges from each consumer, based on water volume they consume. The NWSDB is confronted with many challenges. Limited number of water sources, non-revenue water, pollution, saltwater intrusion into water supply sources and inadequacy of alternative water resources are some of those. In this context, city dwellers, especially the underprivileged, have to face water shortages and water supply interruptions. Communities' responses to water issues are numerous. Government has taken corrective steps and remedial measures to meet water demands. Possible threats to water sector by climate change are an additional problem which must be taken into consideration while finding solutions to water supply issues.

In preparation, Colombo for a water-secured future, urban drought risk should be managed efficiently with proactive approaches, encouraging more community participation in water conservation and bridging the policy and governance gaps.

Acknowledgements The author wishes to acknowledge the support extended by Ms. Ushawarnie Mendis, Waruni Perera and Nalaka Madusanka in editorial support, Eranjnaja Wickramsinghe for comments, Water Specialists Eng. DS Jayasiriwardena, Eng. Badra Kamaladasa and Eng. Danesh Gunathillake for valuable insights.

References

ADB (2016) https://www.adb.org/projects/45148-004/main. Accessed on 1 July 2017

ADB Report (2016) https://www.adb.org/sites/default/files/publication/189411/awdo-2016.pdf

Athukorala PSW, Weerasinghe LS, Jayasooria M, Rajapakshe D, Fernando L, Raffeeze M, Miguntanna NP (2013) Analysis of water quality variation in Kelani river, Sri Lanka using Principle Component Analysis. In: Proceedings of SAITM research symposium on engineering advancements, Colombo

Atwater R (2008) Salinity management issues facing Southern California. Southwest Hydrol 16–17

Bastiaanssen WGM, Harshadeep NR (2005) Managing scarce water resources in Asia: the nature of the problem and can remote sensing help? Irrig Drain Syst 19:269. https://doi.org/10.1007/s10795-005-5188

CCS (2000) First draft report of climate change secretariat. Ministry of Environment, Colombo

CEA (2014) Central environmental authority of Sri Lanka. www.cea.lk. Accessed on 27 June 2017

Changnon SA (2002) Impacts of the midwestern drought forecasts of 2000. J Appl Meteorol 41(10):1042–1052

Cook CB (2012) Water security: debating an emerging paradigm. Glob Environ Change 22(1):94–102

DCS (2012) Annual report of department of census and statistics. Government of Sri Lanka, Colombo

DCS (2015) Annual report of department of census and statistics. Government of Sri Lanka, Colombo

Dharmapala D, Ranasinghe M (2006) Public private partnerships for water supply to urban poor. In: Annual transactions of IESL, Institution of Engineers, Sri Lanka, pp 1–10

DNP (2014) Unstoppable Sri Lanka. Department of National Planning, Sri Lanka

Fabre J, Ruelland D, Dezetter A, Grouillet B (2016) Sustainability of water uses in managed hydrosystems: human- and climate-induced changes for the mid-21st century. Hydrol Earth Syst Sci 20:3129–3147. https://doi.org/10.5194/hess-20-3129-2016

Henley BJ, Thyer MA, Kuczera G (2013) Climate driver informed short-term drought risk evaluation. Water Resour Res 49:2317–2326

Imbulana KAUS, Wijesekara NTS, Neupane BR (eds) (2006) Sri Lanka national water development report. MAI&MD, UN-WWAP, UNESCO and University of Moratuwa, Sri Lanka, Paris and New Delhi

IWA (2016) http://www.iwa-network.org/publications/water-scarcity-and-drought/. Accessed on 7 June 2017

Karunaratne ADM, Pathmarajah S (2002) Groundwater development through introduction of agro-wells and micro-irrigation in Sri Lanka. In: Symposium on use of groundwater for agriculture in Sri Lanka. Post Graduate Institute of Agriculture, Peradeniya, Sri Lanka

Keshavarz M, Karami E, Vanclay F (2013) The social experience of drought in rural Iran. Land Use Policy 30:120–129

Koop SHA, van Leeuwen CJ (2016) The challenges of water, waste and climate change in cities. Environ Dev Sustain. https://doi.org/10.1007/s10668-016-9760-4

Malathi WPP (2016) Impacts of non-revenue water on the national water supply and drainage board and methods to minimize it. In: The 7th international conference on sustainable built environment, Sri Lanka

Mancosu N, Snyder RL, Kyriakakis G, Spano D (2015) Water scarcity and future challenges for food production. Water 2015(7):975–992

McDonald RI, Weber K, Padowski J, Flörke M, Schneider C, Green PA, Montgomery M (2014) Water on an urban planet: urbanization and the reach of urban water infrastructure. Glob Environ Change 27:96–105

ME & NR (2004) National watershed management policy. Ministry of Environment and Natural Resources, Sri Lanka

Mekonnen MM, Hoekstra AY (2016) Four billion people facing severe water scarcity. Sci Adv 2(2):e1500323. https://doi.org/10.1126/sciadv.1500323

NWS & DB (2015) World water day publication. National Water Supply and Drainage Board, Sri Lanka

Parry S, Wilby RL, Prudhomme C, Wood PJ (2016) A systematic assessment of drought termination in the United Kingdom. Hydrol Earth Syst Sci 20:4265–4281. https://doi.org/10.5194/hess-20-4265-2016

Peiris LLA, Dayawansa NDK, De Silva RP (2008) Assessment of non-revenue water in Hantana water supply scheme. Trop Agric Res 20:290–302

Ratnayake NP, Silva KB, Kumara IGIK (2012) Chloride contamination in construction aggregates due to periodic saline water intrusion: a case study in the Kaluganga River Estuary, Sri Lanka. Environ Earth Sci 69(8):2529–2540

Rossi G (2000) Drought mitigation measures: a comprehensive framework. In: Vogt J, Somma F (eds) Drought and drought mitigation in Europe. Springer, Netherlands, pp 233–246

Sevanatha (2003) The case of Colombo, Sri Lanka. Prepared for understanding slums: case studies for the global report on human settlements, Colombo

Srinivasan V, Seto KC, Emerson R, Gorelick SM (2013) The impact of urbanization on water vulnerability: a coupled human–environment system approach for Chennai, India. Glob Environ Change 23(1):229

UN Report (2012a) http://www.un.org/waterforlifedecade/water_at_rio.shtml
UN Report (2012b) http://www.un.org/waterforlifedecade/scarcity.shtml
UNEP (2012) 2012 UN-water status report on the application of integrated approaches to water resource management. United Nations Environment Programme, Nairobi
Wilhite DA (2000) Drought: a global assessment, vol I. In: Wilhite DA (ed) Chap. 1, London, pp 3–18
Wilhite DA, Glant A (1985) Understanding the drought phenomenon: the role of definitions. Water Int 10(3):111–120
Wilhite DA, Svoboda M, Hayes M (2007) Understanding the complex impacts of drought: a key to enhancing drought mitigation and preparedness. Water Resour Manag 21:763–774

Chapter 20
Water Challenges in Ulaanbaatar, Mongolia

Serjmyadag Dalai, Oyunbaatar Dambaravjaa and Gomboluudev Purevjav

Abstract Ulaanbaatar, the capital city of Mongolia, faces an increasing urban population, a growing industrial sector that lacks effective regulation, and a related increase in the demand for residential and commercial water. Exacerbating these challenges is a combination of low precipitation, high evaporation rates and decreasing groundwater supplies, and as a result, the city has predicted water shortage issues in the near future. In recent years, the number of persons living in Ulaanbaatar has been increasing rapidly and 45% of Mongolia's total population has been living in that city. Ulaanbaatar is located in a valley along the Tuul River, and the current and future water supply of the city is heavily dependent on the water resources of Tuul River, namely groundwater resources in the alluvial deposition of the river. Groundwater resources in Ulaanbaatar have been decreasing for last 50 years due to environmental degradation in Tuul River basin, and its usage has been increasing rapidly comparatively with replenishment of water resources. Government research has shown that daily water consumption is 212 m^3 and this volume shall reach 438 m^3 in 2020 and 708 m^3 in 2050. Therefore, Ulaanbaatar is encountering water scarcity caused by rapidly increasing water consumption. For the people who are living in Ulaanbaatar, the most problematic issues will be the water scarcity which is caused by climate change, environmental degradation, and rapid urbanization. This study pointed out the fact of water-related challenges, particularly water scarcity caused by climate changes, urbanization, and water consumption increase in Ulaanbaatar.

Keywords Water scarcity · Water supply · Water demand

S. Dalai (✉)
Institute of Disaster Research, National Emergency Management Agency, Ulaanbaatar, Mongolia
e-mail: mail_serj@yahoo.com; serjee0415@gmail.com

O. Dambaravjaa · G. Purevjav
Information and Research Institute of Meteorology, Hydrology and Environment, Ulaanbaatar, Mongolia
e-mail: oytetuarel@yahoo.com

G. Purevjav
e-mail: p_gombo@hotmail.com

B. Ray and R. Shaw (eds.), *Urban Drought*, Disaster Risk Reduction,
https://doi.org/10.1007/978-981-10-8947-3_20

20.1 Introduction

Mongolia is a landlocked country in East Central Asia. It is bordered by Russia to the north and China to the south, east, and west. It has an area of 1.565 million km^2; Mongolia has a total population of approximately 3.2 million.

The geography of Mongolia is varied, with the Gobi Desert to the south and with cold and mountainous regions to the north and west. The country has an average elevation of 1,580 m. The geographical location is the main cause of an extreme continental climate of Mongolia.

In Mongolia, the average summer temperature is +20 °C and winter is −20 °C. Global warming influences have affected the climate of Mongolia. The average annual temperature in Mongolia has increased by about 2.1 °C over the past 50 years.

Administratively, Mongolia is divided into 21 provinces and the capital city, Ulaanbaatar. Territory of the capital city covers only 0.3% of total territory of Mongolia, with over 1.4 million people.

The Ulaanbaatar is located at about 1350 m above mean sea level, along the Tuul River and at the foot of the mountain ranges or surrounded from all sides by mountains such as Bogd mountain from south side and Chingeltei, Bayanzurkh, and Songino mountains from other sides.

According to the longest observation time series at Buyant-Ukhaa and Ulaanbaatar meteorological stations, annual mean temperature varies from −4.4 to +2.0 °C and annual precipitation ranges 61.1–403.6 mm. Absolute air maximum temperature was reached to 39.5 °C on July 15, 2005, and minimum was −46.7 °C on January 9, 2005, respectively. Maximum daily rainfall was recorded 82.5 mm on June 16, 1967. Summer rainfall is taken 70% and winter's is 3% in the total amount of precipitation. The climate of Ulaanbaatar is harsh and continental, which mainly is characterized by large amplitude of temperature and low precipitation.

Ulaanbaatar has changed rapidly in recent years as a result of rapid economic growth and rural–urban migration. According to the National Statistics Office of Mongolia, the urbanization rate is at 66.4%. Currently, 45% of Mongolia's population lives in Ulaanbaatar.

In addition to its continuing level of urbanization, Ulaanbaatar faces limited regulation in its expanding commercial and industrial sector, both factors contributing to higher demand for water supplies. Further, the city faces challenging weather conditions including low precipitation and high evaporation, thus, in addition to decreasing groundwater supplies, Ulaanbaatar is expected to experience a shortage in water supplies within several years (Altai 2013).

20.2 Climate Change in Ulaanbaatar, Mongolia

20.2.1 Present Climate Change

The climate of Ulaanbaatar is changing due to global climate warming. Annual mean temperature and annual precipitation anomaly compared to 1961–1990 climate at Buyant-Ukhaa and Ulaanbaatar meteorological stations are shown in Fig. 20.1. Their temperature change of dynamic almost similar and relative high-intensity warming is detected from 1990 to up to now (Fig. 20.1a). In terms of precipitation, there are dry and wet periods; however, relative low-precipitation period is continuing still now (Fig. 20.1b).

Climate data at Buyant-Ukhaa meteorological station during last 76 years is used as presenting Ulaanbaatar (UB) current climate change. Annual mean temperature of UB city is increased by 2.6 °C (0.4 °C higher than country average), and annual precipitation is decreased by 5%. If seasonal change is considered, winter change has most intensity such as 3.7 °C, then 2.5 °C in spring, and 2.2 °C in both summer and fall. Winter and spring precipitations are increased by 38 and 57%, respectively, and there are 13% decreasing in summer and 9% in fall as well (Table 20.1). Generally, warming intensity in cold season as well as increasing of precipitation is higher. Oppositely, there is generally decreasing trends in warm season.

20.2.2 Climate Extreme Event Change

Monthly mean climate variables are not enough to assess climate change impact on socioeconomic sectors. Therefore, daily-based climate extreme indices are effective for assessing climate change impact on human and natural systems.

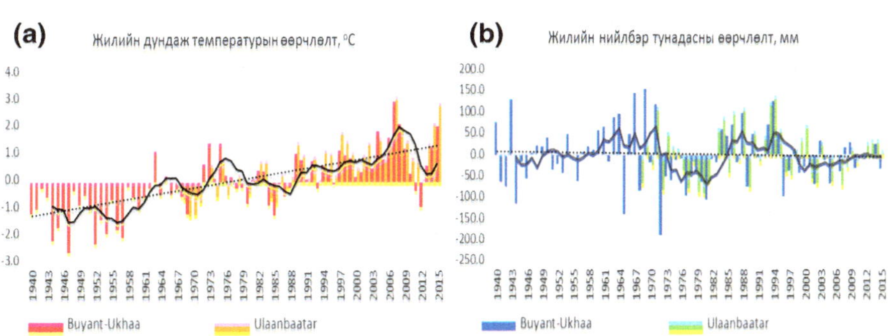

Fig. 20.1 **a** Anomaly of annual mean temperature and **b** precipitation change compared to 1961–1990 climate

Table 20.1 Present seasonal climate change at Ulaanbaatar, 1940–2015

Season	Temperature, °C		Precipitation, mm	
	1961–1990 mean climate	Change	1961–1990 mean climate	Change (%)
Winter	−22.5	3.7	5.2	2.0 (38%)
Spring	−0.2	2.5	24.0	13.6 (57%)
Summer	15.3	2.2	184.0	−24.1 (−13%)
Fall	−2.4	2.2	35.6	−3.1 (−9%)
Annual	−2.5	2.6	248.7	−11.7 (−5%)

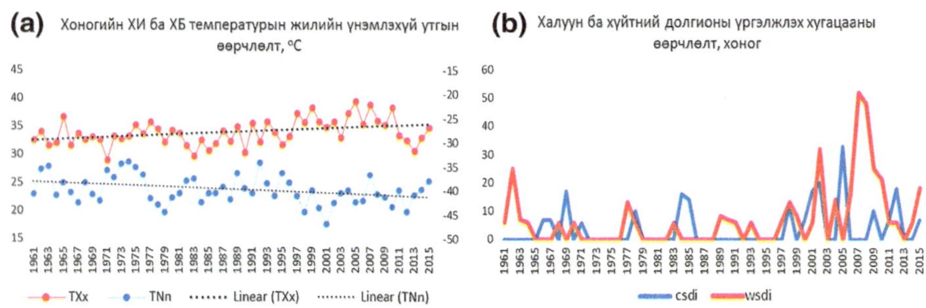

Fig. 20.2 **a** Maximum of daily maximum temperature and minimum of daily minimum temperature, **b** heat and cold wave duration indicators

Maximum of daily maximum and minimum of daily minimum temperature are shown in Fig. 20.2a. According to linear trends, maximum of daily maximum has increased by 3.9 °C and minimum of daily minimum temperature is decreased by 4.8 °C. It means both extreme hot and cold conditions are intensified, while also heat and cold wave duration indicators were increased by 20 and 7 days, respectively (Fig. 20.2b).

Precipitation extreme indices are considered as maximum of 1 and 5 days precipitation amount and maximum number of consecutive precipitation day. Amount maximum of 1 and 5 days precipitation amount is decreasing under general trends of precipitation (Fig. 20.3a). However, inter-annual variability was higher before 1970, and biggest flood disaster in UB city occurred during 1966–1967. Since that time, inter-annual variability is reduced. But consecutive wet days have increased (Fig. 20.3b). Generally, high-intensity rainfall is decreasing, while maximum numbers of consecutive wet days are increasing.

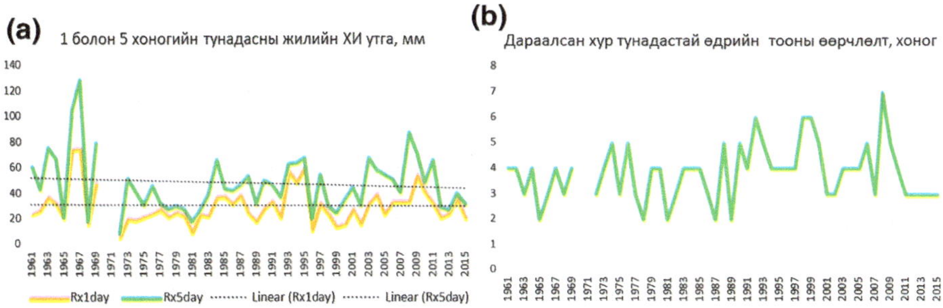

Fig. 20.3 **a** Maximum of 1 and 5 days precipitation. **b** maximum number of consecutive rainy days

20.2.3 Climate Change Projection

Future projection of annual and seasonal temperature and precipitation of UB city are shown in Figs. 20.4 and 20.5 under different greenhouse gas scenarios (GHG) such as low-RCP2.6, mid-RCP4.5, and high-RCP8.5. Generally, temperature change directly depends on the intensity of GHG emission. However, winter temperature change is slightly low and inter-annual variability is higher compared to other seasons.

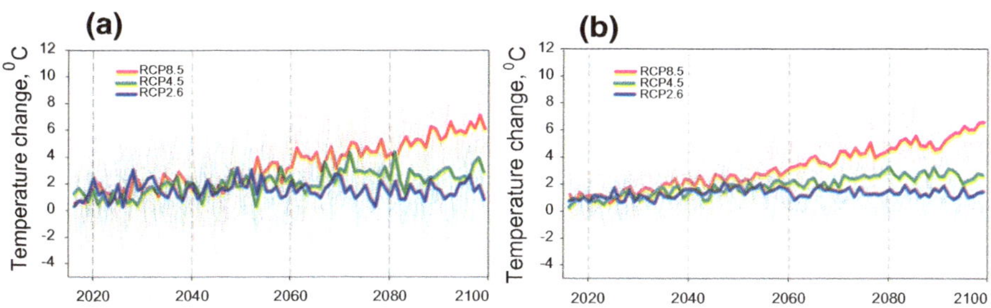

Fig. 20.4 **a** Winter and **b** summer temperature change projection

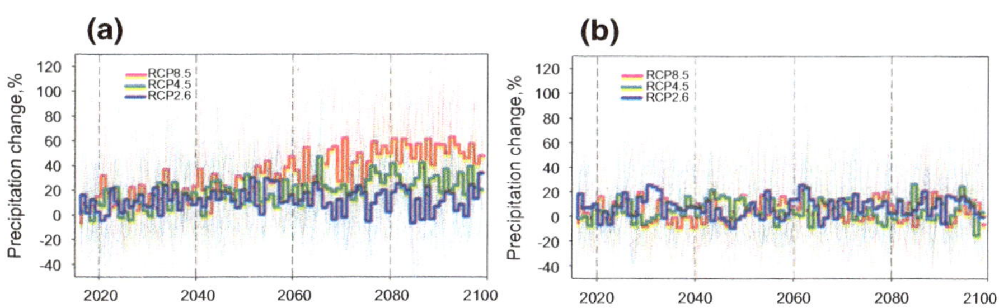

Fig. 20.5 **a** Winter and **b** summer precipitation change projection

Seasonal temperature change will be projected 1.0–1.5 °C in 2016–2035 for GHG emission scenarios. Since that time, it gives different values depending on GHG emissions such as temperature increasing up to 3.1 °C in 2046–2065 and 5.6 °C in 2081–2100. Winter precipitation will be projected to increase by 12.1, 30.7, and 52.4% as corresponding periods. Summer rainfall will increase less than nearly 10% for all periods in Ulaanbaatar.

The impacts of climate change, including global warming, changes in precipitation, and melting glaciers, have started to contribute to water scarcity throughout the world and worsen the impacts of natural hazards (UN-Water 2007; UNDP 2007 in Altai 2013).

Water resources are sensitive to climate changes. Thus, possible climate warming can cause significant changes in the hydrological regime and water resources not only in Ulaanbaatar but also in Mongolia.

20.3 The Demographic Description of Ulaanbaatar, Mongolia

Ulaanbaatar is the capital city of Mongolia and was established in 1639. However, rapid urbanization of Ulaanbaatar has been a recent phenomenon, starting just after the democratic revolution. It has been estimated that Ulaanbaatar's annual population growth rate was less than 1% before 1992, however, increased to between 3 and 4% between 1993 and 2009 (Altai 2013). Migration, especially, toward Ulaanbaatar has increased due to natural disasters such as "drought" and "dzud" affecting herders in rural areas. Also, there are many stable and safe working opportunities in the capital city that mainly attract population and are expanding the urbanization process (UBSO 2016).

Figure 20.6 shows that population of Ulaanbaatar city is increased in the last decade. The majority of the Mongolian population, around 45%, lives in Ulaanbaatar. The city is divided into nine districts (duureg), and each district is subdivided into sub-districts (horoo), of which there are 121.

As of 2016, there are 1396.3 thousand people in Ulaanbaatar, and 44.7% of them are male while 43.3% of them are female.

Figure 20.7 shows that approximately 64% of the population is currently in the working age group, with 30.3% of the population being children between 0 and 15 years old, and 6.2% being 60 years and older (Baldangombo and Puntsagsuren 2012).

The expansion in Ulaanbaatar's population creates the opportunity to expand economic operations, through the increased provision of goods and services. In 2016, there were over half a million persons contributing to Ulaanbaatar's economy, while 70% of all companies in Mongolia are registered in Ulaanbaatar, and 64.2% of the country's businesses operate in the city. Almost two-thirds of Mongolia's gross domestic product (GDP) is generated in the country's capital (UBSO 2016).

Ulaanbaatar's population continues to grow rapidly, creating both health and safety issues for its residents, and placing ever increasing demands the cities water supply and sewage systems (Baldangombo and Puntsagsuren 2012).

20.4 Water Scarcity in Ulaanbaatar

Water scarcity occurs where available water supply is not adequate to meet existing demand (UN 2010 in Altai 2013). Water scarcity can be measured by various kinds of indicators such as water crowding index (Falkenmark 1997), water stress index (Maplecroft 2010), and water poverty index (Lawrence et al. 2003).

In 2000, Mongolia was identified as a water-stressed country, with between 1700 and 1000 m^3 of freshwater available per capita, per year, as measured by Falkenmark indicator (Smakhtin et al. 2000 in Altai 2013), considered to be at the moderate to high level (UNEP 2002 in Altai 2013). Later, in 2010, the country's status was assessed to be in the high to extreme risk range, with Ulaanbaatar highlighted (Altai 2013).

The concept of a water poverty index (WPI) is relatively new compared to others and is defined as "situation where nation or urban center cannot afford the cost of sustainable safe and clean water to all people at all times" (Feitelson and Jonathan 2002 in Altai 2013). This concept infers that water is also available for future generations.

The water poverty index map is indicating that Mongolia was already suffering high water stress (Fig. 20.8).

The Ulaanbaatar Water Supply and Sewerage Authority (USUG) is responsible for supplying water to Ulaanbaatar, from seven groundwater sources (Central, Industrial, Meat complex, Upper, Gatsuurt, Airport, Yarmag), (Baldangombo and Puntsagsuren 2012).

According to 2016 USUG report, there were abstracted 52.5 mln m^3 water and it was increased 0.2 mln m^3 compared to previous year. Water demand of UB city was

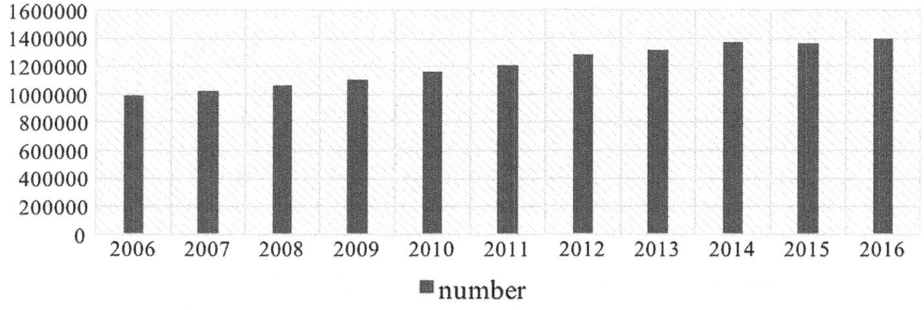

Fig. 20.6 Population of Ulaanbaatar, 2006–2016. *Source* UBSO (2016). Statistical Department of Ulaanbaatar; www.ubstat.mn

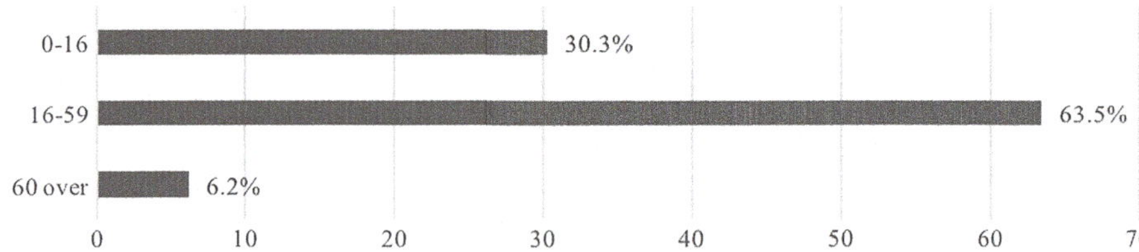

Fig. 20.7 Population by age in Ulaanbaatar. *Source* UBSO (2016). Statistical Department of Ulaanbaatar; www.ubstat.mn

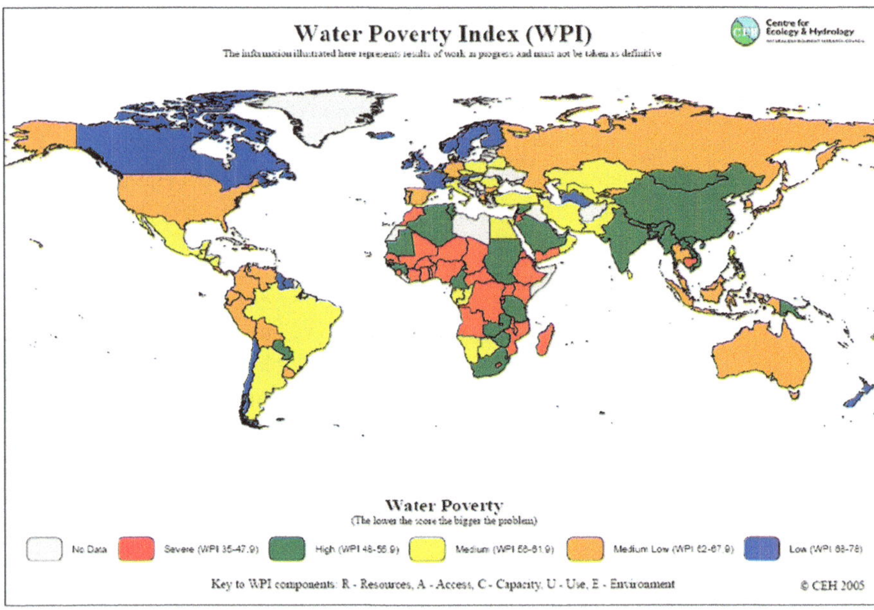

Fig. 20.8 World water poverty index map. *Source* Wisener et al. (2007). Climate change and Human security; www.researchgate.net

44.8 mln m^3 in 2016, which was decreased compared to previous year. In 2016, the water demand and supply ratio of UB was 0.85 (Fig. 20.9).[1]

Forecasts including the Urban Development Master Plan of Ulaanbaatar to 2025, Ulaanbaatar Water and Sewerage Master Plan to 2020 predict that the city will face a critical shortage of water supply within the next ten years, due to supply challenges including climate change impacts and weak infrastructure and demand pressures including an increasing urban population and economic growth (Altai 2013).

[1] The water demand and supply ratio of UB are calculated using data abstracted from USUG's internal statement.

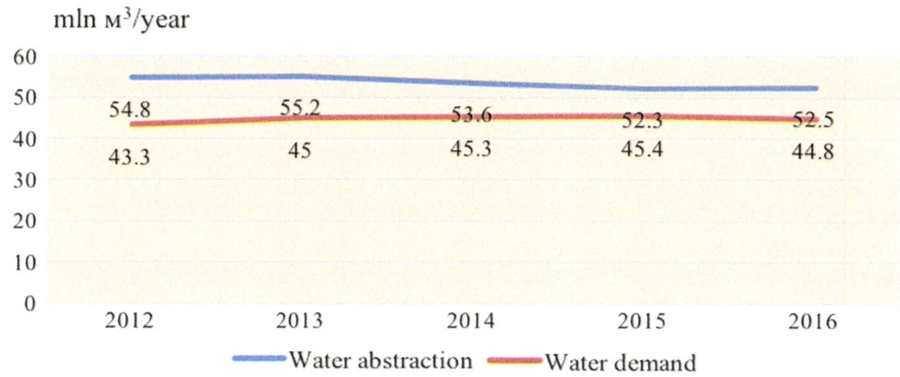

Fig. 20.9 Water demand and supply ratio change in the city. *Source* USUG 2016. USUG's internal statement

20.4.1 Water Supply of Ulaanbaatar

The supply-side issues for water scarcity are caused by a decrease in natural hydrological resources, due to both man-made and natural causes (Altai 2013).

Groundwater resources. Ulaanbaatar is located in a valley along the Tuul River. The current and future water supply of Ulaanbaatar is depending heavily on water resources of the Tuul River, namely groundwater resources in alluvial deposition of the river (Batima et al. 2008). Almost all of the water used by Ulaanbaatar residents is drawn from groundwater sources (JICA 2008 in Altai 2013).

Research of the alluvial deposits of the Tuul River valley has shown a decrease of 3 m since 1950, caused by the drawing of water supplies from this source, with additional research predicting a decrease of 3.7 m in 2020 and 4.6 m in 2050. This suggests that further increases in abstraction from groundwater sources will not be possible in 2050 (Baldangombo and Puntsagsuren 2012) (Table 20.2).

The daily groundwater extraction is increasing due to greater water demands from residential, commercial, and industrial users in Ulaanbaatar (Baldangombo and Puntsagsuren 2012). The governmental research has been showing that daily water consumption is 212 m^3 in 2010 and this volume shall reach 438 m^3 in 2020 and 708 m^3 in 2050. Therefore, Ulaanbaatar may encounter water scarcity caused by rapidly increasing water consumption.

Groundwater resources in Ulaanbaatar have been decreasing for the last 50 years due to environmental degradation in the Tuul River basin as well as its usage has been increasing rapidly comparatively with recharging of water resource.

Surface water resource. Ulaanbaatar lies on the country's Tuul River, which is over 700 km long and flows southwest through the city, and formed by precipitation, snowmelt, and discharge from natural springs. Runoff consists of 25% groundwater, 6% snow/ice, and 69% rainwater. Annual precipitation is between 243 and 402 mm in Ulaanbaatar and surrounding areas, and it is estimated that half of the annual average

Table 20.2 Changes in Tuul River basin groundwater level fluctuation

Time and authors	Groundwater level fluctuation (m)
1946–1948 K.I. Gomaniko (specialist of USSR)	1.6
1959–1961 R.A. Kriger, 1961 (Russian Hydro-project Institute)	Central area 1.0–1.3 Western area 1.8–2.4
1979–1980 PNIIS Institute of Russian Federation	2.7
1997–1998 G. Tserenjav and D. Unurjargal (Geo-Ecology Institute, Mongolia)	3.1
2020 Prognoses by Geo-Ecology Institute, Mongolia	3.7
2050 Prognoses by Geo-Ecology Institute, Mongolia	4.6

Source Ministry of Environment and Green development. Tuul River basin integrated water resource management assessment report (2012)

Table 20.3 Runoff decline in the last 10 years in Tuul River in Ulaanbaatar

	Long-term mean, m^3/sec	Spring IV–VI	Summer VII–IX	Autumn X–XI	Winter XII–III
Long-term mean	25.78	27.47	68.28	10.50	0.28
1996–2007	13.29	15.92	31.81	7.51	0.32
Changes, %	48.4	42.0	53.4	28.5	−14.3

precipitation becomes evapotranspiration, while the other half forms surface water and groundwater recharge. Total annual runoff varies between 25.1 and 327.1 mm and amounts to 127.5 mm on average (Altai 2013; Baldangombo and Puntsagsuren 2012; Davaa et al. 2012; Sarantuya et al. 2002).

Approximately 90% of the water used by residents of Ulaanbaatar is drawn from groundwater sources, and rapid urban development has altered much of the surface water regime (Altai 2013).

Runoff from the Tuul River is reduced in Ulaanbaatar as a result of the infiltration of surface water, caused by the abstraction of groundwater by the city's water supply boreholes. This has been measured to be approximately 2 m^3\sec which is about 10% of the average annual flow (Baldangombo and Puntsagsuren 2012).

The result of recent studies shows that the mean runoff of the Tuul River decreased by 48.4% since mid of the 1990s when low flow years began in the Tuul River basin compared to the long-term mean. Such a decreasing trend was also in the seasonal runoff (Table 20.3).

Impacts of climate change clearly influence surface water regime of the Tuul River. According to the greenhouse gas scenario-A2, runoff is expected to decrease on average by 2 mm in the Tuul River basin areas by 2020. Thus, river basin will continue to dry in the near future (Davaa and Erdenetuya 2007).

Both groundwater resource and surface water resources have been decreasing for the last two decades in Tuul River basin.

20.4.2 Water Demands of Ulaanbaatar

The main contributors to demand-side factors are the increasing population and economic growth and development (Altai 2013).

Population growth. As the world's population continues to grow by 80 million people per year, this indicates an increased demand for freshwater of 64 billion m^3 every year (WWAP et al. 2009; Kjellen and McGranahan 1997 in Altai 2013). Similar research indicates that 1.8 billion people will be affected by water scarcity by 2025, with two-thirds facing water stress (UN-Water 2007 in Altai 2013). The future population growth is an important factor for future water demands.

In recent years, the population of Ulaanbaatar has been growing rapidly, with nearly half of Mongolia's population residing in the city. Figure 20.10 shows that the projection of Ulaanbaatar population in 2021 has been calculated by methodology calculated population growth 2010–2040 by the National Statistical Committee (Baldangombo and Puntsagsuren 2012).

Subsequently, water demand is increasing due to this population growth and due to the expanding number of organizations in the city (Baldangombo and Puntsagsuren 2012). Prior to 1992, Ulaanbaatar's annual population growth rate was less than 1%, before rising from 3 to 4.5% between 1993 and 2009; however, this excludes migration from rural areas (Altai 2013).

This rapid urban population growth continues to add to demands on urban services including the water supply. According to the census (2015), 68% of the Mongolian population lives in urban areas, of which nearly 66% reside in the capital city, Ulaanbaatar (UBSO 2016). Geographically, Ulaanbaatar is divided into two main areas: the city center area and the ger areas. A "ger" is a traditional dwelling designed for the country's rural nomadic lifestyle, consisting of a circular structure comprised of a wooden frame covered with felt. The city center area is connected to the central

Fig. 20.10 Population growth and the prospect of Ulaanbaatar

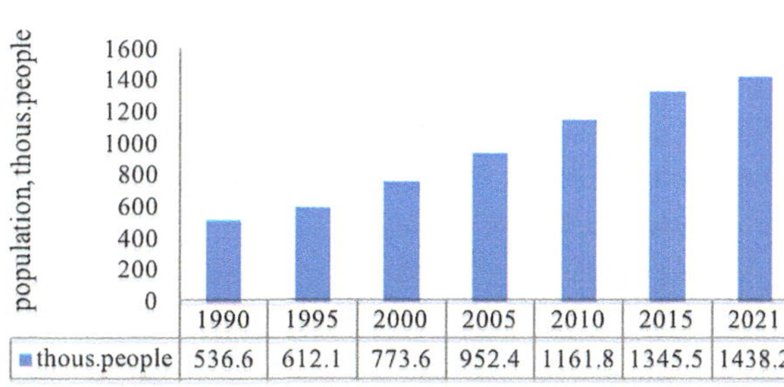

	1990	1995	2000	2005	2010	2015	2021
thous.people	536.6	612.1	773.6	952.4	1161.8	1345.5	1438.2

Table 20.4 Water consumption and water demand of Ulaanbaatar population

	Water supply source and status of supply		Population, *thous. person*			Water consumption, *thous.* m^3*/year*		
			2010	2015	2021	2010	2015	2021
1	Central water supply	With hot and cold water	431.1	636.9	762.7	36,190.0	46,490.9	44,541.3
		From water kiosk	251.0	318.7	370.3	916.0	2908.3	4055.2
2	Non-central water supply	From water kiosk	243.4	213.3	305.2	710.7	1167.9	2227.8
Total			925.5	1168.9	1438.2	37,816.7	50,567.1	50,824.3

Source Ministry of Environment and Green development. Tuul River basin integrated water management plan (2012)

water supply network while the ger areas are not (Altai 2013; Baldangombo and Puntsagsuren 2012).

The city center area contains mostly commercial and administrative buildings and high-rise apartments, however, the most rapidly growing area in Ulaanbaatar is the ger areas, where nearly 60% of the households or 58% of the total city population (NSOM 2017).

The population and future demand of water in Ulaanbaatar have been projected to increase from 2015 to 2021 (Table 20.4).

Economic development. As mentioned above, the major water demand-side factors in urban areas are economic development.

Ulaanbaatar is an important hub for commerce and industry, with approximately two-thirds of the country's GDP produced in the city (NSOM 2016 in Altai 2013). This is comprised of 48% industrial sector, 52% construction, 41% trade, 75% hospitality, and 56% of transportation and communication services (Emerton et al. 2009 in Altai 2013).

Future water demand has been forecast based on expected population growth, estimated to be 8036.6 thousand m^3 in 2021 for the public service sector, based on service sector production growth of 6.9% on average annually. For light and food industries, water demand has been calculated to be 553.0 thousand m^3 in 2021. Future demand has also been forecast for the construction industry and energy and heat industry, refer to Table 20.5 (Baldangombo and Puntsagsuren 2012).

Compared to the water consumption use in 2010 and 2021, water consumption and water demand of municipal services and industries sector in Ulaanbaatar are increased by 53%.

Table 20.5 Water consumption and water demand of municipal services and industries sector in Ulaanbaatar

	Water-consuming and using sector		Water consumption use and water demand, *thous.* m^3/*year*		
			2010	2015	2021
1	Municipal service	Public services (education and health organization)	3241.7	3475.0	3777.3
2		Commercial services (shops, restaurants, hotels, hairdressers, laundry service, etc.)	3590.1	5178.1	8036.6
4	Industry sector	Food industry	2735.0	3810.8	5687.0
5		Light industry	260.2	370.5	553.0
6		Construction and construction material industries	394.3	550.3	821.3
7		Energy and heat industry	22,779.5	30,484.1	43,242.3
Total			33,000.8	43,868.8	62,117.5

Source Ministry of Environment and Green development, Tuul River basin integrated water management plan 2012

20.5 Discussion and Conclusion

Mongolia's capital, Ulaanbaatar, is facing growing long- and short-term problems of water scarcity which are caused by climate change, environmental degradation, and rapid urbanization. Population growth, urbanization, and economic development are ongoing problems which increase water demand in Ulaanbaatar.

The impacts of climate change affect hydrological cycles and have long-term implications including temperature increases and greater land degradation and are significant contributors to supply-side water scarcity.

According to the present climate change of UB city:

- Warming intensity is higher in cold season and also there is increasing precipitation trends, while decreasing trends in summer season;
- Extreme hot and cold conditions are intensified, while also heat and cold wave duration indicators were increased, respectively.

According to projection of global climate output:

- Seasonal temperature will be projected to increase up to 5.6 °C during this century. It has a positive impact to reduce harsh winter conditions; however, it will intensify hot summer conditions.
- Winter snow will increase up to 52.4%, and summer will increase less than 10%. This will contribute to intensify heavy snow and drought as well as dryness.

Water scarcity in urban settings is dependent upon factors of both supply and demand. Demand continues to increase due to population growth and increase economic activity, while supply continues to be affected, and hindered, due to the impacts of climate change (Altai 2013).

In Ulaanbaatar, both residential and commercial users of water rely upon the groundwater of the Tuul River basin, a supply that continues to decline (Altai 2013). Research has shown that the alluvial deposits of the Tuul River valley have decreased by 3 m since 1950 and will continue to fall to 5 m by 2050 (Baldangombo and Puntsagsuren 2012). The result of recent studies indicated that the mean runoff of the Tuul River decreased by 48.4% since mid of the 1990s when low flow years began in the Tuul River basin in comparison with the long-term mean.

Water demand has been rising due to high-intensive urbanization and economic development. Ulaanbaatar is the center of administrative, commercial, and financial activities of the country. The city covers approximately 0.3% of the territory of Mongolia, but almost half of the Mongolian population, with almost two-thirds of Mongolia's GDP produced in the Ulaanbaatar. The population of Ulaanbaatar will reach around 1.5 million in the next 5 years. Future demand of water is strongly related to population growth and rapid urbanization with improved living conditions. Water demand of municipal services and industries sector in Ulaanbaatar has been projected to increase around 50% from 2010 to 2021 (Altai 2013; Baldangombo and Puntsagsuren 2012).

Consequently, a huge imbalance has been generated between the demand and supply of water in Ulaanbaatar. As can be observed from the research outcomes, Ulaanbaatar is likely to encounter water scarcity within the next few years due to the effects of social, economic, and natural factors, such as climate change. The impacts of climate change will be a notable factor, having a discernable impact on water supply. Further, policy makers should concentrate efforts on more effective and efficient uses of existing supply, rather than seeking additional sources.

The combination of a shortage in water resources and limited financial resources for establishing additional water supply requires a shift in policy and behavioral attitudes by water users, in order to achieve greater efficiency in managing existing supply. Such policy activities include education and awareness programs, aimed to inform water users about the issue of water scarcity and communicate methods for users to reduce their water consumption in the long term. For residential users, the government needs to promote activities to raise public awareness of water scarcity and water-use efficiency through professional and civil society communities and media. For non-residential users, the use of advisory and consultant services will be more effective in reducing their water usage. Such services serve to promote activities to reduce water consumption through measures such as water recycling or reuse.

Overall, public information and education activities can be effective in reducing the risk of water scarcity in Ulaanbaatar; however, such activities will be only successful in the longer term if demand continues to be less than available supply.

References

Altai Z (2013) Urban water demand management in Ulaanbaatar, Mongolia, PhD thesis

Baldangombo E, Puntsagsuren Ch (2012) Water supply, water consumption-use, water demand and hydro-construction. In: Tuul River basin integrated water resources management assessment report, Ulaanbaatar

Batima P, Batnasan N, Bolormaa B (2008) Uncertainties in water resource management: causes, technologies and consequences

Davaa G, Erdenetuya M (2007) Hydrological changes in the upper Tuul River basin. World water consultative meeting "water policy and Strategy"

Davaa G, Ouynbaatar D, Tumurchudur S, Wim van der Linden (2012) Surface water resources. In: Tuul River basin integrated water resources management assessment report, Ulaanbaatar

Emerton L, Erdenesaihan N, Veen B, Tsogoo D (2009) The economic value of the upper Tuul river ecosystem, Mongolia

Falkenmark M (1997) Meeting water requirements of an expanding world population. https://www.researchgate.net/publication/25456684_Meting_water_requirements_of_an_expanding_world_popuation

JICA (2008) The study on city master plan and urban development of Ulaanbaatar city

https://www.maplecroft.com/about/news/water-stress.html

Kjellen M, McGranahan G (1997) Comprehensive assessment of the fresh water resources of the world. Urban water towards health and sustainability. In: Altai Z 2013 Urban water demand management in Ulaanbaatar, Mongolia, PhD thesis

Lawrence PR, Meigh J, Sullivan C (2003) The water poverty index: an International comparison. In University K (ed) Keele Economics Research Papers, Keele University, Staffordshire

Ministry of Environment and Green Development (2012) Tuul river basin integrated water management plan (Online). Available http://bic.iwlearn.org/en/documents/documents/proceedings/2012/tuul-river-basin-integrated-water-management-plan-mongolia-english/view

NSOM (2016) National statistical office Mongolia, statistical database. www.1212.mn

Sarantuya Z, Borchuluun U, Munkhtuya Sh (2002) Reserves, consumption and contamination of groundwater in Ulaanbaatar, Mongolia. Atlas of Urban Geology, vol 14. ESCAP, Environment and Sustainable Development Division

Taylor KE, Stouffer RJ, Meehl GA (2012) An overview of CMIP5 and the experiment design. Bull Amer Meteor Soc 93:485–498. https://doi.org/10.1175/bams-d-11-00094.1, http://cmip-pcmdi.llnl.gov/cmip5/availability.html

UBSO (2016) Ulaanbaatar statistic's year book-2015, Mongolia. www.ubstat.mn

UN-Water (2007) Coping with water capacity: challenge of the twenty-first century. UM-water thematic initiatives and world water day

UNDP (2007) Fighting climate change: human solidarity in a divided world. Human development report 2007/2008

UN-Habitat (2010) Cities and climate change initiative. UN-HABITAT Fukuoko, Regional Office for Asia and the Pacific

USUG (2016) USUG's operational report for internal use

Wisner B, Fordham M, Kelman I et al (2007) Climate change and human security. https://www.researchgate.net/publication/237390220_Climate_Change_and_Human_Security

WWAP, UNFP, UNESCO (2009) Water in changing world. World water development report 3

Chapter 21
Water Scarcity in Himalayan Hill Town: A Study of Darjeeling Municipality, India

Tarun Kumar Mondal and Paramita Roychowdhury

Abstract Water scarcity poses serious threat to urbanization process in India. Nature of the scarcity varies across the geographical diversity of this country and is influenced by the management of available water resources. Urban areas in the hill slopes of Himalayas are facing water crisis due to adverse topographical characteristics and unplanned urban expansion. Darjeeling town popularly known as the 'Queen of the Hills' lying in the Eastern Himalaya, one of the most attractive tourist spots in India, is suffering from acute water crisis. Initially, an effective water supply system was installed in Darjeeling town during colonial India (1910–1915) with a capacity to cater 10,000 people. Population of this town has increased more than twelve times, till now; inhabitants of the town mainly depend on this age-old water supply system. Rapid urbanization, expansion of tourism industry, massive deforestation and climate change are worsening the situation fast. Consequently, a huge imbalance is generated between the demand and supply of water in this town, and the gap is widening at a faster rate. The future existence of this world-famous tourist destination, therefore, depends on the ways to overcome this crisis. This chapter deals with the nature of the water scarcity, factors responsible for this crisis and viable management strategies in Darjeeling Municipality. The issues of urban expansion and water scarcity in Himalayan towns and in Darjeeling Municipality have been dealt in the second section of this chapter. The context of water scarcity and hydrogeological, demographic and politico-economic factors responsible for this scarcity have been analysed in the third and fourth sections consecutively. Different water supply systems and their comparative advantages and disadvantages have been discussed in the fifth and sixth sections. The critical factors for water crisis have been analysed in the seventh section. Finally, in the eighth section, the necessary strategies and options to mitigate water scarcity in this municipality have been proposed.

T. K. Mondal
Department of Geography, University of Kalyani, Nadia 741235, West Bengal, India
e-mail: mondaltarunkly@gmail.com

P. Roychowdhury (✉)
Department of Geography, Krishnagar Government College, Nadia 741101, West Bengal, India
e-mail: parorc@gmail.com

B. Ray and R. Shaw (eds.), *Urban Drought*, Disaster Risk Reduction,
https://doi.org/10.1007/978-981-10-8947-3_21

Keywords Darjeeling Municipality · Eastern Himalaya · Himalayan towns Tourism · Urbanization process

21.1 Introduction

Historically, availability of drinking water at its nearest place had been considered as the most critical factor for locating human settlements. Advent of technology has gradually pushed this factor to the back, in spite of the fact that water especially drinking water is not ubiquitous. Urban settlements across the world have grown in response to several factors other than availability of sustainable local source of fresh drinking water; water treatment and transportation are the integral parts of modern urban societies. Urban water crisis is, therefore, ingrained often in their locale and is exhorted by the rise in urban population. The changing pattern of water demand, precisely from the growing urban middle class, intensifies this crisis. The normative estimate of the per capita water consumption of city residents is often far less than the actual water demands. Hence, the city administrations are misled towards a lower amount of target of supply. Moreover, the floating population in the cities is hardly taken into account in calculating water need of the city. Both these practices conceal the real magnitude of the water crisis in any city. Expanding urban India represents a somewhat similar scenario.

21.2 Urban Expansion and Water Scarcity in Himalayan Regions, India

According to Census of India (2011a) estimates, the level of urbanization increased only by 3.35% (from 27.81% in 2001 to 31.16% in 2011), whereas the total urban population increased by around 91 million persons during this period. To augment the water supply at par with the water demand of the particular city is undoubtedly a hard task for the municipal administration, mainly in the fast-growing cities, e.g. Delhi or Mumbai. In addition, the expansion of urban settlements in the regions where water availability is constrained by the hydrogeological conditions induces more complexities and crises. Himalayan hill slopes are example of such a region in India (as well as in Nepal). This region is experiencing rapid urbanization, in spite of difficult terrain and climate. Tourism fuels the process of urbanization on the mountain slopes of Himalaya along with administrative, trade and defence-related activities. Many of the Himalayan hill towns owe their origin to British rulers and gradually have been transformed into regional growth centres. The nature of the water crisis in these towns is completely different from the urban centres of the plain land where easy transportation of water helps to solve the crisis. Darjeeling, located on the magnificent hill slopes of Eastern Himalaya, is a classic example of this type of towns. The *Queen of Hills*[1] as Darjeeling is named (Ganguly-Scrase and Scrase 2015) is experiencing haphazard urbanization and as a result facing acute water crisis. This world-famous tourist destination has now become synonymous with unavailability of water.

Water scarcity in Darjeeling town, therefore, is significant from two aspects: one is the problem of water supply in average medium-sized Indian towns passing through relatively fast process of urbanization during the last three decades. Water crisis in Darjeeling town could be considered as the case of (mis)management or (in)ability of the municipal administration in India to handle the issue of water supply. The second aspect of the water crisis in Darjeeling town is linked to the rationale of process of urbanization on the hill slopes of Himalaya, its viability and sustainability. Being located at the Eastern Himalaya, Darjeeling town receives plenty of monsoon rain and appears to be the water surplus region; crisis of water amidst this rainfall is also a paradox. Moreover, Darjeeling town provides the opportunity to critically examine the nuances of water scarcity in an urban centre at more than 2000 m from MSL.

21.3 Water Scarcity in Darjeeling Municipality: The Context

21.3.1 Darjeeling Municipality: Genesis

Located on the Darjeeling-Jalapahar ridge in the Lower Himalaya (Fig. 21.1), Darjeeling was chosen as a summer sanatorium to relieve the British ruler from the heat of plains of Bengal and for military purposes (O'Malley 1907; Mell and Sturzaker 2014). Darjeeling Municipality, one of the oldest municipalities in India, was established in 1850[2] (Drew and Rai 2016). Darjeeling became the summer capital of the British India and grew like a typical British hill town with mall, church, gardens, exotic buildings and hotels (Ganguly-Scrase and Scrase 2015). The scenic beauty of the lash green Himalayan forests, mystic rivers and jhoras and the grandeur of the Kanchenjunga transformed the town as tourists' paradise. Establishments of tea gardens by British rulers (O'Malley 1907) around the town added spectacular dimension to the beauty of this magnificent hill town. Darjeeling gradually earned her worldwide fame for three 'T's—Tea, Timber and Tourist (Bhattacharya et al. 2001), and these three 'T's have remained as the raisons d'être for Darjeeling.

[1]The name itself indicates the scenic beauty of the town and surrounding areas.

[2]http://darjeelingmunicipality.in/.

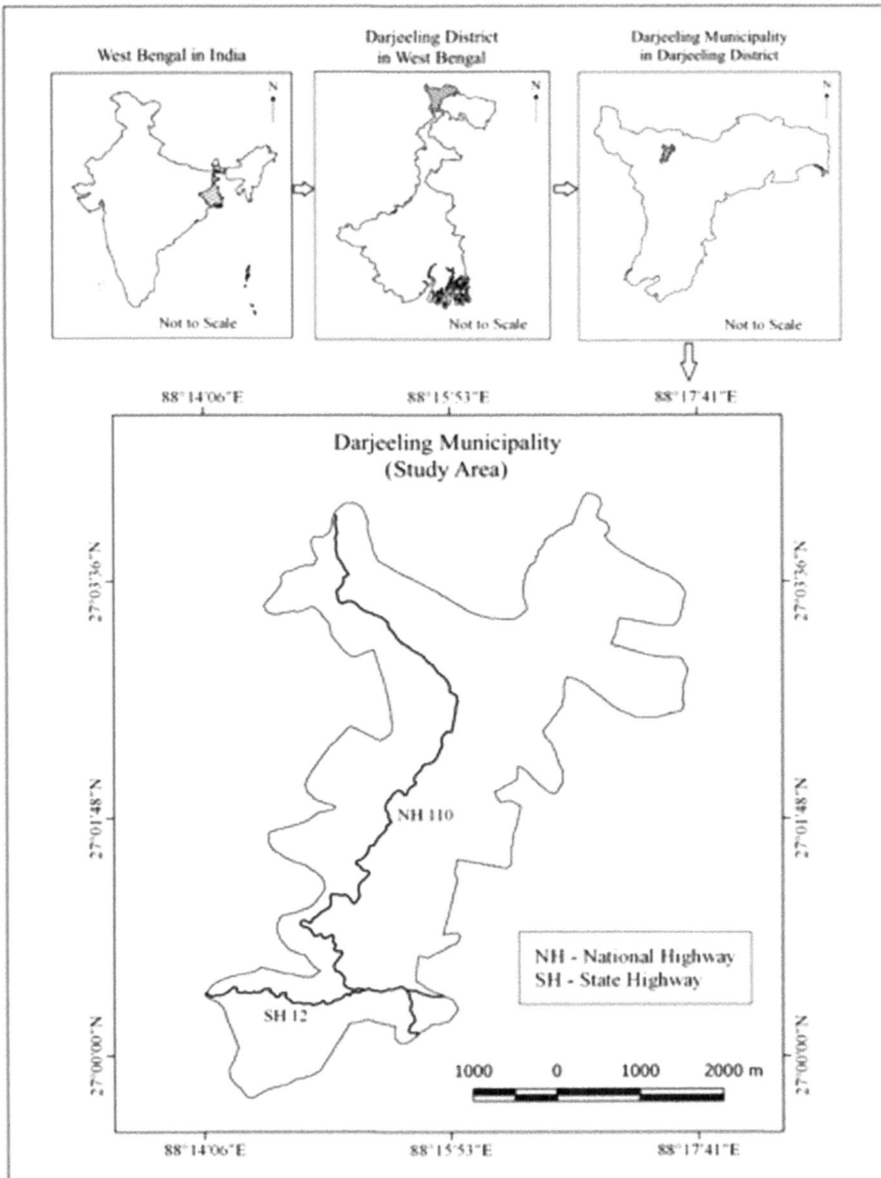

Fig. 21.1 Location of Darjeeling Municipality. *Source* Prepared by authors

21.3.2 Darjeeling Municipality: Evolution

The small sanatorium town started evolving as a trade centre and attracted people from the surrounding areas to settle in. By the end of nineteenth century and beginning of twentieth century, Darjeeling town achieved a cosmopolitan character with British settler, Bengali- and Nepali-speaking people. The Nepali language still is the thread of commonality among the diverse social groups of Lepchas, Bhutias, Tibetans, Bengalis, Marwaris and Biharis (Lama and Rai 2016). Hinduism, Buddhism, Christianity and Islam are the main religious practices of the people of the town. According to Lama and Rai (2016), Darjeeling is a 'complex conglomeration of various castes and ethnic groups, with the ancestry of many of them rooted in tribal and animist traditions'.

The growth of the town followed a somewhat different trajectory during the post-colonial period. Headquarters of Darjeeling district was established in Darjeeling town, and the town gradually assumed the characteristics of an administrative town. A new 'aspirational' middle class emerged in the town (Ganguly-Scrase and Scrase 2015) along with the old-settler of the town. The new middle class mainly belongs to the professionals or service-based people and new business-entrants. Population of the town experienced a rapid growth in the last fifty years, and Census of India (2011b) estimated the total population of the town is 118,805 persons (Tables 21.1 and 21.2). Population-wise, Darjeeling town, therefore, portrays a growth trajectory similar to that of several other district headquarters located in the vast alluvial tract of Bengal plains. Unique to this town is that it receives additional 165 thousand tourists annually. Moreover, Darjeeling is also place with heritage of world-class educational institutions established by the British. The students comprise a significant portion of the population of the town.

21.3.3 Darjeeling Municipality: Water Scarcity

Rapid urbanization with continuous inflow of immigrants and tourist has incapacitated Darjeeling Municipality to keep pace with the ever-increasing water demand. The gap between demand and supply of water is widening particularly during the peak tourist and pre-monsoon season. However, with a total annual rainfall of more than 2900 mm (Table 21.1), Darjeeling Municipality is thriving to meet up the water demand.

In addition, it needs to be mentioned that Darjeeling Municipality has a demarcated boundary and all population accounts and water demand estimation follow this boundary. The process of urbanization is naturally not confined within the municipal boundary of Darjeeling. Urban expansion beyond the municipal boundary depends on the urban facilities including water provided by the municipality. The actual dimension of the water crisis in Darjeeling town, therefore, is inherently linked to the urbanization process centring around Darjeeling Municipality.

Table 21.1 Darjeeling Municipality: at a glance

Year of establishment[a]	01 July 1850
Average elevation from MSL[a]	2123 m
Area[c]	7.43 km^2
Average annual rainfall[b]	2981.8 mm
Mean monthly maximum and mean monthly minimum temperature[b]	19.6 and 1.8 °C
Total population[c]	118,805 persons (59,187 males and 59,618 females)
Population density (persons per km^2)[c]	15,990 persons
Literacy rate[c]	Total literacy rate: 93.85% Male literacy rate: 96.41% Female literacy rate: 91.31%
Work participation rate[c]	34.38%
Percentage of scheduled tribes to total population[c]	22.39%
Percentage of scheduled castes to total population[c]	7.68%
Religious categories[d]	Hindu: 66.51% Buddhist: 23.91% Christian: 5.13% Muslim: 3.94% Other religion: 0.51%
Slum population[d]	25,026 persons (21.06%)
Number of households including houseless households[c]	21,782
Number of wards[a]	32

Sources [a]Darjeeling Municipality (2017), [b]IMD (2017), [c]Census of India (2011b), [d]Census of India (2011c)

21.4 Water Availability and Supply in Darjeeling Municipality: Controlling Factors

The hydrogeological conditions limit the scope of development potential of water resources along with the demographic and politico-economic factors. Water availability in Darjeeling town also depends on these natural endowment and human activities. Naturally, availability and supply of water in Darjeeling Municipality reflect the interplay between all these factors. On the pretext of the management, supply and evolving scarcity of water, an account of these factors in Darjeeling town is required.

Table 21.2 Population and its growth in Darjeeling Municipality

Census year	Total population	Decadal growth[a]		Average annual exponential growth rate (%)[a]
		Absolute	(%)	
1881	7018	–	–	–
1891	14,145	7127	101.55	7.26
1901	16,924	2779	19.65	1.81
1911	19,005	2081	12.30	1.17
1921	22,258	3253	17.12	1.59
1931	21,185	−1073	−4.82	−0.49
1941	27,224	6039	28.51	2.54
1951	33,605	6381	23.44	2.13
1961	40,651	7046	20.97	1.92
1971	42,873	2222	5.47	0.53
1981	57,603	14,730	34.36	3.00
1991	73,062	15,459	26.84	2.41
2001	107,197	34,135	46.72	3.91
2011	118,805	11,608	10.83	1.03

Sources Das (1947) (1881–1891), Census of India (2011a) (1901–2011), [a] calculated by the authors

21.4.1 Hydrogeology of Darjeeling Town

Darjeeling district spreads across the spurs and valleys of the Lower or Outer Himalaya. This region is 'vulnerable to landslide and earthquake being situated at the fault line of Eurasian and Gondwana plates' (Darjeeling Disaster Management Section 2016). The Singalila range running almost south to north divides Darjeeling and Sikkim from Nepal. To the south-east of Singalila range, four hill ranges are observed to spreading from the Ghum.[3] Of these four ranges, on the Darjeeling-Jalapahar range, which extends towards north, Darjeeling Municipality is located along with Jalapahar and Lebong cantonment areas. This range is divided into Lebong spur and Takvar spur beyond the Chowrasta[4] and observatory hill. Darjeeling Municipality is situated on this 'Y'-shaped (Fig. 21.1) spur that encloses the river valley of Rangit (O'Malley 1907). The average elevation of the town is 2123 m above MSL (Darjeeling Municipality 2017). During the period from 1901 to 2000,[5] mean maximum monthly and mean minimum monthly temperatures of Darjeeling station were observed as 19.6 and 1.8 °C, respectively (IMD). During this period, average annual

[3]Ghum is the famous railway station on the Darjeeling Himalayan railway.

[4]The Chowrasta is almost centrally located in Darjeeling Municipality.

[5]97-year records are available during the period 1901–2000.

rainfall of Darjeeling station was recorded as 2981.8 mm and out of which 81.3% rainfall occurred from June–September (IMD).

The total rainfall received by the town gives an impression that the entire region well endowed with water resources. However, the steep slopes (20–48°) running long stretches of 1–2 km (Guha and Kujur 2009) lead to huge surface run-off. Guha and Kujur (2009) mentioned that Darjeeling town receives 19 MCM water per year as net rainwater provided the run-off coefficient is 60%. They also mentioned that located on the 'sequence of overthrust towards south and dipping towards north', Darjeeling town is mainly formed by the deeply weathered metamorphosed rocks, i.e. Darjeeling gneisses, Daling metamorphics, Damuda shales and sandstones. The groundwater is found in 'unconfined condition' in these weathered materials and in 'confined condition' in joints, faults and fractures (Guha and Kujur 2009). The groundwater spurts out as springs in numerous occasions across the hill slopes, which are the primary source of water in the town, and these springs are not always perennial in nature.

21.4.2 Demography of Darjeeling Town

Being situated on the ridge and valley topography overlooking to the snowy peaks of Mt Kanchenjunga, Darjeeling was chosen as a place for British sanatorium. This small hill town later developed as a world-famous tourist destination, a premium place for education and district headquarters. The demographic scenario of the town changed with its changing nature and population sharply increased after 1971 onwards (Table 21.2). According to Census of India (2011b), density of population in Darjeeling Municipality was 15,990 persons per km^2.

The initial population of the town in 1881 and development of infrastructure by British rulers have been far exceeded by this present growth of population (Mell and Sturzaker 2014). The military expenditure, revenue from tea, timber and tourist transformed Darjeeling to a growth centre for the regional economy (Mell and Sturzaker 2014) and location of district administration added momentum to this growth. As Darjeeling Municipality evolved as an administrative and trade centre, connectivity of Darjeeling town increased with the plain lands. Consequently, people from surrounding areas have been migrating to this town. Moreover, the people from hills of neighbouring areas prefer Darjeeling town to settle down as the cold climate suits them than hotter plain lands. Darjeeling town experienced enormous pressure on land as population density increased rapidly. The collective consensus of the people and the migrants of this town promoted an unplanned development in the absence of any legal framework for construction of buildings and master plan for future expansion (Mell and Sturzaker 2014; Ganguly-Scrase and Scrase 2015). Ever-increasing volume of tourist also intensified the demand for more hotels in the town. The hotels, buildings and shops were established in a haphazard manner, and Darjeeling town expanded at the cost of compromising with future sustainability. This unplanned expansion put forward a great challenge in the process of proper design and installation of new water pipes.

The availability of water from Senchal lakes and natural springs has fallen short with the galloping demand. Though the water requirement is less in this cold climate, the changing lifestyle of the growing urban middle class in Darjeeling town asks for more water to flush toilets, use of washing machines and cleaning of houses and cars which has never been a practice earlier. The changing demography of the town therefore is a key determinant of water scarcity.

21.4.3 Politico-Economic Constraints and Water Supply in Darjeeling Town

Water supply is the responsibility of the urban local body in India. Water scarcity in Darjeeling town therefore could not be apprehended if the administrative constraints are not considered. The management of the existing water supply system and future expansion to cope with the increasing demand is the primary duty of the Municipality, and any shortfall automatically has a potential to be an important political agenda for election in the urban local bodies. In Darjeeling Municipality amidst the crisis of water, election agendas were critically not focused towards this. Various reasons for such political silence may be mentioned.

One of the most critical reasons for such silence is the presence of private water vendors, who are also referred as 'water mafia's (Drew and Rai 2016; Lama and Rai 2016). The paid supply of water firstly reduces the grievance of the common people who are able to pay for water as their demand is sufficiently met. Moreover, as many houses, hotels and shops in this town are not constructed with proper permission, automatically their legal right for water demand is denied. Significant portion of such construction though use water from municipal supply are unable to raise their voice for any mismanagement. Direct involvement of the political leaders in this profiteering and rising business of water supply also suppresses the agenda of adequate water supply from municipality.

Secondly, the rampant illegal tapping of water in Darjeeling Municipality also works in favour of the 'power'ful people in the city. In case of scarcity of water during the lean period, whatever amount of water is available for municipal supply is geared towards the home of the leaders. Even if the majority of the residences go dry, the leaders do not feel the actual crisis. The urge to work for better water supply therefore is not dealt empathically by political leadership in Darjeeling town.

Third, significant reason for such political indifference arises out of the complex administrative structure of the Darjeeling. After the Gorkhaland Territorial Administration (GTA), an autonomous body for administration was formed in 2011 (Joshi 2014), infrastructure development is the responsibility of Public Health Engineering Department (PHED) of Government of West Bengal and GTA, whereas the water supply remains the sole responsibility of the Darjeeling Municipality. The divided responsibility among three bodies is also a crucial reason behind the lack of proper management of water supply system in Darjeeling town.

21.5 Sources of Water and Water Supply System in Darjeeling Municipality

The water supply system in Darjeeling Municipality carries its colonial legacy. The water infrastructure that was developed during the colonial period still works as the backbone of the present water supply system. The public water supply system maintained by the Municipal administration and private water vendors are the two main water supply systems in Darjeeling town.

21.5.1 Public Water Supply System

The municipal supply of water in Darjeeling town basically depends on the water sources in its surrounding catchment areas. Water is transported to the town mainly from catchment area of Senchal Wildlife Sanctuary and Rambi catchment area.

21.5.1.1 Catchment Area of Senchal Wildlife Sanctuary

Senchal Wildlife Sanctuary is the main catchment area for water supply system of Darjeeling Municipality. This sanctuary is located around 11 kms. away from the Darjeeling Municipality with an area of 38.97 km^2. and its elevation ranges from 1500 to 2600 m (WWF-India 2012). Twenty-six natural springs within the Senchal Wildlife Sanctuary are the main sources of drinking water supply in Darjeeling Municipality. Water from these springs is collected in an arrestor tank, being transferred through masonry conduit line with around 8 kms. in length and finally get stored in the twin Senchal Lakes following the gravity. The twin lakes are known as North Senchal Lake and South Senchal Lake, and their combined water holding capacity is 33 million gallons. The North Senchal Lake was constructed in 1910 with water holding capacity of 20 million gallons and the South Senchal Lake in 1932 with water holding capacity of 13 million gallons (Drew and Rai 2016). Special attention is given to avoid contamination of these lakes as well as springs.

Water of these lakes is treated by the rapid sand filtration method at Jorebunglow water treatment plant. After purification, water from this treatment plant is transmitted into the main reservoirs, i.e. St. Paul's Tank with capacity of 235,812 gallons, Rockville Tank Iron with capacity of 56,651 gallons and Rockville Tank Masonry with capacity of 58,012 gallons (Darjeeling Municipality 2015). From these reservoirs, water is supplied directly or through subsidiary reservoirs at various places in

Darjeeling Municipality. Excluding the length of public hydrants and service lines, water system of this Municipality comprises of 35 kms. 'transmission main' line, 83 kms. 'distribution main' line, more than 20 masonry bridges sustained the masonry conduit and '14 pipeline bridges along transmission main' (Darjeeling Municipality 2015).

In this water supply system, water is transmitted 'through a series of pipes through a filter house and two distribution tanks' following the principle of gravity (Drew and Rai 2016). During lean period, the water level of the Senchal Lakes comes down due to the decrease in water supply from springs. To sustain the water supply in this Municipality during this crisis period, to augment the water level of Senchal Lakes, pumped water is transmitted to these lakes from the electrical pumping station located at Khong Khola.

To meet the increasing demand of water in Darjeeling Municipality, Sindhap Lake was constructed as the third reservoir in 1984 with 15 million gallons water holding capacity. Water is stored in this lake from the surrounding natural springs and pumping water from the Khong Khola and Bangla Khola water pumping stations (Ghatani 2015). Due to seepage, water holding capacity of Sindhap Lake is never fully utilized (Ghatani 2015).

21.5.1.2 Rambi Catchment Area

Water supply scheme from the Rambi Catchment Area to Darjeeling Municipality was completed into two phases. The first phase was initiated in 1969 and the second phase in 1993 with the aim of pumping of water from this catchment area to Jorebunglow water treatment plant. Water is collected from the tiny streams and eleven natural springs lying within the Rambi Catchment Area, and the provision of 1.5 lakh gallons water per day in Rambi water line is ensured during the second phase of this scheme (Darjeeling Municipality 2015).

21.5.1.3 Balasun River Project

Darjeeling Water Supply Pumping Scheme from Balason River usually known as 'Balasun River Project' was planned in 1995, and the foundation stone of this project was laid in 2006. This project was planned to supply daily additional one million gallons water into the two operational lakes, located around 12 kms distant at Senchal (Drew and Rai 2016). Intake point for raw water collection from Balason River has been fixed near Ceder Tea Garden at an altitude of 806 m (PHED, GoWB 2016). Raw water from this intake point is pumped to the delivery point, i.e. Senchal Lake at an altitude of 2262 m. To cover this high altitude, i.e. 1456 m, water is pumped in two successive stages. In the first stage, from the intake point of Balason River, water is pumped to the intermediate point located at the Kalej Valley Tea Garden with an altitude of 1574 m (PHED, GoWB 2016). In the second stage, water is pumped from this intermediate point to the delivery point Senchal Lakes. For purification and filtration, water from the Senchal Lakes is pumped and transferred to the water treatment plant located close to Sindhap Lake and finally potable water is pumped to St. Paul's reservoir located at an altitude of 2234 m and Rockville reservoir situated at an altitude of 2157 m (PHED, GoWB 2016).

However, this project has failed to fulfil the objective as the mixing impure water from the Balason River to the Senchal Lakes was observed. Water of the Balason River is highly contaminated at the intake point for raw water as the continuous mixing of pesticides, fertilizers from the tea gardens and urban sewage and other pollutants at large scale. At the very beginning during the test run of the project, it was reported by the local people that water supplied from Senchal Lake 'appeared discoloured and unclean' (Drew and Rai 2016).

21.5.2 Private Water Supply System

Apart from the municipal water system, private water supply system plays a significant role to meet the water crisis of the people. Natural springs located within the Darjeeling Municipality are the major source of water controlled privately. In Darjeeling Municipality, more than 32 perennial springs have been identified (Boer 2011). These springs' water is managed by either individually or community based. Individually managed springs are located in the individual's land. 'Traditionally, these springs were common property resources with a non-codified oral tradition of open access' (Lama and Rai 2016). However, as time passed, community-based management systems replaced the earlier traditions. These community organizations are popularly known as *samaj* which reflects unique identity derived from 'histories of colonial neglect' that motivated the villages in the entire Darjeeling to find solutions for social and economic crises in their own way (Drew and Rai 2016).

Due to insufficient supply of water from Darjeeling Municipality, large numbers of people are compelled to purchase water either hand carts vendors or water tanker truck vendors in the open market. The water vendors are considered as a reliable source for water as they are consistent in supplying water. They collect water from the springs located within the municipality as well as bring water from outside as the situation needs.

The public and private water supply systems coexist in Darjeeling town. To be precise, three separate systems, i.e. (1) state owned and managed, (2) market owned and managed and (3) community owned and managed, operate within Darjeeling Municipality. Comparative (dis)advantages among the three systems exist in supply, price, consistency and timing of water.

21.6 Water Supply Systems in Darjeeling: A Comparison

Amidst the debate of water to be considered as a public good or a common property resource as well as the management and distribution to be controlled by state or market, a unique system of water supply and distribution has evolved in Darjeeling town. Though difficult to calculate the exact share of each category in this system, state (Darjeeling Municipality)-, market (private vendors)- and community-based (managed by *samaj*) systems are prominent in their presence in the water landscape of Darjeeling town (Lama and Rai 2016; Drew and Rai 2016; Ganguly-Scrase and Scrase 2015; Ghatani 2015).

21.6.1 State-Owned and State-Managed System

The official system of water supply in Darjeeling town is run by the municipal administration (state run),[6] mainly a piped water supply scheme with household connection. This century-old system still operates with the infrastructure set-up by the colonial ruler. The twin lakes of Senchal catchment area, as already mentioned, have remained as the main source of water as the third lake, i.e. Sindhap Lake (Darjeeling Municipality 2015) stores only half of its estimated capacity due to faulty design and planning (Ghatani 2015). The Rambi Khola catchment area is another source of water for the town though this water collected in this area is now mainly supplied to the army area. The municipal water storage is highly inadequate and insufficient in supply. Municipal water reaches the houses ‘once in four days or even as infrequent as once in 20 days for hardly half an hour’ (Lama and Rai 2016). Inconsistency and insufficiency even in this meagre amount of supply is a regular feature. In a world-famous tourist destination, a regular water supply from municipal connection in such a frequency indicates that water management has not been prioritized in the urban scene of postcolonial Darjeeling.

Besides the storage, the distribution pipeline is mainly based on the old networks established during 1930s. Myriad of water pipelines along roadside is a common scene in Darjeeling town indicating that with advent of time, additional connections were added to the old networks, leading to more chances of leakage. The leakages are not regularly repaired and at times people collect water from the leakage points (Ghatani 2015). According to Drew and Rai (2016), estimates of a water engineer suggested 30–35% of the total water supplied from Senchal catchment area is lost during transmission.

With the rise in water scarcity, manipulation of the distribution of available water is a common practice in Darjeeling. ‘Power’ plays crucial role in determining the flow and amount of water to be supplied in particular household cutting on the ‘common man’s’ share. As the system is maintained by the valve-mans who open

[6]According to the 73rd amendment of Indian Constitution, the urban local bodies are responsible for providing basic services to the citizens including water supply.

and close the innumerable valves in the water supply system are often bribed and manipulated for additional water, illegal tapping is widespread in Darjeeling town (Ghatani 2015). As the demand increases with the arrival of tourists in the town during March–May, availability of water to the residential households declines. Respite from the situation needs additional allocation of resources[7] that will install a more efficient water distribution system.

Massive deforestation in the Senchal catchment area and in the areas around the Darjeeling town is gradually reducing the water availability. According to Basumajumdar (2016), deforestation and global warming have resulted in shortage of water, and 'water yield of the surviving springs' have declined considerably here. Hence, in near future, with increase in water demand and shortage in supply, crisis over water will intensify inviting more corruption.

Considering all these disadvantages, municipal supply remains as the main source of water basically for two reasons—firstly, water is supplied through household connection and second, it is free of cost. Public institutions are compelled to rely on the municipal water supply as the institutions are governed by certain rules and norms. The poor households are bound to depend on this source; water is reused within a household, and it is a common practice in Darjeeling.

21.6.2 *Market-Owned and Market-Managed System*

Given the uncertainties and inadequacies of the state-run system, a parallel system of water supply run by the private vendors is found Darjeeling town. Private water vendors collect water from springs, bring water to the town and sell for commercial and domestic purposes. The water vendors in Darjeeling operate at different scales, from individual water sellers carrying water in jerrycans as head loads to big vendors operating with water tankers in trucks. Individual vendors generally collect water from springs within the town either free of cost or by paying the 'owner'[8] of the springs. Individual vendors use jerrycans (average carrying capacity of 15–20 L) and carry the cans on their head or hand-pulled carts. The elderly persons generally carry water in single can as head load and supply the water to the doorsteps of individual houses. The young vendors carry 16–18 cans or buckets in hand-pulled carts and sell water according to the demand of individual households. On the other hand, the trucks ply with water every day in Darjeeling Municipality to supply water mainly for the hotels and other commercial establishments. They supply water in bulk, not less than 1000 L at one time. Since all these vendors operate in response to the market demand and no legal permission is required from any authority (Ghatani

---

[7]According to the estimates of the Darjeeling Municipality, INR 8.5 million is required annually in order to run the system in present condition. According to Drew and Rai (2016), additional INR 1700 million is required to revamp the system.

[8]In Darjeeling, if any spring is located within any private land, the owner of the land claims full right over the water from the spring.

2015), no official estimate on the quantity, price and frequency of their supply is available. However, in the peak season (March–May), according to Ghatani (2015), 120 trucks ply in the town and according to Lama and Rai (2016) 105 trucks ply everyday carrying 5500–6500 L of water at one trip, often making four–five trips. This number reduces to 60–70 trucks in the non-peak season.

The water vendors are seen all over the city, and dependency on them is on rise during the last three decades. The charges of the water vary according to the demand and availability of water, on an average INR 0.25–0.30 per litre. Houses located up or down from the main road prefer the small vendors as they carry water to the doorsteps of individual houses. The hotels owners on the other hand highly depend on the big vendors as the huge water demand during the tourist season is possible to be met by this system. Any establishment, which runs without proper permission of the administration, also depends on the water vendors, as the municipal supply will never be available to them. The high cost of municipal connection also acts as reason for their existence. The water vendors, therefore, are invincible in the water supply system in Darjeeling Municipality. The tourists and families prefer to pay this assured supply to overcome the water crisis. However, the families who are unable to afford this price and generally depend on springs and jhoras to meet their basic water need are further pushed away by the pressure of the water vendors from these water sources.

21.6.3 Community-Owned and Community-Managed System

Beside the municipal supply and water vendors, a 'decentralized' water supply system (Drew and Rai 2016) managed by *samaj* exists in Darjeeling town. As many as 150 *samaj* are present in Darjeeling town (Lama and Rai 2016). Initially, providing necessary support during any crisis or emergency faced by its members, the *samaj* are now working to address the water need. *Samaj* operate in a not-for-profit mode and intervene in controlling and managing water access to its members and non-members. Thirty-two springs present in the town are now managed by *samaj* (Ghatani 2015), with different types of management strategies. The *samaj* generally set the rule, i.e. amount and time of water collection for the members or non-members during wet and dry seasons, charges for water, cleaning and pollution check in the water source areas and afforestation in the catchment areas. Along with managing their own sources, *samaj* are also active in lobbying with municipality for assuring water supply to their members (Drew and Rai 2016).

The control of the *samaj* over the existing water sources has led to a situation where non-members are getting restricted access to water from the springs they manage. Hence, in Darjeeling town, the taps or springs as public sources of water are gradually disappearing and the overall cost for water either in terms of money or time is increasing for the urban poor (Drew and Rai 2016). *Samaj*, though operates on not-for-profit mode, it cannot be denied that the water access is not always egalitarian. The asymmetry in power relations among the members in *samaj* leads to such inequality.

Table 21.3 Darjeeling town: total water demand and supply

• Total demand of water (per day): 1,970,000 gallons	1,860,000 gallons for 120,000 population + 110,000 gallons for supply to Army, Hospital and St. Paul's School
• Net availability of water (per day): 637,500 gallons	Total water production 850,000 gallons – wastage 212,500 gallons
• Actual water supply (per day) for public: 527,500 gallons	637,500 – 110,000 gallons
• Water deficit (per day): 1,332,500 gallons	1,860,000 – 527,500 gallons
• Ratio of duration of water supplied	4 days

Source Darjeeling Municipality (2015)

21.7 Water Crisis in Darjeeling Municipality

Given the three water supply systems are running parallel, water demand of the people of Darjeeling Municipality is not addressed in totality. The magnitude of water crisis was intense that till 2011, Darjeeling Municipality provided to only 9.85% household's connections and 144 commercial connections (Darjeeling Municipality 2015).

The official estimates of demand and supply indicate 1,332,500 gallons/day (Table 21.3) water deficit per day. While the water mafias in Darjeeling town earn high profit from this gap poor people are compelled to compromise with their minimum water need and proper sanitation. The shortage in water results in unclean living environment and spread of diseases among the low-wage earners in the town.

The situation worsens in Darjeeling during natural calamities, i.e. landslides, earthquakes and prolonged political agitation when water supply system is disrupted. Storing of water in sufficient quantity to meet the demands during any such exigency either by the suppliers or consumers is difficult in Darjeeling due to paucity of storage space. In addition, the weight of stored water appears to be accelerating slope instability. Likewise, rainwater harvesting is not considered as a feasible method to reduce the water scarcity in Darjeeling town.

The water scarcity in Darjeeling Municipality, apart from the population growth and continuous in-migration, is attributed to (i) old supply system of the colonial period, (ii) seasonality of rainfall and (iii) increasing volume of tourism.

21.7.1 Old Supply System of the Colonial Period

The main reason behind the shortage of water supply in Darjeeling Municipality is attributed to the non-upgradation of the old supply system. Increasing pressure on the age-old conduits of the colonial period has made them almost non-functional. After failure of Balasun River Project, no comprehensive plan has been prepared for water supply in Darjeeling town. Deforestation and shrinkage in the existing catchment have also reduced the quantity of water being stored in the twin lakes of Senchal. The easy manipulation of this old system in favour of particular groups has increased the overall crisis of water in Darjeeling Municipality.

21.7.2 Seasonality of Rainfall

Though the water crisis in Darjeeling is observed all through the year, the intensity of the crisis fluctuates over seasons. The natural springs and the catchment areas are fed by monsoon rain, which is mainly concentrated during June–September as already mentioned. During monsoon period, water gushes out of the joints and cracks, the springs are in full flow, and the lakes are also full of water. Consequently, the availability of water from natural sources increases during the monsoon period. During the lean rainfall period, i.e. from November to May, many springs gradually dry up. However, water crisis in Darjeeling Municipality reaches its high during March–May with increase in arrival of tourists.

21.7.3 Increasing Volume of Tourist

Tourism being the main economic activity is promoted for generating more employment in Darjeeling town. The volume of tourist is ever increasing in Darjeeling, and this inflow of tourist is further intensifying the present crisis of water availability. As mentioned earlier, peak tourist seasons in Darjeeling are during the pre-monsoon (March–May) and post-monsoon (October and November) phase when water supply is naturally reduced. According to the estimate of the Darjeeling Municipality (2017), in 2011, around 165,000 tourists visited Darjeeling town, out of which about 10,000 persons are foreign tourists. They have also estimated approximately 30,000 persons visited Darjeeling town per month during the tourist season. Considering the inflow of tourist in Darjeeling town during peak season, it is estimated that Darjeeling Municipality needs to cater the water demand of additional 10,000 tourists per day. As per the consumption norms of WHO, these tourists further augment the water demand by 155,000 gallons/day. However, the unofficial estimates mentioned that Darjeeling receives almost double tourists per day during peak season, simply implying the water demand also raised by 310,000 gallons/day.

21.8 Water Crisis in Darjeeling: Strategies and Options

To address the issues of water scarcity in Darjeeling Municipality, a comprehensive planning for the urban development as well as water supply is urgently needed. In absence of such planned development, Darjeeling Municipality will remain vulnerable to severe water scarcity. Hence, the integration and co-ordination among all the responsible administrative authorities, community-based organizations and Public Health Engineering Department are required for preparation of a master plan to envisage a sustainable future for Darjeeling Municipality. However, long-term vision and preparation of master plan are unlikely to provide immediate respite from present water crisis faced by Darjeeling town. This more than 150-year-old town, therefore, requires an all-encompassing planning with immediate action (short-term), formulation of planning strategies (medium-term) and implementation of planning (long-term).

21.8.1 Immediate Action

Prior to this long-term planning and action, rectification in the existing water supply system is crucial which is possible only in the presence of a strong political desire to run the system equally for everybody. Beside the master plan and establishment of new water supply system, the illegal tapping, manipulation of the system and minimization of the transportation loss of water are needed. Metering of water in households and in commercial establishments may be introduced to monitor the actual amount of water supplied. Water meters will enable the municipal authorities to estimate the actual demand–supply gap and also to introduce user charges for this scarce resource. Introduction of user charges for water, particularly beyond a basic minimum level will restrict misuse of water in Darjeeling town. Revenue earned from water metering will pave the way for raising funds to maintain the existing water system.

As installation of a new water supply system is a long-term process, the provision of water supply through water tanker at a controlled price may be introduced by the competent authorities in Darjeeling. It is evident from the present water supply system in the town that the private water vendors play an inevitable role to reduce the water scarcity mainly in pre-monsoon season. However, the pricing and supply of water are purely demand driven and often unaffordable for the low-income groups. Given this practice, municipal administration in Darjeeling is required to adopt strategies for regulating supply and controlling the prices of water through private vendors. The private vendors, big as well as small, who operate on their own to supply water in the town, may be given due recognition by the municipal authority and monitored for supplying water at a prescribed rate. Option of tax collection from the big vendors supplying water particularly to the hotels and other commercial establishments may be considered by the municipality. In addition, Darjeeling Municipality is proposed to engage itself in supplying water by tankers at reasonable price on regular basis. To involve the private vendors, a pre-designed guidelines need to be implemented by the municipal authority. An effective public–private partnership in water supply has an immense potential to meet the water demand.

Besides, immediate action is needed to introduce the water quality monitoring system in Darjeeling town. The quantity of water supplied has earned great concern in Darjeeling, whereas the quality of water remained a less important issue. It is obvious that the uncertainty of water supply is increasing the use of unsafe water and poor residents of Darjeeling are being exposed towards water-related health hazards.

Afforestation in the catchment area of water sources, proper disposal of the urban waste and removal of waste from the water sources are essential. Awareness generation among the users including tourists to reduce the wastage of water would be emphasized in Darjeeling.

21.8.2 Formulation of Planning Strategies

The immediate action as suggested in the earlier section would be more effective provided the planning strategies are formulated and subsequently implemented in Darjeeling town. To begin with a formal engagement of the existing community-based organizations, professionals, bureaucrats and administrators are required to identify the veracity of the problem and exploration of the alternate sources of water.

The actual water demand considering the floating population comprising of tourists, students and migrant labourers in the peak season needs to be ascertained. Given the present pace of growth of Darjeeling town, quantity of such demand in future is also required to be identified.

The alternate sources of water, possible storage options and transportation of water to the town at an affordable cost have to be considered as the integral part of the planning. Decentralization of the management and distribution of water sources are also required to be incorporated in the plans.

21.8.3 Implementation of Planning

In absence of any master plan or any stringent legal framework, Darjeeling town is growing in response to economic impulses regardless of its environmental viability. Darjeeling town, therefore, requires a blueprint, which is environmentally acceptable and economically feasible for development in future. Long-term planning and formulation of viable options for supplementary and alternative sources of water, construction of additional lakes for storage and maintenance of the existing system in Darjeeling town have to be prioritized. Private operators, it is presumed, would have a greater role to play in long-term planning, in building infrastructure and maintaining the water supply system and collection of revenue.

21.9 Conclusion

Darjeeling today is more of a cosmopolitan administrative town than a small and magnificent hill resort. Water scarcity in Darjeeling Municipality reflects the impact of unplanned urbanization on hill slopes of the Eastern Himalaya. In spite of high annual rainfall, Darjeeling Municipality stands as a classic example of acute water-scarce town. The location of Darjeeling on the steep slopes and natural hydrogeological condition of the town hinder the storing of huge amount of rainfall and consequent supply. However, the absence of strong political will and administrative endeavour to improve the existing supply system are also responsible for worsening the situation further. The shortage in supply of water is resulting into unhygienic sanitation condition, increasing pollution and higher risks of spreading waterborne diseases. Collective cost of water is imposing additional burden to the household budget in this hill town. The urban poor in Darjeeling Municipality are becoming more vulnerable to public health hazards. Future of tourism industry in Darjeeling also depends on the water availability. Conversion of Darjeeling town from water-scarce to water-sufficient region is not unattainable. A long-term vision is essential to initiate this process. Water scarcity in Darjeeling requires to be addressed with utmost attention from all sectors so that a sustainable and affordable water supply system operates in this world-famous tourist destination.

References

Basumajumdar A (2016) Impact of global warming on climate change regarding water supply in the Darjeeling Hills of the Eastern Himalaya and change in mountain ecology. In: Chand R, Leimbruger W (eds) Globalization and marginalization in mountain regions assets and challenges in marginal regions. Springer International Publishing, pp 161–172

Bhattacharya P, Joshi B, Dubey P, Singh SP, Kapoor R (2001) Forests and displaced people: fuel wood collection and trade as a first step survival strategy. RWEDP Report No. 57, FAO Regional Wood Energy Development Programme in Asia Bangkok, Thailand

Boer L (2011) The perennial springs of Darjeeling, a survey to community based conservation. Ashoka Trust for Research in Ecology and the Environment (ATREE), Bangalore

Census of India (2011a) Office of the Registrar General & Census Commissioner, Ministry of Home Affairs, Government of India. http://censusindia.gov.in/

Census of India (2011b) District census handbook: Darjeeling, Village and Town Wise Primary Census Abstract (PCA), Series-20, Part XII-B. Directorate of Census Operations West Bengal, Government of India

Census of India (2011c) District census handbook: Darjiling, Village and Town Directory, Series-20, Part XII-A. Directorate of Census Operations West Bengal, Government of India

Darjeeling Disaster Management Section (2016) Disaster management plan book-2016. Office of the District Magistrate, Government of West Bengal

Darjeeling Municipality (2015) Waterworks. http://darjeelingmunicipality.in/. Accessed 20 June 2017

Darjeeling Municipality (2017) About Darjeeling. http://darjeelingmunicipality.in/. Accessed 20 June 2017

Das AJ (1947) Bengal District Gazetteers: Darjeeling. Bengal Government Press, Alipore, Bengal

Drew G, Rai RP (2016) Water management in post-colonial Darjeeling: the promise and limits of decentralised resource provision. Asian Stud Rev 40(3):321–339. https://doi.org/10.1080/10357823.2016.1192580

Ganguly-Scrase R, Scrase TJ (2015) Darjeeling re-made: the cultural politics of charm and heritage. South Asia J South Asian Stud 38(2):246–262. https://doi.org/10.1080/00856401.2015.1031203

Ghatani S (2015) Sustainable urban water management in Darjeeling. M.Phil. Dissertation, Department of Geography, School of Human Sciences, Sikkim University, Sikkim

Guha RK, Kujur AR (2009) Roof top rainwater conservation in Darjeeling town, West Bengal, an option to mitigate the crisis of water supply—a case study at Raj Bhawan, Darjeeling West Bengal. Bhu-Jal News 24(1):85–90

India Meteorological Department (IMD) (2017) Government of India. http://www.imd.gov.in/section/climate/climateimp.pdf. Accessed 20 June 2017

Joshi D (2014) Feminist solidarity? Women's engagement in politics and the implications for water management in the Darjeeling Himalaya. Mt Res Dev 34(3):243–254. https://doi.org/10.1659/MRD-JOURNAL-D-13-00097.1

Lama MP, Rai RP (2016) Chokho Pani: an interface between religion and environment in Darjeeling Himalaya. J Assoc Nepal Himalayan Stud 36(2):89–98

Mell IC, Sturzaker J (2014) Sustainable urban development in tightly constrained areas: a case study of Darjeeling, India. Int J Urban Sustain Dev 6(1):65–88. https://doi.org/10.1080/19463138.2014.883994

O'Malley LSS (1907) Bengal district gazetteers: Darjeeling. The Bengal Secretariat Book Depot, Calcutta

Public Health Engineering Department, Government of West Bengal (2016) Darjeeling water supply pumping scheme Balason river source. http://www.wbphed.gov.in/main/index.php/darjeeling. Accessed 10 June 2017

WWF-India (2012) Status and distribution of Asiatic black bear and the status of human-bear conflict at Senchal wildlife sanctuary, Darjeeling, West Bengal, India. Khangchendzonga Landscape Programme, Darjeeling and Sikkim and West Bengal Forest Department, Wildlife Division I, Government of West Bengal

Chapter 22
Urban Water Governance: Pricing of Water for the Slum Dwellers of Dhaka Metropolis

Mahfuzul Haque

Abstract Increasing population in the urban areas of Bangladesh takes its toll on the scarce resources like water for drinking, washing and sanitation purposes. Ever expanding Dhaka, the capital of the country, is reeling under the pressure of burgeoning population of around 14.5 million with a density of population of 45,000 per km^2, while the country's population is around 156 million. In the poor urban areas of Bangladesh, availability and the quality of water are fast decreasing. In Dhaka, the percentage of slum population out of total population of the metropolis has marked an increase although, occupying an area of only 4% of the metropolitan area. The paper raises the issue of "water governance" and stated that poor governance in water management has resulted in pilferage of water and caused miseries for the marginal dwellers living in the slums. Crisis of water continues to be one of the major concerns for the slum dwellers of Dhaka. Due to continuous withdrawal of groundwater and gradual loss of recharged areas, the groundwater table of Dhaka has declined at an alarming rate over the last couple of decades. On the top of that mushrooming of slums has generated booming business for the middlemen, corrupt officials and musclemen with political connections posing a great threat to an effective water governance. The study based mainly on se condary sources of the literature, argues that the slum dwellers are paying an exorbitant price than legal connection holders covered by Dhaka water supply and sewerage authority. They spend nearly a quarter of their average monthly income for domestic water supply. The study concluded that no matter that they are poor, the slum dwellers are paying more than that of general water consumers of the city.

Keywords Dhaka slums · Slum dwellers · Water governance · Water pricing

M. Haque (✉)
Department of Development Studies, Adjunct Faculty, University of Dhaka,
House 36/1, Road 4, Dhanmondi R/A, Dhaka 2105, Bangladesh
e-mail: drmahfuzulhaque@yahoo.com

B. Ray and R. Shaw (eds.), *Urban Drought*, Disaster Risk Reduction,
https://doi.org/10.1007/978-981-10-8947-3_22

22.1 Introduction

United Nations General Assembly in its decision in July 2010 stated that access to water and sanitation is a human right (UN 2010). The decision stated that clean drinking water and sanitation are vital for the realization of all human rights. Among the most significant urban services, water and sanitation play a significant role in poverty alleviation and sustainable development. Urban governance has to ensure equitable provision for safe, clean, accessible and affordable drinking water and sanitation, with provisions to accommodate urban expansion and population growth (UN-Habitat and WHO 2016). According to the United Nations Human Settlements Programme (UN-Habitat), approximately 863 million people globally live in urban slums and estimates suggest that this figure may rise to 1.5 billion by 2020 (UN-Habitat 2013). Slums are generally characterized by poor housing, high population density, limited sanitation facilities, lack of security tenure and poor governance. In fact, the slum dwellers are in constant threat of eviction and coercion. Access to health, education, power, gas, water supply, sanitation and waste management is very limited for the urban slum dwellers. This had a telling effect on many of the developing and least developed countries of the world in attaining the Millennium Development Goals (MDGs). Achieving the UN Sustainable Development Goals (SDGs) number six on ensuring access to water and sanitation for all by 2030 would be a big challenge for most of the developing countries of the world having poor urban population in the slums.

According to UN-Habitat, a slum is characterized by (a) overcrowding; (b) poor housing conditions in difficult and inaccessible terrain, tenure insecurity, make-shift structure; and (c) inadequate access to improved water at affordable price; improved sanitation; and other infrastructure services (UN-Habitat 2003; Sabbaraman et al. 2015). UN-Habitat considers slum households as "a group of individuals under the same roof lacking one or more of the following conditions: (i) access to improved water; (ii) access to improved sanitation; (iii) sufficient-living area; (iv) durable housing; and (v) security of tenure" (Adam 2013). The term "slum" has been controversial and is avoided by some researchers on the ground that it represents a highly negative view of the residents (Dagdeviren and Robertson 2011). Slums are also known "informal settlements" or "peri-urban areas". In some countries, these informal settlements are known as "urban ghettos"; "labour lines" or "un-serviced urban neighbourhoods" or very commonly as "shanties". The joint IMF/World Bank Global Monitoring Report (2013) observed that "slums are the urban face of poverty and emerge when cities are unable to meet the demand for basic services and to supply the expected jobs" (Adam 2013).

In slums, due to the absence of legal rights, the dwellers face constant threats of eviction and are denied of formal basic service provisions (UNICEF 2011). They are deprived of services from both public and private sectors. No public funds are allocated for slum development because of tenure insecurity. In most of the urban slums, a parallel structure has emerged, where free education and health services are provided by the NGOs and middlemen are delivering other services at a pretty high cost. In developing countries, slums are the reality as most of the low-income groups engaged in transport, construction, domestic work in urban households and petty business live in these informal settlements to earn a livelihood. The slum dwellers face multiple challenges ranging from insecurity, legality, poverty, unsanitary conditions, diseases and death, although they contribute a major share of the urban workforce.

The overall goal of this chapter is to explore the present state of water governance in the slums of Dhaka metropolis with special emphasis on water pricing. The chapter is structured as follows: section one gives a general introduction on global situation of slums and highlighted some definitional issues raised in the chapter; section two deliberated on issues like water governance, OECD water governance principles, and existing national plans and policies on water sector in Bangladesh; section three discussed the state of water supply situation in Dhaka slums; section four touched upon water pricing based on two survey reports conducted recently; and the last section in conclusion suggested some recommendation measures.

22.2 Water Governance

There is no universally accepted definition of governance. According to Olsson and Head (2015), governance goes beyond regulation, public management and traditional hierarchical state activity. Further to that governance emphasizes the use of novel instruments, such as voluntary and market-based approaches, and cooperative structures between state and non-state actors from various sectors of society. Among the non-state actors, private sector and civil society play a vital role in governing a country. Governance implies some form of self-regulation and cooperation between and among various types of actors. Water governance includes institutions, organizations, policies and practices that shape and manage water resources. It includes delivery of water services for both domestic and industrial uses.

Water governance is as old as human civilization itself. Moss and Newig (2010) stated that emergence of some of the great civilizations was due to their capability to tame floods and manage irrigation through an effective water governance. Clean drinking water has been termed as "blue gold" and ensuring an effective water governance has become a big challenge. According to Water Governance Facility (WGF), water governance refers to existing political, social, economic and administrative systems that provide directions for the management and use of water resources. Water governance delineates the principles of availability and quality of water for the community at an affordable price, which means who will get water, when and how to get water including costs of the services. Water governance determines the equity and efficiency in water resource and services including allocation and distribution and balances water use for various socio-economic activities (UNDP-SIWI 2016). Governing water means formulation, establishment, implementation and monitoring of water policies, legislation and institutions. It explains the obligations of the government, civil society and the private sector bodies enabling the local community in accessing water resources and services at an affordable cost. Water governance observes how the stakeholders comply with the rules and regulations and roles assigned to them in this regard. The sector itself is a part of broader socio-political and economic developments and is influenced by actions taken by leaders of other cross-sectoral bodies.

Tortajada (2010) defined "water governance" as comprising of social, political, economic and administrative organizations and institutions, as well as their relationships to water resource development and management. According to UNDP, water governance includes political, economic and social processes and institutions through which governments, private sector and civil society make decisions about how best to use, allocate, develop and manage water resources (UNDP 2004). Water governance and water management are interdependent. Some of the organizations have developed guidelines/principles concerning water governance in order to set some standards on efficient and effective water use and practices. OECD Principles on water governance have set standards for more effective, efficient and inclusive design and implementation of water policies (OECD 2017). Adopted at the Seventh World Water Forum in April 2015 in Korea through the "Daegu Declaration", Akhmouch and Correia (2016) observed that the Principles provide 12 must-do for governments to design and implement effective, efficient and inclusive water policies. They were developed using a multi-stakeholder approach at the OECD Ministerial Council held in June 2015. The principles clearly allocated and distinguished the roles and responsibilities for water policymakers, policy executives, operational managers, and promoted coordination among the stakeholders.

Some of the salient features of the principles are to encouraging policy coherence through cross-sectoral coordination (e.g., water and environment); producing and sharing water-related data, ensuring allocation of financial resources; mainstreaming transparency practices; promoting stakeholder engagement; and regularly monitoring and evaluating water policy and governance. It is interesting to note that within a short period of adoption, a good number of countries and major stakeholder groups have endorsed the OECD Principles.

22.3 Urban Water Governance

Population in towns and cities are fast increasing accommodating more than half of the world's population. This figure is expected to rise to 60% by 2030 (Mitlin and Satterthwaite 2013; Olsson and Head 2015). Generally, people living in the central areas of big cities have access to water and sanitation. In most cases, the urban poor in fringe areas of mega-cities in developing countries lack such services. Due to limited financial resources and inadequate operational capacities, the urban poor face major constraints in accessing the services. For a sustainable healthy environment in cities, strategic water provision needs to be undertaken. Only by addressing the issues of water scarcity, water accessibility, affordability, and quality, basic human needs could be met.

According to World Bank, urban water governance includes institutions, organizations, policies and practices that shape and manage water resources, including the delivery of water services for diverse populations and industries in cities (World Bank 2008a). Urban water governance ensures effective water use and management in urban areas of the country based on the principle of equity and sustainability. Many developing countries are yet to adopt a policy on water governance, and in most cases the enforcement mechanism is rather weak. Water systems present a unique set of challenges to its stakeholders as the cities grow big and urban areas expand.

Both in the developed and developing countries, some of the challenges facing urban water governance include the range of competing interests among different sectors, inter-agency cooperation, different interpretations of integrated water management, power dynamics and lack of capacity building (Olsson and Head 2015). There are various measures undertaken by city municipalities to address those challenges. There are expensively engineered infrastructure options, serving large populations through centralized systems, i.e. large technical organizations that operate effluent treatment plants, irrigation networks, desalination plants, etc. There are also other low-cost options like low-water-use sanitation, including a range of water harvesting and reuse schemes and dry sanitation options where water is scarce, etc. Some of the cost-effective water management practices in urban areas of the developing world are solar-powered pumps, increased use of surface water by lessening dependence on groundwater, treatment of wastewater at source, treatment of municipal waste through composting, etc.

22.4 Water Governance: Bangladesh Scenario

At a national level, Bangladesh has undertaken a number of policies and strategies targeting improvement water and sanitation sectors. Bangladesh National Water Policy, 1999 and Bangladesh National Water Plan, 2004 made it a provision to ensuring availability of safe and affordable drinking water to all elements of the society including the poor and underprivileged (MoWR 1999). National Policy for Safe Water Supply and Sanitation, 2005 mentioned that the Government's goal was to ensure that all people have access to safe water and sanitation at an affordable cost (MoLGRD 2005). Despite adoption of a number of laudable goals, strategies and legislations on accessible and affordable water supply to its citizen, successive governments in Bangladesh were unable to ensure good water governance through making safe water available to people at affordable price.

On the issue of legality of these informal settlements, so far the policymakers in Bangladesh were of the view that slums are illegal establishments and the dwellers therein are to be evicted by whatever means possible. In the past, there were many abortive attempts (especially during military government rules) to evict the slums through coercive methods without considering humanitarian aspects of the dwellers. However, there are changes taking place in the mindset of the policy planners in Bangladesh. Government of Bangladesh is of the view that there is a need to change attitude towards the slum dwellers considering meeting the SDGs. The Seventh Five-Year Plan (2016–2020) recognized that slums/squatters are an integral part of urban areas and contribute significantly to their economy both through labour market contributions and informal productive activities as a significant number of Dhaka's workforce engaged in transport, construction and garment sectors reside in the slums. It is an approach based on positive attitude and sought to improve the lives of the slum dwellers through slum upgrading/improvement. The plan suggested that the slum dwellers are entitled to receive basic minimum services until proper relocation and resettlement provisions have been made. Understanding the realities and necessities of the urban slums by the policymakers is a major departure from their earlier stand on illegality of these informal settlements.

22.5 State of Dhaka Slums

Bangladesh has had one of the highest rates of growth of urban population over the last three decades at over 6% annually compared to the national population growth rate of about 2.5% (Rashid 2000). UNICEF's "Understanding Urban Inequalities in Bangladesh Report, 2010" stated that around 7 million people are estimated to be currently living in urban slums and this number was rising. The report also challenged the belief that the situation of urban dwellers in Bangladesh is better off than those living in rural areas (UNICEF 2011). Dhaka currently is one of the fastest growing megacities in the world with a population of 15 million and an annual growth rate of 5%. World Bank further observed that the city has more than 4,500 slums and squatter settlements within its territory accommodating more than 35% of its population in approximately 112,670 households. An estimated 0.3–0.4 million migrants mostly poor and landless come to city annually for employment, education and health reasons. Dhaka's population of 15 million is expected to grow to around 20 million in 2020, and Dhaka is projected to be the world's third most populous city (CUS 1996). This rapid growth of urban population is fuelled by migration of the rural poor. Most of the migrants are absorbed in the informal sector due to low level of skill and education (Begum and Moinuddin 2010).

Population of Dhaka metropolis is approaching 15 million with a growth rate of around 5% per year. It has been estimated that around one-quarter of Dhaka's population live in slums (Varis et al. 2006). Dhaka is growing rapidly and much of its growth stems from migration with an estimated 300,000–400,000 new migrants, mostly poor, arriving to the city annually (BBS; Hossain and Khan 2012). The percentage of slum population within the overall population has also increased from 25% in 1996 to current 37.4%, occupying an area of only 4% of the total Dhaka Metropolitan Area (Islam et al. 2005). These slum dwellers face several social, economic and political problems. Crisis or scarcity of fresh water is one of the major concerns for the slum dwellers. Due to increase of Dhaka city population, the gross daily water demand is expected to rise from 2,460 ml/d (million litres per day) to 7,970 ml/d within the next 15 years, assuming a system loss of 20%. DWASA Annual

Fig. 22.1 A young girl pumps water in the Korail Slum in Dhaka. *Credit* © UNICEF/2011/Ahsan Khan

Report (2014–2015) claims its capacity of supplying water to the tune of 2,420 ml/d (accessed in June 2017) and denies any shortage of water, although, it is reported that there is rationing of water in many parts of the city during the summer days. Water supply in Dhaka City is predominantly based on groundwater extraction. More than 75% of the water supply of the city comes from groundwater resources. In the fringe areas, nearly the entire supply is from groundwater. According to a World Bank report (2008a), massive extraction of groundwater has resulted in depletion of groundwater levels by an alarming rate of 2–3 m/year.

Water needs of Dhaka metropolis and Narayanganj are addressed by Dhaka Water Supply and Sewerage Authority (DWASA). It is an autonomous commercial organization in the public sector. It covers more than 360 km^2 service area with 12.5 million people. In the slums of Dhaka, water distribution is informally governed by various groups such as political leaders and middleman with the help of DWASA officials. In some slums, various NGOs were found to be working for a long time. As slums are informal settlements in make-shift houses, there is no provision kept by DWASA for setting up water line on illegal land (Sakib and Islam 2014). In some cases, DWASA supplies water to slum areas through Community Based Organization and street tap water. Water governance is facing enormous challenges in the slum areas and access to freshwater by the slum dwellers remains a major challenge. Moreover, frequent movement of slum dwellers from one slum to another often makes it difficult to assess actual demand and supply of water in poor urban areas of the metropolis.

Fig. 22.2 Unhealthy sanitary conditions of a slum in Dhaka during monsoon. *Credit* World Bank (2008a)

Mostly the slums are serviced by middlemen, who control these slums by utilizing their connections with political power, law enforcing agencies and corrupt officials. These middlemen are also known as urban "water mafias" in some other developing countries. For example, in many parts of the Indian city of Bangalore, where government water supply has failed, or a scarcity has been created artificially, a coalition of middlemen, local politicians and corrupt officials of water department run a parallel private water supply network (Ranganathan 2014). Win-win situation for both the parties (water mafias and slum dwellers) triggers access to water in the slums. Slum people want their daily water supply at any cost and with the help of middlemen they are able to get access to DWASA water supply paying a hefty price. Threats of eviction in the absence of legal land rights prohibit the slum dwellers in accessing formal services. Slum evictions are a constant threat and occur in the absence of a proper resettlement plan. Insecurity of tenure and its implicit potential for politicization serves as a deterrent when it comes to allocating funds for the urban poor. As a result, a parallel structure has emerged where free education and health services are delivered through NGOs and the local *mastaans* (musclemen) with political connections are providing water services at a higher cost (UNICEF 2011) (Figs. 22.1 and 22.2).

22.6 Water Pricing in the Slums

Recently, two studies were conducted on issues like water governance in general and water pricing in particular in some of the slums of Dhaka in 2014 and 2015. The first one was undertaken by two researchers on three selected slums, namely, Agargaon BNP Bazar, Tejgaon Railway and "Ta" Block of Mirpur 12 (Sakib and Islam 2014). They saw a dismal picture of water governance in these slums. The slum dwellers had to depend on the middlemen to access DWASA water pipelines at an exorbitant price. Besides paying a hefty amount for illegal water connections, they pay high monthly charges. In some slums, there are Community-Based Organizations (CBOs) working, who were found to be ineffective and submit to the wills of the local political leaders and middlemen. The study found that a Dhaka city middle class household consumes 7.5–10 times more water than the slum people and pay lesser water bill. Water charges by the middlemen are uncontrolled. In some cases, water was found unsuitable for drinking due to its bad smell. On the issue of water pricing, the study concluded that the slum dwellers pay between Tk 30 and Tk 80 for 1,000 l, whereas Dhaka's domestic users with metered connection pay only Tk 6.34 per 1,000 l. The slum dwellers expressed their dissatisfaction on overall water governance in their slums and looked for a change involving the government, local political leaders and the CBOs (Table 22.1).

The second study was conducted in three other slums, namely, Tejgaon slum, Godown slum and Korail slum to determine current water price (Rahman et al. 2015). The study found that slum dwellers in the study area pay between Tk 50 and Tk 100 for 1,000 l of water against DWASA's rate of Tk 7.33 in 2015 (Table 22.2).

In a recent notification, DWASA rescheduled its water tariff for domestic users from Tk 8.49/1,000 l to Tk 10/1,000 l with effect from 1 November 2016 (Website of DWASA). In summary, the slum dwellers pay between 7 and 14 times more than the water users with metered connection. The study also disclosed that the slum dwellers pay between 13 and 23% of their monthly income for having access to water services.

Although, the slum dwellers pay an exorbitant price compared to the metered users, they often fall sick from drinking this water. As a result, they suffer from diseases like diarrhoea, jaundice, cholera and many water-borne diseases. The illegal

Table 22.1 Comparative scenario of water pricing in slum dwellers and DWASA users

Water price per 1,000 l of DWASA	Water price per 1,000 l in Agargaon BNP Bazar	Water price per 1,000 l in Tejgaon railway slum	Water price per 1,000 l in "TA" Block, Mirpur slum	Comments
Tk 6.34	Tk 30.00	Tk 80.00	Tk 50.00	Slum dwellers pay between 5 and 12 times more than DWASA users

One US $ is equivalent to Tk 79.94 as of 16.8.2017

Table 22.2 Comparative scenario of water pricing in slum dwellers and DWASA users

Water price per 1,000 l of DWASA	Water price per 1,000 l in Tejgaon slum	Water price per 1,000 l in Godown slum	Water price per 1,000 l in Korail slum	Comments
Tk 7.33	Tk 50.00	Tk 50.00	Tk 100.00	Slum dwellers pay between 7 and 14 times more than DWASA users

water connection is to be blamed for as the water runs through a narrow pipe, which easily gets contaminated by dirty slum water. On certain spots, the pipe is broken and taped. Bacteria can grow easily on those spots, and water can easily get polluted. Still this dirty water allows some middlemen earn a good earnings as much as Tk 120 per family a month (DS 2013).

In a recent move, the government appreciating the contribution of these informal settlements decided to provide legal water connections in phases. Terming the slums as "Low-Income Community (LIC)" the DWASA in a major departure decided to provide legal connections to the slums by December 2016. The biggest LIC, the "Korail slums" has already been covered with legal water connection (DWASA website). It is reported that currently, fifty per cent of slum dwellers are getting water through the legal supply network and DWASA hoped that the rest would receive the facilities by the end of 2017. Following legal connections, it is expected that the cost of per unit of water (1,000 l) would come down to Tk 8.49 only, thereby monthly water bill per family would be Tk 50 instead of Tk 120 being paid (DS, 5 Sept 2016; UNB, 5 October 2016). It is interesting to note that the vested interest groups in these slums, the water *mafias*, *mastaan* and some corrupt officials were found opposing the DWASA plan to legalize water connections.

22.7 Conclusion and Recommendations

Based on the study findings of the selected slums of Dhaka, it could be concluded that the slum dwellers in other parts of the country pay no less than that of the regular urban users with metered connection. Although, they pay more, they receive poor quality of contaminated water containing bacteria and parasites. Water-borne diseases keep them abstained from work causing loss of income. The slum dwellers are hostage to local *mastaans*, politicians, middlemen and DWASA corrupt officials and pay exorbitant price starting from water connections to regular supply. On one hand, a Dhaka city middle class household consumes more than that of the slum people and on the other, they pay less. It is heartening to note that the slum dwellers are being recognized as "Low-Income Community", which would give them some legal status with proper connection of essential utility services. Legalizing water services by DWASA would greatly contribute to national economy as the major portion of urban workforce comes from these slums. Furthermore, the following recognition of this informal sector, it is hoped that the proposed "Slum Development Corporation (SDC)" would help improve living standard of the slum dwellers, so far neglected.

It is true that accountability could be increased, if citizens can form social groups to discuss and express their views and if they could establish communications with decision-making authorities (World Bank 2008b). In a well-functioning governance system, many of these channels would ultimately influence the sector-specific service providers. Unfortunately, in most of the developing countries, provider managers are not motivated to do what citizens want or are unable to do what citizens want, because of lack of transparency and political commitments on the part of the government. The following recommendations are suggested in order to alleviate suffering of the slum dwellers:

a. Recognition of the urban slum dwellers as "Low-Income Community" by the government is a welcome move. DWASA should implement its future plan to bring all slums of the country under legal water connections at the earliest. Legalizing water services by this public sector organization would greatly contribute to national economy as the major portion of urban workforce comes from these slums;
b. In order to lessen water crisis in the slum areas, Community-Based Organizations (CBOs) could play a vital role in water supply and distribution. They are to be allowed to work free from the influence of local politicians, rent seekers and water *mafias*. "Local Water Committees" are to be formed by local representatives, local elites and other local stakeholders. Such committees are to ensure a healthy water governance in these slums so long they are relocated to a proper place with basic amenities;
c. "Ward Sanitation Task Force" existing in these slums to monitor CBO's activities is to be strengthened. Combined efforts by this organization, local government and DWASA could help providing quality water services in the slums;
d. Involving the private sector in providing safe drinking water in the slums could be thought of and introduction of private–public partnership could be a meaning way in resolving existing water crisis in the slums;
e. Growth of slums in Dhaka City is huge and it may not possible to find a solution for water problems within a short period of time due to chronic social problems being faced by these slums. National and local efforts are to be directed initiating feasible and effective plans and strategies;
f. Since the existing slums are on illegal lands and due to their very volatile nature, policymakers in the government find it very difficult to develop a long-term strategy for these slums. The "Slum Development Corporation (SDC)" set-up to help improve living standard of the slum dwellers could play an important role if provided with proper patronage by the local government authorities;
g. Quality of water being provided by DWASA needs to be improved. Regular monitoring could help ensure quality water services by this organization. Estab-

lishment of the office of "Water Ombudsman" could be a good idea in order to ensure regular supply of safe drinking water to the subscribers;

As the citizens of the country, the slum dwellers have every right to access safe drinking water as enshrined in the constitution of Bangladesh. They had to migrate to these unhospitable lands devoid of basic facilities like fresh air, water, ventilation, hygiene and a healthy environment. Mainly due to poverty, land erosion and river erosion, they landed up in urban ghettos for a living. They contribute greatly to urban workforce in transport, construction, ready-made garment and domestic work. Terming the slum dwellers as the low-income community is a laudable policy decision. Understanding contributions of these slum dwellers by the government would certainly pave the way for economic progress and prosperity, as the country is striving to achieve the SDGs by ensuring access to water and sanitation for all by 2030.

References

Adam A-W (2013) Perceptions of slum dwellers and municipal officials on factors impacting the provision of basic slum services in Accra, Ghana (Monograph). International Institute of Social Studies-Erasmus, The Hague, The Netherlands

Akhmouch A, Correia FN (2016) The 12 OECD principles on water governance—when science meets policy. Utilities policy, vol 43, Part A. Elsevier, December 2016, pp 14–20

Begum H, Moinuddin G (2010) Spatial dimension of social exclusion, an imperial investigation into the relationship of housing and social exclusion in the slums of Dhaka City. Manage Res Pract 2(3):314–328

Dagdeviren H, Robertson SA (2011) Access to water in the slums of Sub-Saharan Africa. Dev Policy Rev 29(4):485–505

DWASA Annual Report (2014–2015) Internet: https://dwasa.org.bd. Accessed on 27 June 2017

Hossain MA, Khan TH (2012) Poverty and public services: social exclusion of urban poor. Decelerated Decline, State of Poverty in Bangladesh, 2012, Unnayan Onneshan, Dhaka

Islam N, Mahbub AQM, Nazem NI, Angeles G, Lance P (2005) Slums of urban Bangladesh: mapping and census. Center for Urban Studies, National Institute of Population Research and Training, Dhaka, pp 25–26

Ministry of Water Resources (MoWR) (1999) National Water Policy 1999, Government of Bangladesh

Mitlin D, Satterthwaite D (2013) Urban poverty in the global south: scale and nature. Routledge, Oxon, UK

MoLGRD (2005) National sanitation strategy, February 2005, Local Government Division, Ministry of Local Government, Rural Development and Cooperatives, Government of Bangladesh

Moss T, Newig J (2010) Multi-level water governance and problems of scale setting the stage for a broader debate. Environ Manage 46:1–6. Available online at www.springerlink.com

OECD (2017) Principles on water governance. www.oecd.org. Accessed in June 2017

Olsson L, Head BW (2015) Urban water governance in times of multiple stressors: an editorial. Ecol Soc 20(1):27. https://doi.org/10.5751/ES-07300-200127

Rahman MM, Ahmed TS, Al-Hadi A (2015) Water pricing for slum dwellers in Dhaka Metropolitan area: is it affordable? (Monograph). University of Asia Pacific, Dhaka

Ranganathan M (2014) "Mafias" in the waterscape: urban informality and everyday public authority in Bangalore. Water Altern 7(1):89–105

Rashid SF (2000) The urban poor in Dhaka City: their struggles and coping strategies during the floods of 1998, BRAC. Disasters 24(3):240–253
Sakib HN, Islam TM (2014) State of water governance in Dhaka Metropolitan city of Bangladesh: evidence from three selected slums. Int J Interdisc Multidisc Stud 1(2):19–38. Available online at http://www.ijims.com
Subbaraman R, Laura N, Kiran S, Shrutika S, Tejal S, Mahesh N, Anita P, David EB (2015) Multidimensional measurement of household water poverty in a Mumbai Slum: looking beyond water quality. PLoS ONE 10(7):e0133241. https://doi.org/10.1371/journal.pone.0133241, http://dx.doi.org/10.1371/journal.pone.0133241
The Daily Star (2013) Dhaka's biggest slum to finally get legal water supply, Dhaka, 27 February 2013
The Daily Star (2016) Legal water for Dhaka slums next year, Dhaka, 5 September 2016
Tortajada C (2010) Water governance: some critical issues. Int J Water Resour Dev 26(2):297–307. https://doi.org/10.1080/07900621003683298
UNB (2016) Dhaka slums to have legal water supply by 2017, Dhaka, 5 October 2016
UNDP (2004) Water governance for poverty reduction: key issues and the UNDP response to millennium development goals. UNDP, New York, NY
UNDP-SIWI (2016) Water governance facility. Issue sheet, UNDP—Stockholm International Water Institute (SIWI), Sweden, August 2016
UN-Habitat (2003) The challenge of slums: global report on human settlements. Earthscan, London
UN-Habitat (2013) State of the world's cities 2012–13: prosperity of cities. United Nations Human Settlements Programme, Nairobi
UN-Habitat and WHO (2016) *Global Report on Urban Health: Equitable, Healthier cities for Sustainable Development.* WHO, Geneva
UNICEF (2011) Urban water challenges in Bangladesh, March 2011
United Nations (2010) Resolution A/RES/64/292. The human right to water and sanitation. In: Sixty-fourth general assembly of the United Nations, New York, 16 July 2010. United Nations, New York, 2010 (internet accessed on 27 June 2017)
Varis O, Biswas AK, Cecila T, Lundqvist J (2006) Megacities and water management. Water Resour Dev 22(2):377–494
World Bank (2008a) Bangladesh policies for mainstreaming strategic environmental assessment in the urban development of Greater Dhaka, No. 68831
World Bank (2008b) Deterring corruption and improving governance in the urban water supply & sanitation sector. A sourcebook, water working notes, December 2008, No. 46829

Chapter 23
Urban Water Management Issues and Challenges After the 2004 Indian Ocean Tsunami Recovery: Lessons Learned from Banda Aceh City, Indonesia

Alfi Rahman, Shimpei Iwasaki, Stephen Anthony Sutton, Aiko Sakurai and Parmakope

Abstract Banda Aceh city faced an urban water problem after the 2004 Indian Ocean tsunami devastated and damaged most of the city. Over 60% of the city area was destroyed, including the entire loss of some areas of land and massive impacts on infrastructure. Water supply infrastructure in particular was badly hit, with only

A. Rahman (✉)
Department of International Environment and Resources Policy, International Post Graduate Program in Human Security, Graduate School of International Cultural Studies, Tohoku University, 41 Kawauchi, Aoba-Ku, Sendai, Miyagi 980-8579, Japan
e-mail: alfi.rahman@unsyiah.ac.id

S. Iwasaki
Department of Environmental Science, International College of Arts and Sciences, Fukuoka Women's University, 1-1-1, Kasumigaoka, Higashi-Ku, Fukuoka 813-8529, Japan

S. A. Sutton
Department of Psychological and Clinical Sciences, Charles Darwin University, Darwin 226-0015, Australia

A. Sakurai
Department of Social Sciences, Toyo Eiwa University, Tokyo 226-0015, Japan

Parmakope
Urban Development Planning Agency, Banda Aceh, Indonesia

A. Rahman · S. A. Sutton
Tsunami and Disaster Mitigation Research Center (TDMRC), Syiah Kuala University, Banda Aceh 23233, Indonesia

S. A. Sutton
Bushfire and Natural Hazards Cooperative Research Centre, East Melbourne, Australia

A. Rahman
Faculty of Social and Political Sciences of Syiah Kuala University, Banda Aceh 23111, Indonesia

B. Ray and R. Shaw (eds.), *Urban Drought*, Disaster Risk Reduction,
https://doi.org/10.1007/978-981-10-8947-3_23

20% of the clean water available immediately following the event. Contamination of the groundwater by seawater was a crucial factor as the tainted seawater disrupted the clean water balance, which in turn damaged the facilities of the local public water utility company (Perusahaan Daerah Air Minum, PDAM). This loss of capacity came at a critical time, with a large number of disaster victims having a pressing need for water, for crucial hygiene as well as consumption. The reconstruction and rehabilitation phases of the disaster recovery sought to resolve the water resource problem for Banda Aceh citizens, with the restoration of PDAM becoming a top priority. This chapter highlights the urban water management issues and challenges of this situation and the problem of providing the clean water to Banda Aceh's citizens following the 2004 Indian Ocean tsunami.

Keywords Banda Aceh urban water management
The 2004 Indian Ocean tsunami · Clean water recovery

23.1 Introduction

Indonesia is located in Southeast Asia and is the world's fourth most populous country with more than 255 million inhabitants (BPS Pusat 2016). Indonesia is also recognized as the world's largest archipelagic country. According to Indonesian Agency for Meteorological, Climatological, and Geophysics, there are currently 13,466 islands that have been registered with valid coordinates (BMKG n.d.).

Indonesia's climate tends to be relatively even all year round with no extreme weather due to its position along the equator. The tropical equatorial climate dominated by high rainfall can be found on every major island in Indonesia, with an average of 2000 mm on most islands except Sunda. In terms of water resources, Indonesia has over 5590 rivers, but most of them short and steep. The combination of rainfall and streams makes Indonesia a country with relatively abundant water resources.

Across Indonesia, water utilities are supplied from rivers and lakes (60%), springs (25%), groundwater (15%), and rainfall harvesting (10%). The provision of fresh water supply services in Indonesia is dominated by the local company known as Perusahaan Daerah Air Minum (PDAM).

Even though Indonesia is rich in terms of water resource availability, most major cities still find difficulty in providing adequate supplies of clean and fresh water for their citizens. Some studies have indicated those factors which contribute to the national deficit in fresh and clean water country. These include rapid growth of population, the changes in land use, and industrialization (Hadipuro 2010). Environmental degradation including that caused by natural disasters has also contributed significantly to problems in the supply of clean water. One major example is found in the impact of the earthquake of M_w of 9.2 near Banda Aceh that triggered a tsunami, which destroyed a large proportion of the province's population and infrastructure.

This chapter uses the Banda Aceh examples as a case study to examine urban water management issues and challenges and identifies the main problems in the water management system for local water companies tasked with providing the clean and fresh water to the community. The case study reviews relevant literature to highlight the critical issues that arose from the tsunami event and the subsequent water supply challenges.

The primary data for this study is derived from publicly available information and data, which has been adapted and remodeled. A field study was also conducted comprising interviews with the relevant agencies focusing on their involvement after the 2004 Indian Ocean tsunami seeking to identify the critical issues in reconstructing of public water utility company.

23.2 Profile of the Case Study

Banda Aceh is located at the northern tip of Sumatra Island, Indonesia (BPS Kota Banda Aceh 2016). Astronomically, Banda Aceh is located between 5° 16′–5° 36′ north latitude and 95° 16′–95° 22′ east longitude; it is in the northern hemisphere (BPS Kota Banda Aceh 2016). The province of Aceh lies at the northwest tip of Indonesia. It has boundaries in north through the Strait of Malacca with Malaysia, to the south and the east with District of Aceh Besar and to the west with Indian Ocean. Banda Aceh is as the capital city of Aceh Province since hundred years ago (see Fig. 23.1).

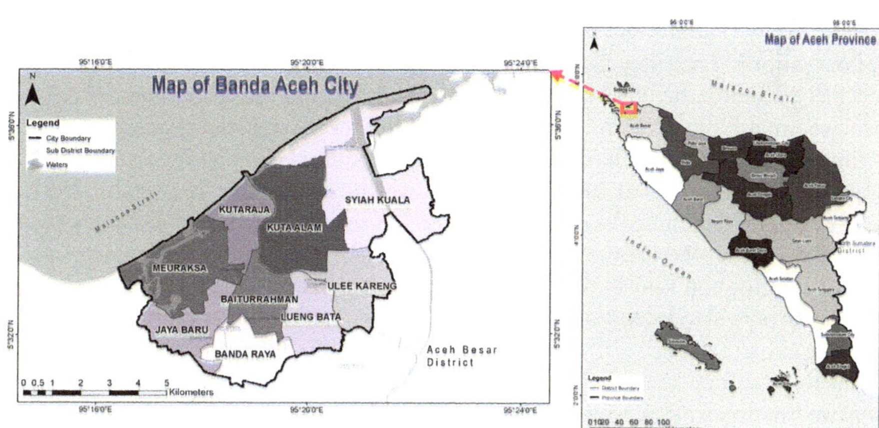

Fig. 23.1 Map of Banda Aceh city where the study was conducted. Adapted from BAPPEDA Kota Banda Aceh (2014)

According to 2016 statistics data released by BPS Banda Aceh, Banda Aceh city has a total population of 250,303 with 61,641 households (BPS Kota Banda Aceh 2016).

Prior to the 2004 Indian Ocean tsunami, Aceh had suffered a conflict between the Free Aceh Movement (FAM) who want to separate Aceh as new independent country and Indonesian Armed Military for more than 30 years. The long time conflict had ultimately cost at least 1258–2000 lives (Schulze 2004). As a consequence, the Indonesian government declared martial law and restricted travel. The violence overwhelmed the stability and security of the province. Delivery of health and education services was disrupted, and the economy languished behind the rest of the nation. There was also a direct and indirect impacts on infrastructure, first as roads, bridges, and water supplies became a military target and then as the conflict depressed the capacity to build and maintain essential services. Further, only limited research into the water supply was undertaken such that at the time of the tsunami and despite the heavy reliance on groundwater, only a limited understanding of the aquifers was available.

The combined result of the Aceh civil war was a significant lack of civil works infrastructure development in Aceh compared to other Indonesian regions, including water supply facilities. So, despite the fact that the province has a good natural supply of surface water, by 2004, most of the people of Aceh relied upon groundwater to supply their daily water needs.

This difficult situation was exacerbated by the tsunami on December 26, 2004; however, in addition to the physical impacts, the 2004 Indian Ocean tsunami also had a significant impact on the political situation in Aceh. The Former President of the Republic of Indonesia, Susilo Bambang Yudhoyono, himself a Former Head of the Indonesian Army declared the 2004 Indian Ocean tsunami a national disaster. Subsequently, 15,000 additional military personals deployed to Aceh for humanitarian relief operations. Their key tasks were to hasten recovery and burial of dead bodies as well as generally the clearing up debris (Masyrafah and Mckeon 2008). Not long after the tsunami, the FAM and Government of Indonesia signed a peace pact which was finally accepted by both parties and has successfully been continued to this day.

As noted above, 60% of Banda Aceh city was destroyed by the tsunami. Fortunately, this did not include the city's main access road which, heading east, was away from the coast and the area of tsunami impact. The road therefore remained open, but, as it ran through ranges that were contested by rebels, was not in good condition. The airport also inland of the city remained functional. Within two days—by December 28, 2004, access to Banda Aceh was formally opened by the government to allow relief efforts to enter. The tsunami attracted huge international attention and a large international effort to aid recovery and rehabilitation. Immediately access was opened to international non-government organizations (NGOs), and foreign government relief teams streamed in, together with local and Indonesian government and volunteers. It was expected that Banda Aceh should provide the logistical and leadership coordination for the relief of the province, but the widespread destruction and loss of life made it difficult for the city to fulfill these roles.

Any possibility of Banda Aceh functioning as the capital city relied on the provision of clean and safe drinking water to the inhabitants and incoming aid services. These services were stretched prior to the tsunami, and the provision of potable water became therefore the most important concern in the emergency and recovery process. While a general need existed for the supply of clean water for the daily needs of Banda Aceh's surviving citizens, there was a real and emerging threat posed by lack of water in the disaster zone on human health (Ferretti et al. 2010). This concern arises in most large-scale disasters but was exacerbated in Banda Aceh in 2004 by the fact that the tsunami directly impacted upon and dramatically reduced the already limited sources of clean water. As already stated, most of the citizens of Banda Aceh relied on groundwater. Most groundwater was accessed from local wells. This water supply was now contaminated, first by seawater and then by runoff containing sewage, debris, and bodies. The focus on the provision of clean water at a suitable scale to prevent a major secondary health disaster became urgent.

23.2.1 *Impacts of the 2004 Indian Ocean Tsunami*

The western coastal area of Aceh, including the city of Banda Aceh, was hardest hit by the 2004 Indian Ocean tsunami. The tsunami left enormous casualties and a trail of destruction across the province as well as other regions including the Nias Islands and North Sumatra, Indonesia (BRR 2009b). For practical purposes, Banda Aceh city was essentially totally destroyed, and the infrastructure turned into garbage heaps with bodies strewn everywhere. Situated on a broad plain, the tsunami had rushed into the city, inundating and smashing everything up to 6 km inland (see Fig. 23.2).

23.3 The Emergency and Recovery Phases

In April 2005, not long after the tsunami event, the Indonesian government established Bureau of Reconstruction and Rehabilitation for Aceh and Nias (BRR) (BRR 2009b). The head of BRR was appointed with ministerial level authority and mandated to coordinate the entire relief effort. According to BRR (2009b), the original role of BRR was to be a coordinator, where it would apply an overview of the entire reconstruction landscape to match demand for aid to the available resources. This remit was complicated by the vast area, the multiplicity of activities, and the sheer number of actors (BRR 2009a). The BRR was also tasked with the role of rebuilding and strengthening local government—a responsibility that was phased out in 2009.

Immediately after the event, the local government and NGOs implemented range strategies to respond to the clean water issue. Water was either processed on-site or brought in from a remote processing site, and it took times (Oliver 2008). The water that was processed on-site used numerous filters and technologies such as the reverse osmosis system in providing clean water for the victims (Oliver 2008). Unfortunately, this method was not able to sustain the volumes of water required consistently.

The wells within the immediate area of inundation, which covered 60% of the land area around Banda Aceh, was contaminated with the saline water from the tsunami (see Fig. 23.2). It was possible to restore wells, but this required them to be sunk to greater depths, between 30 and 150 m (Oliver 2008).

While some of the contamination was of a temporary nature, it nevertheless took time for the water quality to return to normal, with obvious important consequences to the inhabitants who require water daily for survival. The time frame for the flushing of the saltwater contamination was uncertain due to the limited understanding of the local aquifers and the physical properties of the city's groundwater wells. During the emergency phase, no guidelines existed to help people understand or make decisions about how, where, or when to harvest clean water from their normal water supply sources.

According to Jaskolowski (2008), the major factor affecting drinking water quality in most the tsunami-inundated areas was the high salinity and consequent high sodium and chloride concentrations. In shallow groundwater, the contamination was attributed to the saltwater inundation directly caused by the tsunami, while in the deep groundwater cation exchange was occurring (Jaskolowski 2008).

Villholth et al. (2005) also conducted a study related to the impacts of the 2004 Indian Ocean tsunami to the groundwater in the east coast of Sri Lanka. This study

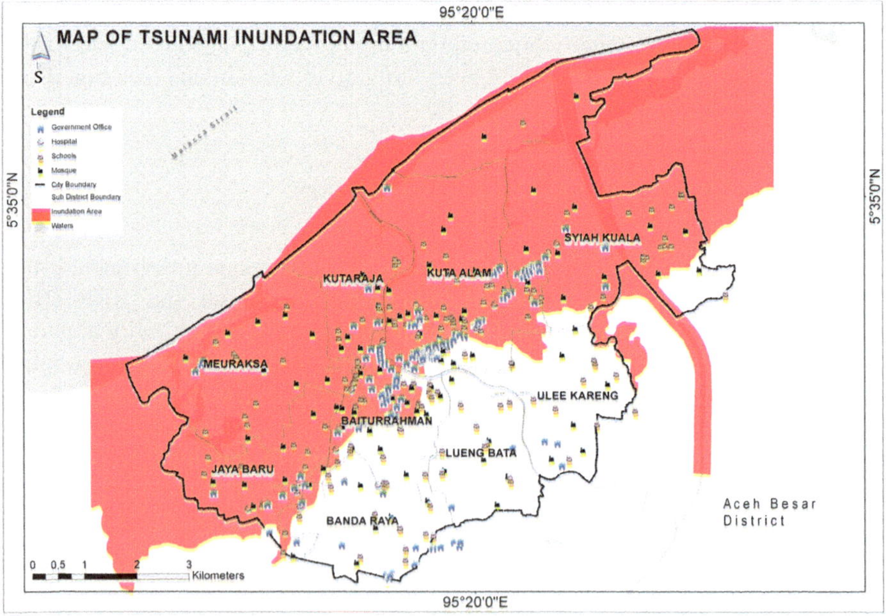

Fig. 23.2 Map of tsunami inundation area in Banda Aceh city. Adapted and modified from BAPPEDA Kota Banda Aceh (2014)

indicated that the well water which was contaminated by seawater varied significantly between inundated areas due to the different inundation patterns, soils, and the characteristics of well itself. The study also described the importance of a monitoring period indicating the wells which could still function for domestic water supply. In some cases, the salinity in the inundated wells persisted up to 7 months after the tsunami (Villholth et al. 2005). Only limited studies were undertaken to determine the impact of the 2004 Indian Ocean tsunami in the Banda Aceh area. This situation contributed to the uncertainty of the status of contaminated groundwater area and whether it was safe to consume or not.

23.4 Issues and Challenges on Water Supply Management System Through Build Back Better Effort

The water supply management system in Banda Aceh city is operated by a local Water Utilities Company named PDAM Tirta Daroy. The name of PDAM Tirta Daroy is derived from the name of the Krueng Daroy River, which originated from the Mata-ie mountains and flows through Banda Aceh city to the Indian Ocean.

PDAM Tirta Daroy was established by law No. 2/1975 by the Governor of Aceh and has been operating as the local company for the provision of clean water to Banda Aceh since 1975. Law number 32/2004 of the central government also mandated that the local water utility company manage and provide clean water for citizens to increase their prosperity. Even though the company has been established for more than 40 years, the lack of infrastructure improvements described above meant that the company faced some serious issues and challenges.

Generally, the long history of conflict between FAM and the Indonesia military impeded development opportunities in Banda Aceh city and made it difficult for the government to invest in the development of water infrastructure, both in absolute terms and compared to other Indonesian cities.

The legal framework of Indonesia drinking water regulation is set out in law No. 7/2004 relating to the water sources. More specifically, the drinking water system is controlled by Government Regulation No. 16/2005. The Indonesian drinking water regulation is implemented by several related ministries: The Ministry of Public Works handles the strategy and technical matters, while the Ministry of Health ensures drinking water quality. The Ministry of Home affairs provides guidance on the management of the resources, and, finally, the Ministry of Finance oversees the financing aspects of water supply (see Fig. 23.3). This complex framework of ministries occurs in the central government, while local governments are actually responsible for the running and supplying of drinking water. Despite this complexity, it is possible for a proper coordination system for technical tasks to occur at the local level. The interesting thing about the framework is the opportunity it provides for the participation of experts, academics, and stakeholders in supervising and monitoring the water supply. While this opportunity exists, it has not yet been

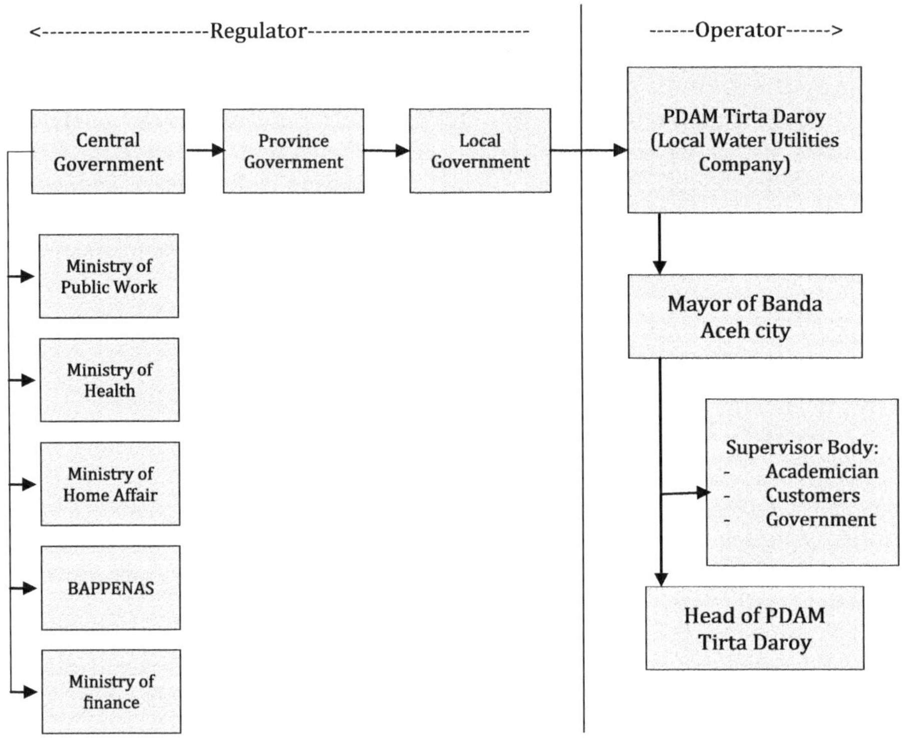

Fig. 23.3 Framework of PDAM Tirta Daroy as local water utility company in Banda Aceh city. Adapted from Subekti (n.d.)

capitalized upon, and the framework has not worked properly, especially in the area of participatory involvement of academic experts, customers, and others.

The Banda Aceh city and BRR paid special attention to the recovery of PDAM Tirta Daroy as the peak utilities water company, hoping that it could provide the city with clean water but also maintain the water balance for the maintenance of the environment. The rehabilitation of PDAM Tirta Daroy was also hoped to build capacity to adapt to the predicted changes brought about by both the development of the Banda Aceh city as an urban area and climate change.

The increasing number of inhabitants and changing of water supply needs in Banda Aceh as industries and trade develop in the area has placed additional requirements on the water supply. From this point of view, PDAM Tirta Daroy is one of the main actors in the future development of the city and its economy, with water supply establishing a fundamental limit on the potential for growth in Banda Aceh city.

Figure 23.4 shows the state of water supply pipeline infrastructure in 2007 and subsequent construction of new pipelines in 2010 and 2015. Figure 23.4a sets out the pipelines in existence in 2007. Comparison with Fig. 23.4b, c shows that a significant amount of new pipeline was installed during the reconstruction and rehabilitation

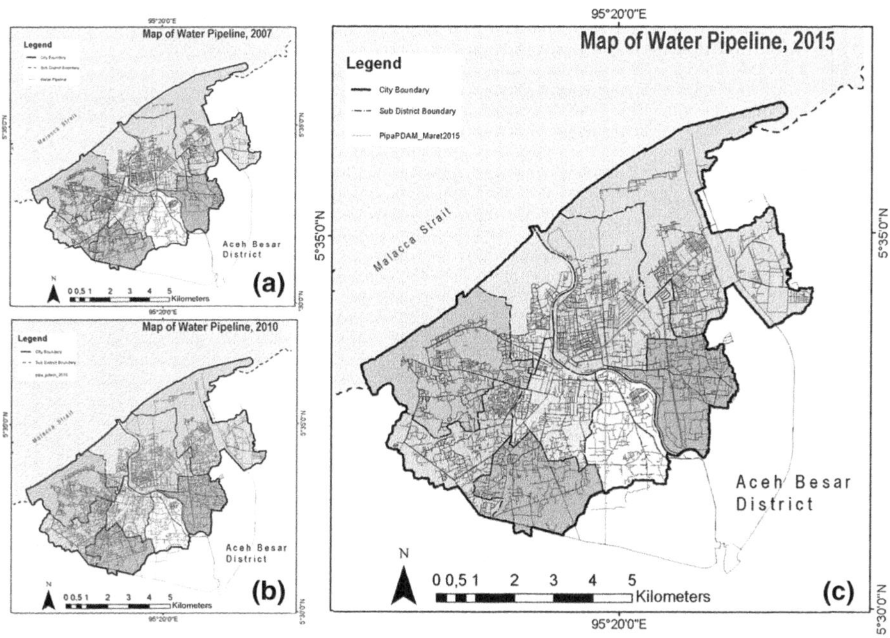

Fig. 23.4 Development of Banda Aceh water pipeline after the 2004 Indian Ocean tsunami, **a** 2007, **b** 2010, and **c** 2015. Adapted and modified from BAPPEDA Kota Banda Aceh (2014)

phase (up to 2010). This is clear in areas such as Syiah Kuala sub-district in the north of Banda Aceh. The pace of infrastructure development slowed between 2010 and 2015 with Fig. 23.4c portraying a pipeline network largely unchanged from 2010.

This increase in pipeline infrastructure during the reconstruction phase provides the basis for data released by PDAM in 2014 which shows the service coverage of PDAM Tirta Daroy reached 94.26% of the population. This exceeded the standard of Millennium Development Goals (MDGs) of 68.87% which benchmarks every person's minimum requirement for clean water per day at 60 L or 0.06 m^3, a figure mandated in the law of number 14/2010 by Ministerial Decree.

Even though, in terms of the infrastructure, the water supply coverage area has improved significantly, data released by BPS Banda Aceh 2016, somewhat surprisingly, shows only 59.38% of the households identified to use the PDAM Tirta Daroy services, while 37.56% use wells, 2.88% rely on deep ground-water pumps and 0.18% on commercial refill water kiosks. This discrepancy (between PDAM data and BPS data) in part arises from the fact that the local water utility did not pay adequate attention to the actual quality of the pipelines installed. BAPPEDA Banda Aceh city suggests that this and a low level of competent staff involved in the installation, recording, administration, payment, technology, and systems impede the effectiveness of the water supply system, and staff integrity may also be a problem (BAPPEDA Kota Banda Aceh 2015).

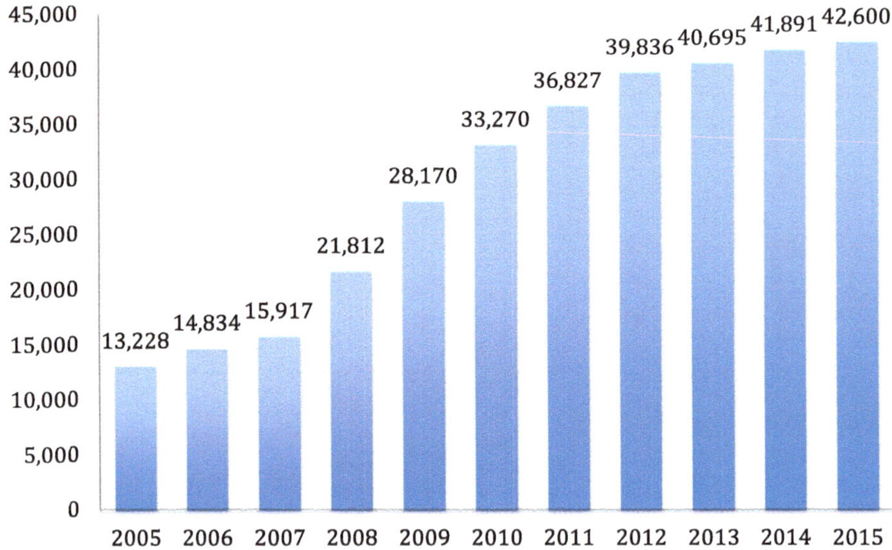

Fig. 23.5 Number of customers of PDAM Tirta Daroy by year. Adapted from PDAM Tirta Daroy (2015)

Figure 23.5 presents data about the number of PDAM Tirta Daroy subscribed customers. There has been increasing number of subscribed customers after the 2004 Indian Ocean tsunami during the reconstruction and rehabilitation phase. In 2015, there were 42,600 households subscribed as PDAM Tirta Daroy's customers, approximately 75% of the total number of households in the city. This data also shows that not all of the subscribed customers are using the PDAM services continuously.

According to a PDAM Tirta Daroy 2013 report, the percentage of water lost from the supply system reached 44.4% (PDAM Tirta Daroy 2016). This situation was exacerbated by a low level of billing efficiency due to customers who had initially received water assistance after the 2004 earthquake and tsunami remaining unregistered as subscribers nine years later. Further, nearly ten years after the disaster, many water meters remained broken resulting in the capacity to accurately bill those consumers who were actually registered remaining very low. It is not clear to what extent damaged pipes and consequent leaks contribute to the total loss of water.

Since 2013, the management system of PDAM Tirta Daroy has focused on solving the water loss problems by conducting data collection and moving to comprehensive control measures to ensure that both official and illegal water supply connections by PDAM water meet their payment obligations. Other efforts to reduce water losses include efforts to re-engage with the wider customer base through increased promotion, establishing branch offices and increasing the number of counterparts to facilitate registration and completion of contributions for customers as well as the imposition of a system of penalties abuses of the water supply system.

The data in Fig. 23.5 suggests that this has not been entirely successful and may in part contribute to the 25% 'gap' between households and subscribers. For example, there was a rapid increase in the number of subscribers from 15,917 in 2008 to 33,270 in 2010, that is, a doubling of subscribers in three years. In contrast, during the last four years from 2012 to 2015, the number of subscribers has only increased by 7% from 39,836 to 42,600. Despite this decreased rate of capture of new subscribers, it is technically possible for PDAM Tirta Daroy to connect all households in Banda Aceh within the next ten years. However, it seems clear that the increase in subscribers relates to the increase in pipeline installation that occurred at the same time, and the subsequent slowing of the roll-out of new infrastructure is likely to mean increasingly fewer new subscribers in future years. In addition, the cost structure for water supply remains a disincentive to connect all households, particularly the poorest inhabitants of the city.

According to the review report released by BAPPEDA Banda Aceh city in 2011 discussing factors affecting poverty levels in Banda Aceh, the lack access to clean water was a significant contributor to an increase in poverty (BAPPEDA Kota Banda Aceh 2011). There is a 'vicious cycle' where the poorest in society have to pay high prices for water to third-party suppliers in order to meet their daily needs for potable water. The implications for sanitation and health affect other services provided by the government and limit the growth potential of the city and its economy as a whole.

23.5 Other Issues and Challenges

Surprisingly, the increase in the water supply business has not only been dominated by the local government company such as PDAMs or the big bottled water companies but has also seen significant growth in the use of refilled bottled water kiosks. These can easily be found in all areas and cities in Indonesia, including in Banda Aceh. Most of the small-scale water providers rely on groundwater or shallow wells as sources in their effort to fill the gaps in PDAM coverage (Hadipuro 2010). These issues and their temporal context are set out in Table 23.1.

As stated, the current water supply situation does not appear to be sustainable. The lack of the study of the totality of the water supply in Banda Aceh combined with poor definition of a strategy encompassing clear roles and responsibilities means that the water supply is unlikely to be able to support any significant growth in the Banda Aceh economy.

Some problems arise from these various attempt to fill the water supply gap. One important consideration is the inadequate monitoring of water quality from the refill kiosks. Another is the lack of consideration of water balances for the future development of Banda Aceh city and in particular the challenge of maintaining groundwater sustainability.

A priority strategy for action should be integrated to ensure the future availability of water resources for humanitarian as well as economic reasons. Any long-term strategy for sustainability of Banda Aceh's water supply must take into consideration

Table 23.1 Critical issues on the provision of clean water after the 2004 Indian Ocean Tsunami

Emergency	Rehabilitation and reconstruction phase	Current issues
2004–2005	2005–2009	2009–now
• The emergency operation paid considerable attention to the need for providing the clean water • Cleaning and renovating local wells which were contaminated with the seawater • Installation of emergency osmotic filter treatment plants	• "Build back better" concept introduced by BRR improved water pipeline infrastructure • The lack of integration of the Banda Aceh blueprint after the 2004 Indian Ocean tsunami to the water management issues • The detailed assessment of water supply system identifies serious shortcomings in infrastructure and management	• Water management and decision-making do not effectively implement findings of reviews or take into account the needs customers • Disincentives for the water supply utility to connect poor households • Regular tests and certification on the quality of PDAM water • Regular test and evaluation for all of small-scale water supply providers

environmental water balances. Regulations need to be strengthened to ensure the clean water framework which includes the right of access to clean water for all citizens; this must include a system of monitoring and assessing water quality and safety from all sources.

The human resource capacity of PDAMs needs to improve. This might assist in understanding the sources of water losses (currently nearly half the capacity) and contribute to an improvement in the performance of the overall business. Improvements need to be made in the monitoring of the overall water business, to improve understanding of the situation by both policy-makers and the general public. Better data is required with greater promotion of information through public awareness campaigns and groundwater balance plans.

23.6 Conclusions

The situation with urban water supply prior to the 2004 Indian ocean tsunami was unstable and unsustainable, in large part due to lack of investment in infrastructure and maintenance because of the civil war. The catastrophic tsunami led to a major focus on rehabilitation of the utility company, PDAM Tirta Daroy, and the water supply itself. Significant progress was made during the reconstruction phase, but progress has slowed or stalled, threatening the social wellbeing of the citizens of Banda Aceh, particularly the poor, as well as the potential for economic growth in the city.

In summary, the urban water management issues and challenges of after the 2004 Indian Ocean Tsunami in Banda Aceh city are:

1. Before the 2004 Indian Ocean tsunami, the long-term conflict between Free Aceh Movement (FAM) and Indonesian armed military had positioned Banda Aceh city with poor water utilities infrastructure and management compared to other cities in Indonesia;
2. The 2004 Indian Ocean tsunami was in many ways revelatory of structural problems and dysfunction in Banda Aceh society (Solway 1994). This was nowhere more evident than in Banda Aceh's water supply system;
3. After the 2004 Indian Ocean tsunami, BRR immediately focused on the provision of clean water to the victims, but these efforts were not fully integrated (in part due to the extent of tsunami impact itself) and because each agency or donors interpreted the form of water supply issue based on the capacity interests of the institution;
4. In the emergency phase, a lack of coordination among international agencies and NGOs and local NGOs was evident. Donors interpreted or chose the methods they adopted. There were no manuals for handling and provision of clean water in the emergency phase, and these are still not yet available;
5. In the rehabilitation and reconstruction process, several approaches were adopted to improve the capacity of water resource management and supply. However, the actual assessment of the water supply situation in Banda Aceh uncovered deeper problems with infrastructure and some even more severe concerns around the lack of financial and managerial capacity;
6. There were no clear mechanisms or definitions of roles and responsibilities for or by local government for the provision of water to the community;
7. Notably, access to clean water by the poor was significantly lacking, threatening social disruption as well as potential health epidemics and reinforcing poverty.

References

BAPPEDA Kota Banda Aceh (2011) Analisis survey validasi TN2PK Kota Banda Aceh, Banda Aceh

BAPPEDA Kota Banda Aceh (2014) Tsunami data project. BAPPEDA, Banda Aceh

BAPPEDA Kota Banda Aceh (2015) Hasil pendataan sarana sambungan rumah PDAM

BMKG (n.d.) Indonesia Memiliki 13.466 Pulau yang Terdaftar dan Berkoordinat. Retrieved 4 Apr 2017 from http://www.bakosurtanal.go.id/berita-surta/show/indonesia-memiliki-13-466-pulau-yang-terdaftar-dan-berkoordinat

BPS Kota Banda Aceh (2016) Banda Aceh dalam Angka 2016. BPS Kota Banda Aceh

BPS Pusat (2016) Statistical Yearbook of Indonesia 2016. In: Sub-directorate of statistical compilation and publication, Ed. BPS-Statistics Indonesia

BRR (2009a) 10 management lessons for host governments coordinating post-disaster reconstruction. BRR NAD-NIAS

BRR (2009b) Tsunami: from disaster to the emergence of light. In: Supit H, Agusta M (eds) BRR Book S. BRR NAD-NIAS

Ferretti E, Bonadonna L, Lucentini L, Della Libera S, Semproni M, Ottaviani M (2010) A case study of sanitary survey on community drinking water supplies after a severe (post-Tsunami) flooding event. Annali Dell'Istituto Superiore Di Sanita 46(3):236–241

Hadipuro W (2010) Indonesia's water supply regulatory framework: between commercialisation and public service? Water Altern 3(3):475–491. Retrieved from http://www.scopus.com/inward/record.url?eid=2-s2.0-78649470842&partnerID=40&md5=94f803089b22323c980449ff24044883
Jaskolowski B (2008) Hydrogeochemistry of groundwater along the tsunami affected coast between Sigli and Bireuen, Northern Sumatra, Indonesia. Institut für Geologie
Masyrafah H, Mckeon JMJA (2008) Post-tsunami aid effectiveness in Aceh, proliferation and coordination in reconstruction. Retrieved from http://www.brookings.edu/~/media/research/files/papers/2008/11/acehaidmasyrafah/11_aceh_aid_masyrafah.pdf
Oliver V (2008) After the tsunami: a project on design action in Banda Aceh. RMIT University. Retrieved from http://researchbank.rmit.edu.au/view/rmit:13377
PDAM Tirta Daroy (2015) PDAM Tirta Daroy company profile, Banda Aceh
PDAM Tirta Daroy (2016) Peningkatan pelayanan air minum kota Banda Aceh, Banda Aceh
Schulze KE (2004) The free Aceh movement (GAM): anatomy of a separatist organization. Policy Stud (2):I
Solway JS (1994) Drought as a revelatory crisis: an exploration of shifting entitlement and hierarchies in the Kalahari, Botswana. Dev Change 25(3):471–495
Subekti (n.d.) Challenges & opportunities water supply sector in Indonesia. Indonesian Water Supply Association
Villholth KG, Amerasinghe PH, Jeyakumar P, Panabokke CR, Woolley O, Weerasinghe MD, Pathirana SRK (2005) Tsunami impacts on shallow groundwater and associated water supply on the east coast of Sri Lanka. International Waste Management Institute, Nov 2005. https://doi.org/10.1201/9780203894569

Chapter 24
Implications of Water Insecurity and Future Prospects in Asian Cities

Bhaswati Ray and Rajib Shaw

Abstract The world has witnessed unprecedented urban growth since the beginning of the twenty-first century, concentrated mostly in the less developed countries of Asia. Cities are particularly vulnerable to water scarcity and water stress, being the centres of economic growth and population concentration. There is also an increased demand for water in order to maintain the intense pace of activities in urban areas. This chapter intends to focus on the problems of water insecurity in the cities of Asia and to highlight some of the best practices adopted in urban water governance. As cities grow in terms of population and diversity of urban functions, demands on the urban water systems rise and hence the existing water systems and available water resources are faced with daunting and multifarious challenges as they are often exploited beyond their sustainable limit. Communities in urban areas are hence seeking resilience in the existing urban water systems and to future uncertainties in water supply because of climate change and population growth. An attempt is hence made for integrated water management approaches involving different stakeholders instead of compartmentalized management of urban water systems. It is already evident and documented that the existing water management approaches are inadequate to address water insecurity and water stress. Transforming urban water systems into more resilient and hence more sustainable systems would require innovative approaches. These approaches have also been discussed under two broad categories—the infrastructure-based approaches and the system-based approaches.

Keywords Resilience · Innovative · Infrastructure based · System based
Water management

B. Ray (✉)
Sivanath Sastri College, University of Calcutta, Kolkata, India
e-mail: bhaswati173@gmail.com

R. Shaw
Graduate School of Media and Governance, Keio University, Fujisawa, Japan

B. Ray and R. Shaw (eds.), *Urban Drought*, Disaster Risk Reduction,
https://doi.org/10.1007/978-981-10-8947-3_24

24.1 Introduction to Water Insecurity in Asian Cities

The world has witnessed unprecedented urban growth since the beginning of the twenty-first century. This growth is concentrated mostly in the less developed countries of Asia. Despite its low level of urbanization, 54% of the world's urban population is found in Asia and the percentage share is expected to reach 66% by 2050. Nearly 93% of the growth is expected to be concentrated in the less developed countries with more than 60% in the cities of Asia. It is also the million-plus cities that are witnessing the most phenomenal urban growth under the impact of forced or voluntary migration from rural areas to urban areas in search of better employment opportunities, better living conditions and better access to education and health services. Along with climate change impacts, the rapid pace of mass urbanization or pseudo urbanization is putting cities under increasing water insecurity and water stress. The megacities have also grown in numbers from 10 in 1990 to 28 in 2014, accounting for nearly 12% of the world's urban dwellers. Sixteen of these megacities are located in the countries of Asia.

Cities are particularly vulnerable to water shortages under the twin impact of rapid urban growth and reduced availability of fresh water. There is also an increased demand for water in order to maintain the intense pace of activities in urban areas. With the subsequent growth of cities, the demands for water also increase, driven on the one hand by the rapid increase in urban population and on the other hand by the increased percentage of urban dwellers using municipal water services. Urbanization and changing lifestyles because of a higher standard of living further increases per capita water use, as the use of modern amenities in the form of showers, washing machines and dishwashers leads to a substantial increase the residential use of water. Cities also concentrate the sources of greenhouse gases caused by anthropogenic carbon emissions that accelerate global warming and impact the hydrological cycle thereby further increasing the water stress in cities. Increase in impervious surface and consequent depletion of groundwater reserves, increased water pollution from city waste and industrial affluent decrease the availability of finite water resources. Choked water bodies and urban drainage systems increase the vulnerability of cities to floods. Cities thus spatially concentrate the water demand of millions within a limited area and also contribute to increased frequency and intensity of water stress, urban flood, global warming and droughts. Urban areas also account for huge water footprint because of the consumption of water-intensive food and high-energy nexus. Climate change impacts are further expected to increase water insecurity by altering rainfall patterns while making cities increasingly vulnerable to major disasters including frequent floods, severe droughts, hurricane, storm surges and landslides that interfere with sustainable development, poverty alleviation, water security and other prominent goals in the region.

The book deals with the various aspects of urban water insecurity, concepts and relevance with special reference to urban water insecurity in Asian cities. The first section discusses the concepts of urban water insecurity and the implications of climate change on urban drought. This section would also explore the parameters for developing urban water security index. The second section deals with various case studies from Asia covering varied dimensions of urban water insecurity and the good practices and adaptive policy responses to manage urban water crisis.

24.2 Urban Drought and Implications of Climate Change

Urban areas concentrate population, economic activities and built environment over a relatively small area and use a disproportionate share of resources making the cities key drivers of global environmental degradation. Urban areas contribute to global warming and heat island effect with anthropogenic carbon dioxide emissions when fossil fuels are burnt for heating and cooling, in industries and for transportation. Clearing of land for the growth and expansion of cities, for the development of transport network and increased intensity of urban activities since historic times are responsible for regional land use change, deforestation and the reduction of natural sinks like wetlands. Urban areas also alter the hydrological cycle because of increased surface run-off and reduced infiltration. It results in groundwater depletion and increased water stress, aggravated by a higher frequency of extreme climatic events, greater variability of monsoon rains, floods and endemic drought due to erratic precipitation pattern and the increased chances of water contamination during such events, sea-level rise and saltwater intrusions. More frequent and high-intensity rainfall caused by global warming is already apparent in streamflow records over many decades. Behavioural and lifestyle changes imposed upon by climate change in urban areas including the increased demand for artificial heating and cooling and increased mechanization that result in increasing water use and overexploitation of groundwater also impact urban water use along with unplanned urban growth in vulnerable areas, impaired ecosystem services and altered hydrological cycle.

Hence urban drought or urban water scarcity is caused by climate change impacts resulting in physical water scarcity or a decline in water resource availability and socio-economic scarcity brought about by inadequate access to water resources and inequitable power distribution (Table 24.1). While policy experts often focus on the issue of physical scarcity and estimates of water availability to explain water insecurity, it is often more than a technocratic response. Both geographical limitations like location in dry climates or far from water sources as well as financial limitations, with the poor cities unable to construct robust urban water infrastructure increases their vulnerability to urban water insecurity, at a time when the rationale for delivery of water services in a non-discriminatory manner is well documented. Institutional, operational and financial causes of water insecurity shape the response patterns that differ in the rural and urban context.

Urban areas are seen to battle urban water insecurity through controlled urbanization, using alternate water sources like rainwater harvesting, desalinization, groundwater recharge, introduction of sewage treatment plants, improving water infrastructure by replacing old worn-out pipelines and extending the water supply network up to the fringe areas, improved water governance and cost recovery through water

Table 24.1 Urban water insecurity: issues and challenges

Urban water insecurity	Issues and challenges	Implications
Physical scarcity	• Low rainfall and arid climate • Water pollution • Reduction in water bodies • Altered hydrological cycle • Saline intrusions	Physical water scarcity or water stress
Climate induced	• Climatic variability • Extreme events • Altered precipitation pattern • Climatic events like El Nino • Heat waves and droughts	Physical water scarcity or water stress
Water governance	• Outdated infrastructure • Leakage loss • Competing water use • Inequitable distribution of water • Inadequate accessibility and affordability	Socio-economic water scarcity or water insecurity
Urbanization	• Rapid population growth • Increased per capita water use • Land-use changes • Choked water bodies • Soil sealing	Physical and socio-economic water scarcity
Disaster risk	• Increased frequency of floods and droughts • Contamination of groundwater • Groundwater depletion • Sea-level rise and coastal flooding • Water footprint	Physical and socio-economic water scarcity

Source By authors

metering, reduced greenhouse gas emissions, green infrastructure like rooftop gardens, permeable pavements, underground water detention systems and improved disaster preparedness. Examples are plenty even in Asian cities.

24.3 Assessing Urban Water Insecurity: Defining the Urban Water Security Index

Water scarcity is the inability to meet the demand for water in an area due to the unavailability of sufficient amount of water. Water insecurity, on the other hand, refers to the denial to use adequate quantities of safe pollution-free water to meet the basic human functions and local ecosystem services as well as an increased risk of water-linked disasters. The two main challenges facing urban areas are the inadequate access to water supply and sanitation facilities and increased frequency and intensity of extreme weather events with adverse consequences on economic growth, health and well-being. Water insecurity implies the denial of access by the marginalised sections of the population and is often the result of poor water governance not necessarily linked to severe scarcity or lack of availability. Ensuring water security is thus a defining global challenge in the twenty-first century, and involves, in addition to having enough water, the mitigation of environmental risks like floods and droughts caused by excess or lack of water, addressing conflicts over shared water systems, reducing stress amongst the different stakeholders and competing uses of water. It is thus embedded in various development issues including food security, social equity and environmental sustainability.

Various attempts have been made to measure and define urban water insecurity. The Falkenmark indicator developed in 1989 is one of the most accepted methods of water scarcity and water stress. It is defined as the fraction of the total annual run-off available for human use (Charkhestani 2015), using a per capita estimate of water requirement. No stress conditions prevail when the annual per capita availability of water is more than 1700 m^3 (Water Strategy Man 2004). Water stress occurs when the amount of water available per head is between 1000 and 1700 m^3 per annum. Water scarcity implies an annual per capita availability between 500 and 1000 m^3. When the available amount of water is less than 500 m^3 per person per year water conditions are one of absolute scarcity. The indicator, however, does not take into account the socio-economic and institutional dimensions that aggravate the water scarcity caused by climate variability and physical water scarcity. An integrated approach to water scarcity parameters was suggested by Vorosmarty et al. (2010) bringing together the physical and socio-economic dimensions. The International Water Management Institute (IWMI) has also assessed both physical water scarcity and economic water scarcity. The Water Poverty Index (WPI) (Sullivan 2002, Lawrence et al. 2002) developed by the Centre for Ecology and Hydrology (CEH), Wallingford further tries to assess the connection between water scarcity issues and socio-economic conditions (Water Strategy Man 2004). It ranks countries according to the provision of water, combining five components (Water Strategy Man 2004). The components include resource availability, access, use, capacity and environment. Each of these components is derived from two to five indicators which are normalized to a scale from 0 to 1 (Water Strategy Man 2004). Vörösmarty et al. (2010) and the Asian Development Bank (2003) assessed global threats to human water security and biodiversity based on twenty-three indicators. The SIPE approach to water scarcity was developed by Abedin and Shaw (2014) covering the socio-economical, institutional, physicochemical and environmental dimensions of water scarcity. Each dimension consisted of five primary indicators (Abedin and Shaw 2014). Each primary indicator was made up of 5 secondary indicators amounting to a total of 20 primary and 100 secondary indicators. Authors have demonstrated that the inclusion of five dimensions (physical, social, economic, institutional and natural) (Abedin and Shaw 2014) is essential to measure resilience including urban water resilience (Joerin and Shaw 2011). The Arcadis Sustainable Cities Water Index assessed 50 global cities on issues impacting the resiliency, efficiency and quality of the urban water sector and their impact on long-term sustainability. Indicators were chosen for the three categories encompassing physical availability, socio-economic parameters and water governance and included among others fresh water withdrawn as percentage of total available water, number of water-related disasters like flood, droughts, percentage of city area covered by green spaces, water charges, proportion of water lost in transit, households having access to safe drinking water, households with metered supply, hours of supply, wastewater reuse and the incidence of water-related diseases. The water security dimensions proposed by the Asian Development Bank 2016 to determine water security status cover five dimensions of water security index (WSI): WSI1or basic water (renewable, supply, sanitation), WSI2 or sufficient water (water supply, consumption, agricultural water), WSI3 or development water (irrigation area, industrial water use, water for energy, water for aquaculture), WSI4: water disaster (loss from floods and drought) and WSI5 or water for future (population growth, urban population growth, water footprint) (Asian Development Bank 2016).

Based on the concepts and indicators that define and measure water insecurity, an attempt has been made in this book to identify the parameters and indicators that may be used to assess water security in urban areas and resilience in the urban water system (Ray and Shaw 2016). Apart from the physical scarcity of water, cities in Asia also suffer from water insecurity, often magnified by improper water management. Hence in the listing of indicators, emphasis has been given to institutional dimensions of urban water security. Fifty indicators have been chosen under the physical, socio-economic and institutional dimensions of urban water security. These indicators are to be normalized on a scale from 1 to 5, ranging from very poor or not available score of 1 to a best score of 5. In addition, all five variables within a parameter would have to be ranked (W_1, W_2, W_3, W_4, W_5) between each other in the range of not important (1) to very important (5) in order to give a particular variable a higher or lower weightage in the calculation of the aggregate scores (Joerin and Shaw 2011). The constant use of five choices, ranks and weights allows the adoption of a formula based on weighted mean (Joerin and Shaw 2011) (Eq. 24.1, Formula for weighted) to calculate the urban water security for each indicator, parameter and dimension (Abedin and Shaw 2014). Cities may then be ranked according to the aggregate-weighted mean index combining all the components. The dimensions and indicators used may be modified to suit local conditions to ensure global applicability of the dimensions and parameter/indicators.

$$\frac{\sum_i^{n=1} W_i X_i}{\sum_i^{n=1} W_i X_i} = \frac{W_1 X_1 + W_2 X_2 + W_3 X_3 + W_4 X_4 + W_5 X_5}{W_1 + W_2 + W_3 + W_4 + W_5} \tag{24.1}$$

24.4 Learning from Asian Cases

Already identified as the world's driest continent in terms of per capita water availability, severe droughts are common over vast regions of Asia, extending from southern Vietnam to central India. With the annual per capita availability at 3920 m^3, the availability of freshwater is less than half the global annual average of 6380 m^3 per inhabitant. Asian cities are the worst hit under the impact of rapid and mass urbanization. Increased demand because of a growing urban population and lifestyle transformations along with climate change-induced water shortage is already putting pressure on the water management systems in the water-scarce urban areas. The region is also vulnerable to disasters, often water linked disasters like endemic floods and droughts, hurricane, storm surges and landslides but continues to be inadequately prepared. Hence the Asian cities are embracing new approaches to urban planning for a water-secure future. This chapter intends to focus on the problems of water insecurity in the cities of Asia and to highlight some of the best practices adopted in urban water governance. The Asia-Pacific Water Forum anticipates that climate change impacts will impose additional threat on the already vulnerable countries in the Asia and the Pacific region, challenging the very concepts of sustainable development, poverty alleviation and improved water security. Water-related natural disasters are also common in the Asia-Pacific region.

The case studies have been selected to cover all aspects of urban water insecurity and thus highlight key-learning experiences (Table 24.2). The selected cities range from megacities with a population above 10 million that include Delhi and Kolkata in India, Dhaka in Bangladesh and Metro Manila in the Philippines. There are million-plus cities with a population exceeding 1 million. Such cities include Tehran in Iran, Darjeeling, Nagpur and Thane in India, Kathmandu in Nepal, Peshawar in Pakistan and Ulaanbatar in Mongolia. The medium-sized towns include those towns that have a population 1 lakh and above. Colombo, Kandy, Galle in Sri Lanka, Banda Aceh city, Indonesia, Iloilo and Cebu in the Philippines, Udon Thani in Thailand fall in this category. Small towns with a population of less than a lakh include Yanagawa in Japan and Galle in Sri Lanka.

Many of the chapters highlight the various aspects of urban water insecurity and water stress. Water stress and urban drought resulting from physical scarcity of water and climate-induced uncertainties and water stress have been dealt with in the context of the megacity of Delhi as well as the million city of Tehran. Darjeeling and Kathmandu valley focus on the urban water challenges in mountain environment. Climate change and its impact have been discussed with reference to the city of Delhi while impact of El Nino occurrences on water availability and consequent urban drought has been dealt within the context of Metro Manila, Iloilo and Cebu in the Philippines. The socio-economic conditions of water insecurity in urban areas have been considered for the cities of Kolkata, Delhi and Dhaka, the three important and densely populated megacities in the Indian subcontinent. Old worn-out infrastructure, inadequate and inequitable distribution of water result in acute water insecurity amongst the urban poor and in the fringe areas of these cities. These issues thus further highlight poor urban water governance and inefficient water management. The key issues of urban water governance have also been dealt with separately for Colombo city of Sri Lanka. Health risks associated with poor drinking water availability is evident from the case study of Hong Kong. Rapid urban growth is also responsible for soil sealing that aggravates the already evident water scarcity by reducing infiltration and groundwater recharge when most Asian cities are seen to supplement municipal

Table 24.2 Key learning from different Asian cities

Key learning from urban experience	Case studies from Asia
Physical scarcity of water, climate-induced uncertainties and water stress	Delhi, India Tehran, Iran
Urban water scarcity in mountain environment	Darjeeling, India Kathmandu, Nepal
Soil sealing and urban drought	Peshawar, Pakistan
Water insecurity in the formal water supply and management system and community resilience in parallel water supply systems	Kolkata, India
Socio-economic water insecurity	Kolkata and Delhi, India Dhaka, Bangladesh
Water insecurity in urban slums	Dhaka, Bangladesh
Urban water governance issues	Colombo, Sri Lanka
Health risks associated with water resources management	Hong Kong, China
Risks and vulnerabilities to water-related disasters	Dhaka, Bangladesh
Institutional obstacles to climate resilience and vulnerability to floods and droughts	Udon Thani, Thailand
Flash floods and water scarcity	Ulaanbatar, Mongolia
El Nino, urban drought and risk governance	Metro Manila, Iloilo and Cebu, Philippines
Impacts of tsunami on the availability of clean drinking water	Banda Aceh city, Indonesia
Traditional water management through canal restoration	Yanagawa, Japan
Role of urban infrastructure in building resilience	Nagpur, India Dhaka, Bangladesh
Studies of policy initiatives for mitigation and management of urban drought	Colombo, Gale and Kandy, Sri Lanka
Water conservation measures	Thane, India

Source By authors

supply with groundwater. Impacts of soil sealing on urban drought have been dealt with for Peshawar in Pakistan. On mitigation and management of urban drought, studies of policy initiatives have also been done with case studies of Colombo, Gale and Kandy. The twin problem of flash floods and water scarcity of Ulaanbatar city and the impacts of tsunami on the availability of clean drinking water in Banda Aceh city, Indonesia cover the disaster aspect of urban water insecurity. Apart from the analysis of urban drought and its causes, water conservation measures that may be adopted for tackling urban drought have also been studied. Analysis of existing water conservation throws light on the issues and challenges in the existing measures while evaluating traditional water management schemes in Yanagawa city of

Japan. The chapter describes the process of the canal restoration to revitalize the water environment in urban areas by bridging the distances between far water and near water. Resilience in the existing urban water system has been evaluated through urban infrastructure in Nagpur and Dhaka.

Urban water insecurity issues discussed in the previous chapters thus focus on some of the targets specified in the Sustainable Development Goals (SDGs), including goals 6, 9, 11, 13 and 15. Goal 6 is devoted to universal access to safe water and improved sanitation, sustainable management of water resources and capacity building through innovations. Goal 9 highlights on innovations for improved infrastructure while Goal 11 emphasizes upon making cities inclusive, safe, resilient and sustainable. Goal 13 focuses on climate change and hence on urban drought while Goal 15 promises to combat desertification, reverse land degradation and loss of biodiversity.

24.5 Future Prospects in Asian Cities

It is quite evident that urban water systems develop and function as complex socio-ecological system. Water management strategies need to adopt a multidisciplinary-integrated approach including all the different stakeholders and must address issues evolving from changes in regional climate as well as other socio-economic factors like population increase, economic development and urbanization. Water is one of the most vital resources for the functioning of urban water systems with competing water use across commercial, residential and industrial sectors. With industrial and domestic water consumption levels expected to increase by twice the present level by 2050 (UNDESA 2007), competition for water use is expected to increase significantly between urban, peri-urban and rural areas. To make matters worse, impact of climate variability including the increased frequency of extreme weather events has altered the quality, quantity and the seasonal availability of water for most urban centres (Global Water Partnership 2000). In the context of changing expectations, urban water systems are confronted with increasingly complex and multi-faceted challenges, in particular, when the existing and available resources reach the limits of sustainable exploitation (Brown et al. 2009; Pubmed). As cities grow in terms of population and diversity of urban functions, demands on the urban water systems rise and hence the existing water systems are often exploited beyond their sustainable limit. It is already evident that existing water management approaches are inadequate to address water insecurity and water stress. An attempt is hence made for integrated water management approaches involving different stakeholders instead of compartmentalized and isolated management of urban water systems. Integrated urban water management encompasses all forms of water sources including blue water (surface water, groundwater, desalinated water), green water (rainwater), black and grey water (Global Water Partnership 2000). The targets for urban water management include assured access to water and sanitation facilities and services, proper management of rainwater, wastewater treatment, stormwater drainage, reduced levels of pollution of water resources, control over waterborne diseases, reduced risk of water-linked hazards of floods, droughts and saline intrusions as well the preservation and proper functioning of wetlands. Integrated urban water management (IUWM) offers a set of principles that ensure better coordinated, responsive and sustainable water management practices (Global Water Partnership 2000). It recognizes alternate water sources for enhanced water security, differentiates and recognizes water quality parameters and the use potential of different sources of water, manages water storage, distribution, treatment, recycling and disposal as part of the same resource management cycle (Global Water Partnership 2000) rather than as discrete activities. Urban water must align formal institutions (organizations, legislation and policies) with informal practices (norms and conventions) (Global Water Partnership 2000), simultaneously pursue economic efficiency, social equity, and environmental sustainability and encourage participation by all stakeholders (Furlong et al. 2017). It must also plan for the protection, conservation and exploitation of water resources (Global Water Partnership 2000). The existing urban water systems must be more efficient, reduce leakage loss to the tune of 32 billion cubic metres per year (Kingdom et al. 2006; Global Water Partnership 2000) and reduce the volume of non-revenue water. Apart from technological solutions to improve water security, urban water management also includes behavioural changes, institutional capacity building and water resource assessment or water audits. The aim of urban water management is to create cities and towns that are resilient, liveable, productive and sustainable (Government of Western Australia). Urban water management thus takes into consideration the total water cycle, facilitates the integration of water factors in the planning process, and encourages all levels of government and industry to adopt water management and urban-planning practices that benefit the community, the economy and the environment (Government of Western Australia).

Urban areas in the less developed countries of Asia, crippled by severe financial and institutional constraints, and poor water governance, need to better understand and manage water security issues including adaptation options. Global urban communities are striving for increased resilience within the existing urban water systems and to future uncertainties in water supply because of climate change and population growth. It is already well documented that the conventional water systems are often inadequate to address water insecurity and water stress in urban areas. Transforming urban water systems into more resilient and hence more sustainable systems would require innovative approaches. These approaches may be categorized into infrastructure-based approaches and system-based approaches (Fig. 24.1).

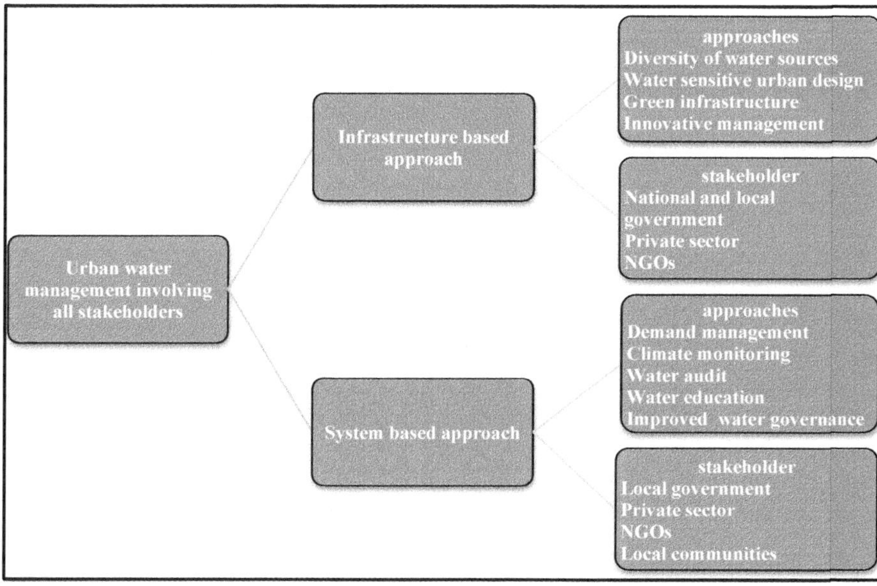

Fig. 24.1 Innovations in urban water management

24.5.1 Infrastructure-Based Approaches

- Diversity of water sources

It is pertinent to note that there are a wide range of water sources that the cities can access to supplement the existing ones including rainwater harvesting, construction of new storage facilities, appropriate and sustainable extraction of groundwater, groundwater recharge, urban stormwater use, treated wastewater and desalination options. Appropriate policy initiatives would enable cities greater flexibility in identifying the alternate sources for improved supply and delivery options.

- Water-sensitive urban design

Water-sensitive urban designs allow cities the option to utilize grey water for non-potable uses in toilets, laundry, gardens and car-washing stations. Such use of water may be made compulsory for residential areas and the building bylaws amended accordingly. Provisioning of green infrastructure in new buildings and urban areas like rooftop gardens, permeable pavements, underground water detention systems may also be explored. Green retrofitting of existing buildings like the introduction of rainwater harvesting at institutional level needs to be promoted through legal binding if necessary.

Strategies to mitigate urban warming by changing the materials used for individual buildings and improved building design and layout would help to reduce energy demand. Increased number of open spaces may also be adopted as a strategy to reduce urban heat island effect while the introduction of detention areas and ponds would reduce urban run-off to increase infiltration for groundwater recharge.

Drainage pumping stations can significantly reduce the chances and intensity of urban flooding. The old worn-out water supply system in many cities needs to be replaced to reduce leakage loss and increase the availability of water.

- Innovations in water management and water usage control devices

The private sector may be engaged in innovative water management strategies like water usage control devices to reduce the flow of water from taps to reduce wastage. The saved water may be reallocated to vulnerable areas of the city and also to the vulnerable sections of the population. Such devices could include the installation of an isolating ball valve or a flow restrictor to monitor the flow of water. With a tap aerator, the design of the tap nozzle allows air to mix with water giving the appearance of increase in water flow. These devices are easy to retrofit to existing taps (Nibusinessinfo). An ‘eco-tap’ or ‘eco-brake’ cartridge stops an user from using the full flow that the tap can provide by stopping the tap lever from moving more than halfway (Nibusinessinfo). It offers substantial savings as the potential water flow is halved (Nibusinessinfo). The other option could be the use of a spray tap. In a spray tap, the nozzle found in normal taps gets replaced with a snout that gives out water in the form of mist or spray. Spray taps can save up to ten litres of water per minute and are hence highly effective at saving water.

- Mobile-based water flow monitoring system

An online or mobile-based application may be introduced that monitors and controls the flow of water through taps (Kumar and Mahmoud 2017). The water flow monitoring system is an interactive android-based mobile application (Kumar and Mahmoud 2017), fitted with external hardware, where registered users can view their water usage and flow on the screen. The rate of flow of water gets updated every second to a database available online. An user’s on/off instruction is set within the database; the hardware synchronizes this instruction and performs the desired action (Kumar and Mahmoud 2017). Set Timer and Set Schedule are two additional functions provided to users (Kumar and Mahmoud 2017). The Set Timer function enables the users to set a timer to turn on/off a water supply tap while the Set Schedule function enables the users to set a timer to turn on/off a water supply tap on a desired date (Kumar and Mahmoud 2017). This can be made available for the centralized potable water schemes at the city level as well as for localized water supply schemes like simple rainwater harvesting systems at household or community levels.

24.5.2 System-Based Approaches

- Increased urban water resilience

Urban water systems are also facing an increasing need to improve water resources management and build efficient infrastructural facilities to meet the requirements of an escalating demand and diminished water availability. Climate monitoring, assessment of water stress, demand management, land-use planning and resource mobilization, urban landscaping will ensure better urban water systems and reduce climate-change-related disaster risks. Institutional capacity building, promoting scientific research and awareness, new and innovative approaches to urban water governance and community involvement aimed at reducing water insecurity and removing water stress is expected to increase urban water resilience.

- Stakeholder participation

The role of urban communities needs to be recognized in identifying urban water stress and their participation should be encouraged in developing water-sensitive strategies. Brown et al. (2008) argue that unless new technologies are socially embedded into the local institutional context, their development in isolation is insufficient to ensure their successful implementation in practice (Wong and Brown 2009; Brown et al. 2008). Participation of the key stakeholders is warranted for understanding and balancing the interests and needs of different stakeholders and for strengthening mutual cooperation.

- Demand management

Managing water demand is considered as an effective tool to ensure urban water security. It promotes water conservation through changes in the attitude of the end-users and through the use of conventional practices for the conservation of water resources. Demand management for urban water security could employ communication tools, public awareness campaigns for behavioural change and to conserve scarce water resources, social media campaigns on water conservation practices and public workshops on water-use efficiency.

- Water education

Water-focused school education and innovative practices like the maintenance of water diaries of domestic water usage per week or month and presentation of water certificates for reduction in water usage may be considered as a system-based approach to save water.

- Water pricing and water audit

Country-specific water pricing based on institutional and cultural framework may be thought of. Water audit may be carried out at institutional or subcity/ward level for effective demand-supply system. Automated water systems are best avoided as they result in inefficient and excessive water use leading to stress in the water supply systems. On occasions when automated water systems cannot be avoided, they need to be fitted with timers to ensure controlled use.

- Water conservation measures

Regulatory tools may also be adopted for forced water conservation and may include such measures as setting water-use limits, slab-based pricing of water use based on local culture and water-use requirements, special concessions for the use of water-efficient technology by the city corporations and legal bindings for water conservation. Coordination between the formal and informal water sectors, improvements in water storage and management, community participation would also lower water insecurity. Rainwater harvesting techniques may be used to divert rainwater captured in rooftops and gutters for groundwater recharge or for gardening. Aiming for water security through diversity and optimum use of all water sources, matching water quality with the purpose of use are some of the other interventions suggested.

24.6 Conclusion

The cities in Asia are faced with multifarious problems and include inadequate availability of fresh water, leakage loss from distribution networks, intermittent supply, poor water quality in municipal taps and poor cost recovery. Excessive use of groundwater has caused rapid depletion of groundwater reserves and severe environmental damages. Inadequately developed sewerage systems and the absence of sewage treatment plants also often result in water contamination, making it exceedingly difficult for cities to supply clean water. In addition, heavy seasonal monsoon rain and frequent floods are common in Asian cities.

Preparing the urban water systems in cities of Asia for climate change impacts requires innovations in the traditional approaches practiced by the local communities. Technological improvements and institutional capacity building are needed to develop resilient and sustainable urban water systems. Many urban centres and countries of Asia have already adopted innovative measures in managing water stress and drought in urban areas. The densely populated city of Singapore, for example, with no source of fresh water supply, is amongst the most highly water-stressed countries with its demand exceeding the naturally occurring supply. Yet the city is able to provide enough water for its industrial, agricultural, commercial and domestic use through investment in technology, international agreements and effective water management. Advanced rainwater detention systems account for 20% of the water supplied in Singapore. 40% is given by Malaysia. Use of grey contributes another 30% while desalination process provides the remaining 10% of the supply. However, in order to make any water management system resilient and sustainable, an integrated water management strategy must be formulated and implemented. Interventions are required at planning and policy level that would have some legal binding on the citizens. Such bindings would ensure the proper implementation of innovations in the urban water system.

References

Abedin MA, Shaw R (2014) Safe water adaptability for salinity, arsenic and drought risks in Southwest of Bangladesh. Risk Hazards Publ Policy 4(2):62–82

Asian Development Bank (2003) Water for all: the water policy of the Asian Development Bank. Asian Development Bank, Philippines

Asian Development Bank (2016) Asian water development outlook 2016: strengthening water security in Asia and the Pacific. Asian Development Bank, Philippines

Brown RR, Keath N, Wong THF (2008) Transitioning to water sensitive cities: historical, current and future transition states. In: 11th international conference on urban drainage. Edinburgh, United Kingdom

Brown RR, Keath N, Wong THF (2009) Urban water management in cities: historical, current and future regimes. Water sci Technol 59:847–855

Charkhestani A, Ziri MS, Rad HA (2015) Wastewater reuse: potential for expanding Iran's water supply to survive from absolute scarcity in future. J Water Reuse and Desal 06(3):437–444

Falkenmark M (1989) The massive water scarcity threatening Africa—why isn't it being addressed. Ambio 18(2):112–118

Furlong C, Brotchie R, Considine R, Finlayson G, Guthrie L (2017) Key concepts for integrated urban water management infrastructure planning: lessons from Melbourne. Util Policy 45(C):84–96

Global Water Partnership (2000) IWRM toolbox integrated water resource management. http://www.gwptoolbox.org. Accessed 10 Apr 2013

Government of Western Australia. http://www.water.wa.gov.au/urban-water/urban-development/urban-water-management

Joerin J, Shaw R (2011) Mapping climate and disaster resilience of cities. In: Shaw R, Sharma A (eds) Climate and disaster resilience of cities, 1st edn. Emerald Publisher, United Kingdom

Kumar K, Mahmoud AM (2017) Monitoring and controlling tap water flow at homes using android mobile applications. Am J Softw Eng Appl 6(6):128–136. http://www.sciencepublishinggroup.com/pdf/10.11648.j.ajsea.20170606.11. Accessed 28 Nov 2011

Lawrence P, Meigh J, Sullivan C (2002) The water poverty index: an international comparison. Keele economics research papers: 1–17. http://ideas.repec.org/p/kee. Accessed 24 Aug 2008

Kingdom B, Liemberger R, Marin P (2006) The challenge of reducing non-revenue water (NRW) in developing countries, how the private sector can help: a look at performance—based service contracting. Water and Sanitation Sector Board Discussion Paper Series No. 8. World Bank, Washington, DC

Nibusinessinfo. https://www.nibusinessinfo.co.uk/content/reduce-water-use-taps

Pubmed. https://www.ncbi.nlm.nih.gov/pubmed/19273883

Ray B, Shaw R (2016) Water stress in the megacity of Kolkata, India and its implications for urban resilience. In: Shaw R, Rahman A, Surjan A, Parvin GA (eds) Urban disasters and resilience in Asia, 1st edn. Elsevier, United Kingdom

Sullivan C (2002) Calculating a water poverty index. World Dev 30(7):1195–1210

United Nations Department of Economics and Social Affairs (2007) World population prospects: the 2006 Revision, Highlights, Working paper no. ESA/P/WP.202. United Nations, New York

Vorosmarty CJ, McIntyre PB, Gessner MO, Dudgeon D, Prusevich A, Green P, Glidden S, Bunn SE, Sullivan CA, Reddy LCR, Davies PM (2010) Global threats to human water scarcity and river biodiversity. Nature 467:555–561

Water Strategy Man (2004) Indicators and Indices for decision making in water resources management. Newsletter 4. http://environ.chemeng.ntua.gr/WSM/Newsletters/Issue4/Indicators_Appendix.htm. Accessed 22 July 2010

Wong THF, Brown RR (2009) The water sensitive city: principles for practice. Water Sci Technol 60(3):673–682

Printed by Printforce, the Netherlands